Multi-Agent Systems

Platoon Control and Non-Fragile Quantized Consensus

Automation and Control Engineering
Series Editors - Frank L. Lewis, Shuzhi Sam Ge, and Stjepan Bogdan

Networked Control Systems with Intermittent Feedback
Domagoj Tolic and Sandra Hirche

Analysis and Synthesis of Fuzzy Control Systems: A Model-Based Approach
Gang Feng

Subspace Learning of Neural Networks
Jian Cheng Lv, Zhang Yi, Jiliu Zhou

Synchronization and Control of Multiagent Systems
Dong Sun

System Modeling and Control with Resource-Oriented Petri Nets
MengChu Zhou, Naiqi Wu

Deterministic Learning Theory for Identification, Recognition, and Control
Cong Wang and David J. Hill

Optimal and Robust Scheduling for Networked Control Systems
Stefano Longo, Tingli Su, Guido Herrmann, and Phil Barber

Electric and Plug-in Hybrid Vehicle Networks
Optimization and Control
Emanuele Crisostomi, Robert Shorten, Sonja Stüdli, and Fabian Wirth

Adaptive and Fault-Tolerant Control of Underactuated Nonlinear Systems
Jiangshuai Huang, Yong-Duan Song

Discrete-Time Recurrent Neural Control
Analysis and Application
Edgar N. Sánchez

Control of Nonlinear Systems via PI, PD and PID
Stability and Performance
Yong-Duan Song

Multi-Agent Systems
Platoon Control and Non-Fragile Quantized Consensus
Xiang-Gui Guo, Jian-Liang Wang, Fang Liao, Rodney Teo

For more information about this series, please visit: https://www.crcpress.com/Automation-and-Control-Engineering/book-series/CRCAUTCONENG

Multi-Agent Systems

Platoon Control and Non-Fragile Quantized Consensus

Xiang-Gui Guo
Jian-Liang Wang
Fang Liao
Rodney Teo

CRC Press is an imprint of the
Taylor & Francis Group, an **informa** business

CRC Press
Taylor & Francis Group
6000 Broken Sound Parkway NW, Suite 300
Boca Raton, FL 33487-2742

First issued in paperback 2022

ISBN-13: 978-0-367-25432-2 (hbk)
ISBN-13: 978-1-03-233831-6 (pbk)
DOI: 10.1201/9780429287770

Publisher's Note

The publisher has gone to great lengths to ensure the quality of this reprint but points out that some imperfections in the original copies may be apparent.

Library of Congress Cataloging-in-Publication Data

Names: Guo, Xiang-Gui, author. | Wang, Jian-Liang, author. | Liao, Fang, author. |
Teo, Rodney Swee Huat, author.
Title: Multi-agent systems : platoon control and non-fragile quantized consensus /
Xiang-Gui Guo, Jian-Liang Wang, Fang Liao and Rodney Swee Huat Teo.
Description: First edition. | Boca Raton, FL : CRC Press/Taylor & Francis Group, 2020. |
Series: Automation and control engineering | Includes bibliographical references and index.
Identifiers: LCCN 2019014816| ISBN 9780367254322 (hardback : acid-free paper) |
ISBN 9780429287770 (ebook)
Subjects: LCSH: Multiagent systems. | Motor vehicle fleets--Data processing. |
Neural networks (Computer science) | Autonomous vehicles.
Classification: LCC QA76.76.I58 G87 2020 | DDC 006.3/0285436--dc23
LC record available at https://lccn.loc.gov/2019014816

Visit the Taylor & Francis Web site at
http://www.taylorandfrancis.com

and the CRC Press Web site at
http://www.crcpress.com

Contents

Preface

Multi-agent systems have emerged as an inexpensive and robust way of addressing a wide variety of tasks, ranging from transportation, exploration, search and rescue, and reconnaissance to cooperative construction and manipulation. In particular, researchers have studied extensively the simple case of a 1-D platoon of vehicles, considering diverse alternatives to achieve coordinated movement of the platoon. The main goals of *platoon control* are to maintain a desired and safe distance between neighboring vehicle in a platoon and to make the following agents reach consensus with the leader's state. The success of these systems relies on efficient information exchange and coordination among members of the team. However, when digital communications are adopted, due to the finite channel capacity and the widespread of analog-to-digital (AD) and digital-to-analog (DA) converters in sensors and actuators, quantization plays an important role in information exchange among agents. Therefore, quantized consensus or consensus with quantized communications becomes interesting and more meaningful, and has drawn the attention of more and more researchers from various perspectives. Besides, inaccuracies and uncertainties do occur in controller implementation due to finite word length (FWL) and numerical roundoff errors, etc. Consequently, how to design a controller non-fragile to some amount of error with respect to its coefficients is of practical significance for the implementation of digital controllers because even vanishingly small perturbations in the controller coefficients may destabilize the closed-loop systems.

In order to solve *platoon control* problem, distributed adaptive *sliding mode control* (SMC) algorithms are proposed to guarantee *strong string stability* based on *modified constant time headway* (MCTH) policy. *String stability* is a measure of how position errors propagate from one vehicle to another in a platoon. The MCTH policy is used to remove the unrealistic assumption in the most existing literature that initial spacing, velocity and acceleration errors are zero. Moreover, by using the leader's information, this policy can also increase the traffic density to an extent that is nearly equal to that of constant-spacing policy. It is known that SMC has been well investigated as a so-called robust control technique to control dynamical systems with parameter perturbations, modeling uncertainties, and external disturbances, while neural networks (NNs) and fuzzy logic systems (FLSs) have the capability to approximate any continuous functions over a compact set to arbitrary accuracy. Since unmodeled nonlinear dynamics and external disturbances often exist in vehicle platoon, this book is interested in investigating the *platoon control* issue by combining SMC technique with NN or FLS approximation methods. Moreover, it is well known that non-smooth nonlinear characteristics such as fault, quantization, deadzone and saturation are common in actuators. Since these non-smooth nonlinearities can severely limit system performance, this book also proposed some methods to attenuate the negative effects of these actuator nonlinearities. Furthermore, some existing results have pointed out that the *string stability* does not provide a formal warranty of collision avoidance between consecutive vehicles. In order to avoid collision between consecutive vehicles as well as the connectivity breaks owing to limited sensing capabilities, a symmetric barrier Lyapunov function is employed.

On the other hand, the neighbors of a given agent may change with time, which may result in a mismatch between the encoder and decoder. Quantization parameter mismatch might induce instability of the closed-loop system. Meanwhile, many effective methods such as structured vertex separator method, randomized algorithm, the sparse structure method and transfer function sensitivity approach have been proposed to deal with the numerical problem caused by interval-bounded controller coefficient variations. However, the number of linear matrix inequality (LMI) constraints involved in the design conditions by using the first three methods is still large, while the fourth method involves the closed-loop system's transfer function and cannot be applied to multi-agent systems with quantization information and nonlinearities. In addition, many real world coupled dynamical systems can be modeled as multi-agent systems with high-order dynamics, such as distributed unmanned air vehicles and coupled manipulators. Therefore, this book also investigates the distributed non-fragile quantized H_∞ consensus problems for general linear or *Lipschitz* nonlinear high-order multi-agent systems with input quantization mismatch and interval controller coefficient variations.

In this monograph, the aim is to present our recent research results in investigating *platoon control* and non-fragile quantized consensus for multi-agent systems. The main feature of this book is that distributed adaptive SMC methods are successfully introduced to solve platoon control for vehicle platoons. Meanwhile, barrier Lyapunov function combined with terminal SMC technique is proposed to achieve collision avoidance between consecutive vehicles. Moreover, based on LMI technique, a non-fragile quantized consensus problem for high-order multi-agent systems is investigated. This monograph provides a coherent approach and contains valuable reference materials for researchers wishing to explore the area of multi-agent systems. Researchers, graduate students and engineers in the fields of vehicular technology, electrical engineering, control, applied mathematics, computer science and others will benefit from this book. Its contents are also suitable for a one-semester graduate course.

We are grateful to Nanyang Technological University, University of Science and Technology Beijing and National University of Singapore for providing resources for our research work. We also gratefully acknowledge the support of our research by National Natural Science Foundation of China (Grant No. 61773056), Fundamental Research Funds for the Central Universities of USTB (No. 230201606500061, FRF-BD-16-005A), China Postdoctoral Science Foundation (No. 2018T110047, 2017M610046), Beijing Key Discipline Development Program (No. XK100080537), the National Natural Science Foundation of China (Grant No. 61673055, 61673056, and 61603274), and National University of Singapore (No. RCA-14/123).

Finally, we would like to thank the Series Editors Frank Lewis and Shuzhi Sam Ge for their encouragement, and the entire team of Taylor & Francis Group for their cooperation in bringing out the work in the form of a monograph.

Xiang-Gui Guo
University of Science and Technology Beijing, China
Key Laboratory of Knowledge Automation for Industrial Processes, China
Beijing Engineering Research Center of Industrial Spectrum Imaging, China

Jian-Liang Wang
Nanyang Technological University, Singapore

Fang Liao
National University of Singapore, Singapore

Rodney Teo
National University of Singapore, Singapore

Authors

Xiang-Gui Guo received a B.S. degree at the College of Electrical Engineering in 2005 from Northwest University for Nationalities, China, and an M.S. degree at the College of Electrical Engineering and Automation in 2008 from Fuzhou University, China. In 2012, he received a Ph.D. degree in Control Science and Engineering from Northeastern University, China. He is currently an Associate Professor with the School of Automation and Electrical Engineering, University of Science and Technology Beijing, Beijing, China. He was also a Postdoctoral Fellow at the School of Electrical and Electronic Engineering, Nanyang Technological University, Singapore. His research interest includes string stability and their applications to flight control system design.

Jian-Liang Wang received the B.E. degree in electrical engineering from Beijing Institute of Technology, Beijing, China, in 1982, and the M.S.E. and Ph.D. degrees in electrical engineering from Johns Hopkins University, MD, USA, in 1985 and 1988, respectively, specializing in nonlinear systems and control theory. From 1988 to 1990, he was a lecturer with the Department of Automatic Control at Beijing University of Aeronautics and Astronautics, Beijing, China. Since 1990, he has been with the School of Electrical and Electronic Engineering at Nanyang Technological University, Singapore, where he is currently an Associate Professor. His research interests include multi-agent systems, vision-based control systems, path planning and trajectory generation, flight control systems, reliable and fault tolerant control, and fault detection and identification.

Fang Liao received the B.E. and M.E. degrees in control and navigation from Beijing University of Aeronautics and Astronautics, Beijing, China, in 1992 and 1995, respectively, and the Ph.D. degree from Nanyang Technological University, Singapore, in 2003. From 1995 to 1999, she was an engineer with the Research Institute of Unmanned Air Vehicles, Beijing University of Aeronautics and Astronautics. From 2002 to 2004, she was a Research Associate and then Research Fellow with the School of Electrical and Electronic Engineering, Nanyang Technological University, Singapore. Since 2004, she has been with Temasek Laboratories, National University of Singapore, Singapore, where she is currently a Senior Research Scientist. Her research interests include robust and adaptive control theories and application, fault tolerant control, multi-agent systems, motion planning, and constrained optimization methods.

Rodney Teo received the B.Eng. in mechanical engineering from National University of Singapore, Singapore, in 1990 and the M.S. and Ph.D. degrees in aeronautics engineering from Stanford University, Stanford, CA, USA, in 1998 and 2004, respectively. From 1990 to 1995, he was a Project Engineer and then, from 1996 to 1997, a Project Manager on helicopter acquisition and system integration projects in the Defence Materiel Organisation of Singapore. He is currently a Principal Member of Technical Staff of the DSO National Laboratories, Singapore, and a Senior Research Scientist with Temasek Laboratories, National University of Singapore. His current research interests include in research and development in the area of manned-unmanned systems.

1

Introduction

There have been great efforts in investigating *platoon control* and *non-fragile quantized consensus* for *multi-agent systems*. Distributed coordination of multiple vehicle agents, including *unmanned aerial vehicles* (UAVs), *high speed trains* (HSTs), *intelligent vehicle highway systems* (IVHSs) and *intelligent transportation systems* (ITSs), has been a very active research topic in the systems and control society. The recent research results about *platoon control* and non-fragile quantized consensus are reviewed in this chapter.

1.1 Platoon Control Problem

One control objective in the field of coordinated systems is formation control. In formation control, a group of vehicles should follow a given group trajectory and in addition every vehicle needs to maintain a prescribed distance to the surrounding vehicles. Increasing commercial and private vehicle traffic motivates a growing interest in the one-dimensional version of this problem which is often called 'platooning,' and some practical examples are shown as in Figure 1.1. *Platoon control* of multiple vehicles has applications in areas such as IVHSs [2, 55, 118], HSTs [36, 52, 81, 101, 194], ITSs [37, 204], aircraft [22, 130] and spacecraft formation flight [158], closely spaced parallel approaches [170], and separation control of arrival stream aircraft [173].

The main goals of *platoon control* are to maintain a desired and safe distance between consecutive vehicles in a platoon [27, 93, 142, 143, 164, 165] and increase the capacity of traffic flow by reducing the neighboring-vehicle spacing [45, 56, 57, 58, 59, 126, 207]. Since vehicles in a platoon are dynamically coupled, disturbances acting on one vehicle may affect other vehicles. Therefore, an important concept regarding vehicle platoons is *string stability*, i.e., *spacing errors* do not amplify as they propagate upstream from one vehicle to another vehicle [57, 93]. If the spacing errors amplify as they propagate upstream (i.e., *string instability*), it not only will likely provide poor ride quality but also could result in collision [57]. An example of *string instability* can be found by considering a formation of piloted aircraft in low-visibility conditions such as fog or clouds. With this limited visibility, each pilot can only see the aircraft directly ahead and attempts to track a position relative to that aircraft. Typically, any position change to the first aircraft is reacted to by the second aircraft with slight overshoot. Each aircraft overshoots the motion of the previous aircraft. This can cause unacceptable motion of the last aircraft in the string. Therefore, the field of *string stability* research has been active for decades due to its broad applications in various areas, such as autonomous formation flying [7, 151, 155], IVHSs [43, 45, 46, 186], etc. In these applications, each vehicle in the platoon should be controlled to maintain a *desired safe spacing* from its leading vehicle and *avoid collision* [27, 142, 143, 164]. In the following, we overview recent research results in platoon control problems.

FIGURE 1.1
Vehicular platoon examples.

1.1.1 Spacing Policies

The relationship between *string stability* and spacing policies has been an important issue [43]. Generally, there are two major strategies to maintain a desired spacing, i.e., the constant-spacing (CS) policy and the variable-spacing (VS) policy, depending on whether the required spacing in-between vehicles is dependent on vehicle speed or not [46, 164, 165]. Research should mainly focus on three main questions: (1) What is the relationship between *string stability* and spacing policy? (2) What are the substantive differences between *CS policy* and *VS policy*? (3) What are the advantages and disadvantages of *CS policy* and *VS policy*?

Currently, most existing work in this area is based on *CS policy* [38, 42, 43, 45, 46, 93, 169, 212]. However, it is known that for high capacity, (i.e., small vehicle to vehicle spacing), the *CS policy* achieves *strong string stability* at the cost of neighboring-vehicle communication by requiring preceding vehicle's acceleration information [164, 169]. Furthermore, such communication may result in time delay and package dropout and thus may cause *string instability* [73, 96, 169]. Therefore, *VS policy* is used to replace *CS policy* in many existing research such as [2, 26, 44, 157, 164] because it can ensure *string stability* just by using on-board information [2, 164]. Nevertheless, the steady-state neighboring-vehicle spacings of *constant time headway* (CTH) policy are very large at high speed, and thus the traffic density is low. It is worth mentioning that *CTH policy* is the simplest and most common *VS policy* [2, 14, 164]. The *CTH policy* does not require a lot of data from other vehicles, and it can ensure *strong string stability* using on-board information only and thus improving *string stability* [164], which is necessary to *avoid collisions.* In addition, since using variable spacing, this policy is more tolerant of *external disturbances* than the *CS policy* [92, 154]. The properties of these two types of policies are shown in Table 1.1. Furthermore, it is worth mentioning that the vehicles in the platoon can have different communication topologies, either radar-based or communication-based. Figure 1.2 shows six commonly used topologies for $N+1$ vehicles, including [212]:

(a) Predecessor following (PF) topology;

(b) Predecessor-leader following (PLF) topology;

(c) Bidirectional (BD) topology;

(d) Bidirectional-leader (BDL) topology;

(e) Two-predecessors following (TPF) topology;

(f) Two-predecessors-leader following (TPLF) topology.

For conciseness, many other topologies are not exhibited here, but they all can be analyzed using similar approaches.

TABLE 1.1
Comparison of constant-spacing policy and variable-spacing policy

	Constant-Spacing Policy	Variable-Spacing Policy
Pros	High traffic density	Not requiring a lot of data from other vehicles; Using on-board information only; Having more safety; Being more robust to disturbances; Collision avoidance
Cons	Inter-vehicle communications; Time delay; Package drop; Preceding vehicle acceleration	Low traffic density

1.1.2 Actuator Nonlinearities

During the past several years, many interesting and important approaches have been developed to guarantee *string stability*, such as *sliding mode control* (SMC) [93, 116, 164], *linear matrix inequality* (LMI) approach [45, 46], backstepping approach [138], adaptive sliding mode approach [93, 116], and distributed receding

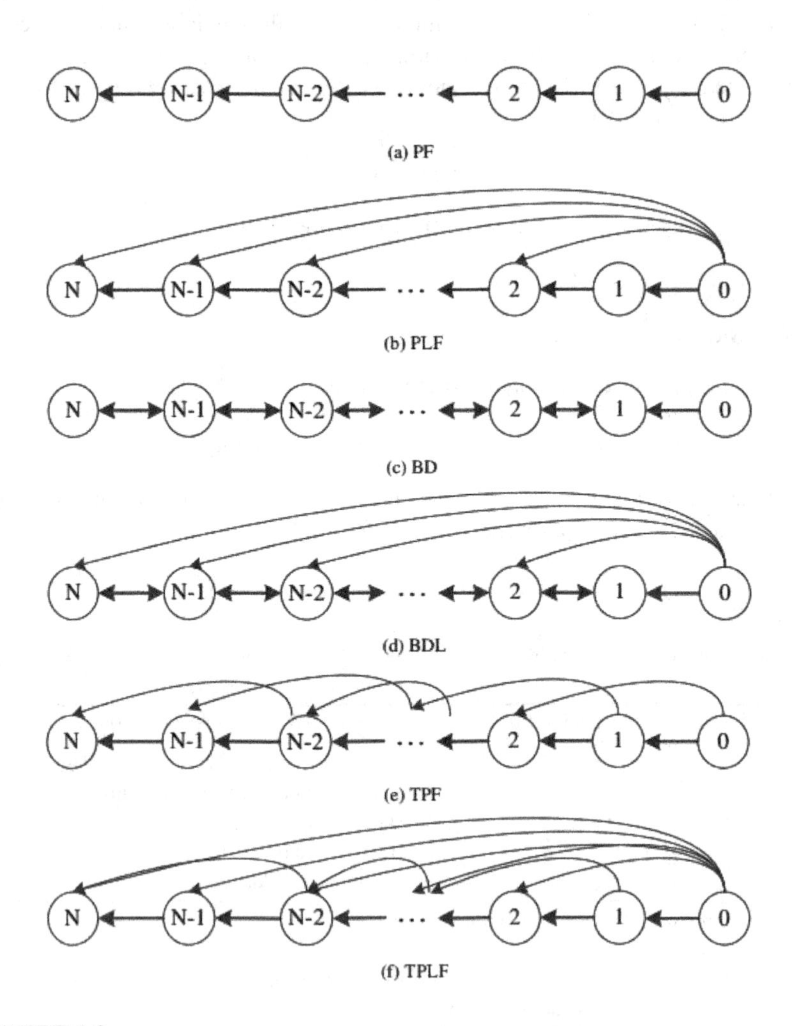

FIGURE 1.2
Typical information flow topologies for platoons.

horizon control approach [34]. Generally, platoon control of vehicle platoons is a difficult task because of high nonlinearity, coupling dynamics effects, time-varying parameters, unknown disturbances, modeling uncertainties and actuator faults. Such undesirable factors may result in inaccuracy in control performance of the whole vehicular platoon and finally lead to instability. Adaptive control system design for uncertain nonlinear systems has received much attention using universal function approximators such as *Chebyshev neural network* (CNN) [135, 217], *radial basis function neural network* (RBFNN) [46, 116, 138] and *fuzzy logic systems* (FLS) [52] to approximate the unknown nonlinearities. Nonsmooth, nonlinear characteristics such as *deadzone*, backlash, hysteresis, *saturation* and *quantization* are common in actuators and sensors such as mechanical connections, hydraulic actuators and electric servomotors. The characteristic nonlinearities of actuators severely limit system performance, and its study has been drawing much interest in the control community for a long time. In the existing literature on the issue of *string stability*, the characteristic nonlinearities of actuators are commonly ignored. Therefore, how to deal with this problem has been a task of major practical interest as well as theoretical significance.

1.1.3 Communications/Sensing Restricted Applications

Although *CS policy* and *VS policy* have been proposed to address the problem of *string stability*, one common feature of the aforementioned policies is that the desired separation distance between neighboring vehicles is given *a priori* and does not evolve according to the communication/sensing quality of environments [60]. In addition, there lacks a comprehensive performance index evaluating both communication/sensing and *string stability*.

1.2 Non-Fragile Quantized Consensus

During the last decade, the consensus problem of *multi-agent systems* has received increasing attention in the control community due to its wide applications in many areas, such as cooperative UAVs, IVHSs, air traffic control and sensor networks, etc. Consensus control often refers to the situation where network-connected subsystems are controlled to achieve the same or very similar control objectives [31]. The success of objectives relies on efficient information exchange and coordination among the agents. However, when information communications are adopted, due to the finite channel capacity and the widespread application of analog-to-digital (AD) and digital-to-analog (DA) converters in sensors and actuators, *quantization* plays an important role in information exchange among agents [39, 49, 54]. Therefore, *quantized consensus*, or consensus with quantized communications, becomes interesting and more meaningful and has drawn the attention of more and more researchers from various perspectives. Besides, inaccuracies and uncertainties do occur in controller implementation due to finite word length (FWL) and numerical roundoff errors, etc.

[48, 50, 53, 54, 61, 64, 66, 67, 68, 191, 192]. Consequently, how to design a controller *non-fragile* to some amount of error with respect to its coefficients is of practical significance for the implementation of digital controllers because even vanishingly small perturbations in the controller coefficients may destabilize the closed-loop systems [61, 64, 191]. In the following, we will overview the recent research results in *quantized consensus* and *non-fragile* control problems.

1.2.1 Quantized Consensus

Consensus under *quantization* has been studied recently with the motivation from the fact that the control signal to a plant is transmitted through a network medium with limited communication capacity [39, 49, 51, 53, 54, 117, 119, 185] and finite sampling rates [98, 128, 197, 198]. Due to the impact of capacity constraints in communication, and the widespread application of AD and DA converters in sensors and actuators, *quantization* plays an important role in information exchange among agents [39, 54, 51, 174]. Since many real world coupled dynamical systems can be modeled as *multi-agent systems* with high-order dynamics, such as distributed UAVs and coupled manipulators, *quantized consensus* control for high-order *multi-agent systems* was further investigated [40, 49, 188]. In addition, the difference of the quantization parameters between the encoder and decoder side is not investigated in the above work. However, the neighbors of a given agent may change with time, which may result in a *mismatch* between the encoder and decoder [103]. When quantization parameter mismatch is ignored in the controller design, it might induce instability of the closed-loop systems [210].

1.2.2 Non-Fragile Control Problems

Because inaccuracies and uncertainties do occur in controller implementation due to FWL and numerical roundoff errors, etc. [61], *non-fragile* control has received significant attention recently. There are two main types of coefficient uncertainties considered in the existing literature, i.e., norm-bounded uncertainties and *interval-bounded variations*. Furthermore, these two types can further be divided to *additive* case and *multiplicative* case [192]. Because the type of norm-bounded uncertainty cannot exactly reflect the uncertain information due to implementation imprecision, *non-fragile* control problems with respect to interval-bounded coefficient variations for linear systems have received more and more attention [48, 50, 53, 54, 64, 66, 67, 68, 191, 192]. Although many effective methods such as structured vertex separator method [190], randomized algorithm [29], the sparse structure method [13] and transfer function sensitivity approach [50, 66, 67, 68, 191] have been proposed to deal with the numerical problem caused by interval-bounded coefficient variations, the number of LMI constraints involved in the design conditions by using the first three methods is still large, while the fourth method involves the closed-loop system's transfer function and cannot be applied to *multi-agent systems* with quantization information. Furthermore, as yet, few results are available for considering the non-fragile consensus control problems for *multi-agent systems*.

1.3 Preview of Chapters

This book is divided into 12 chapters. The background of platoon control and non-fragile *quantized consensus* control for *multi-agent systems* is introduced in Chapter 1, while some basic theories and some technical tools are reviewed in Chapter 2. The other 10 chapters contain new contributions about platoon control and non-fragile quantized consensus control for *multi-agent systems*. We now preview Chapters 3-12 and outline the research contents as follows:

Chapter 3 presents a new distributed adaptive integral SMC (ISMC) protocol to achieve the *finite time stability* of each vehicle and the *strong string stability* of the whole vehicle platoon system in presence of *external disturbances*. The adaptive control scheme does not require the exact knowledge of maximum and minimum values of the *external disturbances*, which is desirable for practical applications. A *modified constant time headway* (MCTH) policy is introduced to remove the unrealized assumption that the initial spacing/velocity errors are zero.

Based on the *MCTH policy* proposed in Chapter 3, a new *MCTH policy* is proposed in Chapter 4 to not only guarantee smooth transient response for non-zero initial spacing and velocity deviations, but also increase the *traffic density* to nearly equal to that of *CS policy*. At the same time, unlike most existing work which assumes that the acceleration of the leader is zero or constant, both the leader and the followers in this chapter may be subject to some unknown bounded nonlinear uncertainties. Moreover, *adaptive compensation* terms without requiring *a prior* knowledge of upper bound of the uncertainties are constructed to effectively reduce the effect of the unknown bounded nonlinear acceleration uncertainties.

Chapter 5 explores a simple and straightforward method by adjusting only one parameter to attenuate the negative effect of input *saturation*. Unlike the previous result in Chapter 4, where the nonlinearities of the consecutive vehicles are required to satisfy matching conditions, this constraint is removed by introducing *Chebyshev neural network*.

Based on Chapters 3-5, Chapter 6 developed a novel adaptive fuzzy *fault-tolerant* control method for multiple HSTs in presence of actuator faults, unknown nonlinear dynamics and *external disturbances*. The FLS is used to approximate the unknown functions. Based on proportional and integral (PI)-based sliding mode technique, two adaptive fuzzy control schemes are then proposed to achieve *strong string stability*. By using Lyapunov analysis, all signals of the closed-loop system are shown to be uniformly ultimately bounded and all HSTs track the reference trajectory to a small compact set around zero.

Chapter 7 investigates the adaptive platoon control for nonlinear vehicular systems with asymmetric nonlinear input deadzone and neighboring-vehicular spacing constraints. Unlike Chapters 3-6, this chapter does not require to guarantee *strong string stability*, but uses a symmetric barrier Lyapunov function to avoid collisions between consecutive vehicles as well as the connectivity breaks owing to limited sensing capabilities. At the same time, a neural-network-based terminal *sliding mode control* (TSMC) scheme with minimal learning parameters is developed to maintain neighboring-vehicles connectivity and simultaneously *avoid collisions*. The uniform ultimate boundedness of all signals in the whole vehicular platoon control system is proven via Lyapunov analysis. Compared with Chapters 3-6, less information is required for feedback control.

Chapter 8 proposes a novel *MCTH policy* for third-order vehicle dynamics platoon with improving road capacity, and simultaneously removing the common restriction of zero initial spacing, velocity and acceleration errors. Compared with the existing results, our proposed method does not *feedback linearize* the system model, and the negative effects of mismatch input quantization and unknown asymmetric *deadzone* are considered simultaneously. In addition, an optimized adaptation method is incorporated into the adaptation mechanism. Then, the number of adaptive parameters will not increase as the number of NN nodes increases and thus the online computation burden can be greatly alleviated.

Based on the *MCTH policy* proposed in Chapter 8, Chapter 9 developed an approximation-free adaptive proportional-integral-derivative-based SMC (*PIDSMC*) scheme for nonlinear *third-order* vehicle platoon subject to *asymmetric actuator saturation*. The scheme is capable of guaranteeing, for any initial system condition, *strong string stability* of the whole vehicular platoon despite *asymmetric actuator saturation*. Moreover, *adaptive compensation* instead of approximation approach such as neuro-network and fuzzy logic approaches is adopted to attenuate the negative effects caused by *asymmetric actuator saturation* and *unmodeled dynamic nonlinearities*.

Chapter 10 concerns the quantization effect on a non-fragile control law for *Lipschitz* nonlinear *multi-agent systems* with quantization information under undirected communication graphs. The design controllers are non-fragile to *additive* interval-bounded controller coefficient variations. The *incidence matrix* instead of *adjacency matrix* is applied to achieve *quantized consensus*. Based on LMI techniques, explicit convergence conditions for both *uniform* and *logarithmic quantizers* are obtained.

Chapter 11 studies the distributed non-fragile H_∞ consensus problems for linear multi-agent systems with *external disturbances* and unknown initial disturbances under switching weighted balanced directed topologies. Unlike Chapter 10, the type of *multiplicative* controller coefficient variations is considered in this chapter. Sufficient conditions for the existence of the proposed control strategy are also obtained by using LMI techniques. Instead of requiring the coupling strength among neighboring agents to be larger than a threshold value as in previous literature, the coupling strength can be determined by solving some LMIs.

Chapter 12 investigates the distributed quantized H_∞ consensus problems for general linear and *Lipschitz* nonlinear multi-agent systems with input quantization mismatch and *external disturbances*. Switching weighted undirected or switching weighted balanced directed topologies as in Chapter 11 are considered. By using LMI techniques, a new distributed quantized H_∞ *consensus* protocol is developed to achieve satisfactory performance. Besides guaranteeing satisfactory performance, complete consensus instead of practical consensus is also established.

2

Preliminaries

This chapter introduces notations used in the book, some useful definitions and lemmas, algebraic graph theory background, algebra and matrix theory background, linear and nonlinear system theory background, and non-smooth analysis background.

2.1 Notations

$\mathbf{R}$	Real number
$\mathbf{R}^n$	The real n vector
$\mathbf{R}^{n\times m}$	The real $n\times$ m matrices
$\|\cdot\|$	The Euclidean norm of a vector
$|\cdot|$	The absolute value of real numbers
$tr\{X\}$	The trace of the matrix X
The superscript T	Transpose for real matrices
I_N	The $N \times N$ identity matrix
$\mathbf{1}_p$	$p \times 1$ column vector of all ones
$\mathbf{0}_p$	$p \times 1$ column vector of all zeros
$\|x\|$	2-norm of a vector x
max	Maximum
min	Minimum
sup	Supremum, the least upper bound
inf	Infimum, the greatest lower bound
sin	sin function
cos	cos function
tanh	The tangent hyperbolic function
$P = P^T > 0$	Symmetric positive-definite matrix
$P = P^T \geq 0$	Symmetric non-negative matrix
$\otimes$	The Kronecker product
$\lambda_i(A)$	The eigenvalue set of A
$\lambda_{max}(A)$	The maximum eigenvalue of a real symmetric matrix A
$\lambda_{min}(A)$	The minimum eigenvalue of a real symmetric matrix A
$\|A\|$	The induced 2-norm of a real matrix A
rank(A)	The rank of a matrix A
$diag\{A_1,\cdots,A_n\}$	A block-diagonal matrix with matrices $A_i, i = 1,\cdots,n$
$\det(A)$	The determinant of a matrix A

$He\{M\}$	$He\{M\} \triangleq M + M^T$
$\text{sgn}(\cdot)$	The signum function
$0_{m\times n}$	$m \times n$ zero matrix
$\|x(t)\|_p, p \geq 1$	$\|x(t)\|_p \triangleq \sup\limits_{t\in[0,\infty)} (\|x_1(t)\|^p + \|x_2(t)\|^p + \cdots + \|x_n(t)\|^p)^{\frac{1}{p}}$
$\|x(t)\|_\infty$	$\|x(t)\|_\infty \triangleq \max\limits_{1\leq i\leq n} \|x_i(t)\|$for all$t \in [0,\infty)$
$\|X(t)\|_p$	$\|X(t)\|_p = \sup\limits_{t\in[0,\infty),x(t)\neq 0} \frac{\|X(t)x(t)\|_p}{\|x(t)\|_p}$
$\mathscr{V}_N = \{1,2,\cdots,N\}$	The integer set from 1 to N containing 1 and N
$\mathscr{V}_N \setminus \{N\}$	The relative complement of $\{N\}$ in $\mathscr{V}_N$
$\mathscr{V}_N \cup \{0\}$	The union of $\mathscr{V}_N$ and $\{0\}$
$\ln(x)$	The natural logarithm of x

2.2 Acronyms

AD	Analog-to-Digital
BD	Bidirectional
BDL	Bidirectional-Leader
CNN	Chebyshev Neural Network
CP	Chebyshev Polynomial
CS policy	Constant-Spacing policy
CTH	Constant Time Headway
DA	Digital-to-Analog
FLS	Fuzzy Logic System
FWL	Finite Word Length
HST	High Speed Train
ISM	Integral Sliding Mode
ISMC	Integral SMC
ITS	Intelligent Transportation System
IVHS	Intelligent Vehicle Highway System
LDCP	Linear Distributed Consensus Protocol
LMI	Linear Matrix Inequality
MCTH	Modified Constant Time Headway
NDCP	Nonlinear Distributed Consensus Protocol
NN	Neural Network
PI	Proportional and Integral
PID	Proportional-Integral-Derivative
PIDSMC	Proportional-Integral-Derivative SMC
PISMC	Proportional and Integral SMC
PLF	Predecessor-Leader Following
RBFNN	Radial Basis Function Neural Network
SBLF	Symmetric Barrier Lyapunov Function
SMC	Sliding Mode Control

TCTH	Traditional Constant Time Headway
TPF	Two-Predecessors Following
TPLF	Two-Predecessors-Leader Following
TSMC	Terminal SMC
UAV	Unmanned Aerial Vehicle
VS policy	Variable-Spacing policy

2.3 String Stability Theory

String stability implies that the spacing error between two neighboring vehicles does not amplify as it propagates downstream the platoon, i.e., to vehicles with higher indices [55, 93, 164, 165]. Consider N follower vehicles and a leader that are organised into a platoon running in a straight line. Let $x_i(t)$, $v_i(t)$ and $a_i(t)$ $(i = 0, 1, \cdots, N)$ denote respectively the position, velocity and acceleration of the ith vehicle in the platoon with $i = 0$ standing for the leader vehicle and the others being followers. It is also defined in the following forms with $e_i^x(t)$ representing the spacing errors.

Definition 2.1 *[93] (String Stability): Suppose $e_i^x(0) = 0$, a platoon is string stable if, given any $\varepsilon > 0$, there exists $\sigma > 0$ such that*

$$\|e_i^x(0)\|_\infty < \varepsilon \Rightarrow \sup_i \|e_i^x(\cdot)\|_\infty < \sigma \ .$$

Definition 2.2 *[46, 93] (Strong String Stability): Suppose $e_i^x(0) = 0$, a platoon is string stable in strong sense if the error propagation transfer function $G_i(s) := E_{i+1}^x(s)/E_i^x(s)$ satisfies $|G_i(s)| \leq 1$, where $E_i^x(s)$ is the Laplace transform of $e_i^x(t)$.*

2.4 Basic Algebraic Graph Theory

Suppose that a team of agents interacts with each other via communication or sensing network or a combination of both to achieve collective objectives. It is natural to model the interaction among agents by directed or undirected graphs. A *directed graph* $\mathscr{G}$ is a pair $(\mathscr{V}, \mathscr{E})$, where $\mathscr{V} = \{1, 2, \cdots, N\}$ is a non-empty finite node set and $\mathscr{E} \subset \mathscr{V} \times \mathscr{V}$ is an edge set of ordered pairs of nodes, called edges. A weighted graph associates a weight with every edge in the graph. The edge (i, j) in the edge set $\mathscr{E}$ denotes that agent v_j can obtain information from agent i, but not necessarily vice versa. For an edge (i, j), node i is called the parent node, j is the child node, and i is a neighbor of j. The set of neighbors of i is denoted by N_i, whose cardinality is called the in-degree of node i.

A graph is defined as being balanced when it has the same number of ingoing and outgoing edges for all the nodes and is said to be outgoing with respect to node i and incoming with respect to j. A graph with the property that $(i, j) \in \mathscr{E}$ implies

$(i,j) \in \mathscr{E}$ for any $(i,j) \in \mathscr{V}$ is said to be undirected, where the edge (i,j) denotes that agents i and j can obtain information from each other. Obviously, an *undirected graph* is a special *balanced graph*.

The *adjacency matrix* $\mathscr{A} \triangleq [a_i j] \in \mathbf{R}^{N\times N}$ of a *directed graph* $(\mathscr{V},\mathscr{E})$ is defined such that a_{ij} is a positive weight if $(i,j) \in \mathscr{E}$, and $a_{ij} = 0$ if $(i,j) \notin \mathscr{E}$. Moreover, we assume $a_{ii} = 0$ for all $i \in \mathscr{V}$. Self-loops in the form of (i,i) are excluded in this book. The *adjacency matrix* of an *undirected graph* is defined analogously except that $a_{ij} = a_{ji}$ for all $i \neq j$ because $(i,j) \notin \mathscr{E}$ implies $(i,j) \notin \mathscr{E}$. Noted that a_{ij} are selected to model the strength of the interaction between nodes. If the magnitude of the weight is not relevant, then a_{ij} is set equal to 1 if $(i,j) \in \mathscr{E}$.

The *Laplacian matrix* $\mathscr{L} = [l_{ij}]_{N\times N}$ is defined as $l_{ij} = -a_{ij}, i \neq j$, and $l_{ii} = \sum_{j=1}^{N} a_{ij}$. For an *undirected graph*, the *Laplacian matrix* $\mathscr{L}$ is symmetric, while for a *directed graph*, $\mathscr{L}$ is not necessarily symmetric and is sometimes called the non-symmetric *Laplacian matrix* [1] or directed *Laplacian matrix* [94].

Define one end of each edge in $\mathscr{E}$ by a positive sign, and the other one is a negative sign. Now, consider the kth edge in $\mathscr{E}$, with $k \in \{1,2,\cdots,M\}$, where i,j are the two nodes connected by the edge. Then, the $N \times M$ *incidence matrix* $\mathscr{D} = [d_{ik}]_{N\times M}$ associated with the graph $\mathscr{G}$ is defined as

$$d_{ik} = \begin{cases} +1 & \text{if } i\text{th node is the positive end of the } k\text{th link} \\ -1 & \text{if } i\text{th node is the negative end of the } k\text{th link} \\ 0 & \text{otherwise.} \end{cases}$$

It should be pointed out that the choice of orientation does not change the results because the graph is bidirectional [3].

It is easy to show that $\mathbf{1}_N^T \mathscr{D} = \mathbf{0}$, and $\mathscr{D}\mathscr{D}^T = \mathscr{L}$, where $\mathscr{L}$ is the *Laplacian matrix* of the graph $\mathscr{G} = (\mathscr{V},\mathscr{E})$ defined as follows

$$\mathscr{L} = [l_{ij}]_{N\times N},\ l_{ij} = \begin{cases} -1 & (i,j) \in \mathscr{E} \\ -\sum_{j=1, j\neq i}^{N} l_{ij} & i = j \\ 0 & \text{otherwise.} \end{cases}$$

To facilitate the consensus protocol design and stability analysis, several properties of the *Laplacian matrix* $\mathscr{L}$ are recalled in the following lemma.

Lemma 2.1 *[114, 149] (1) Zero is an eigenvalue of $\mathscr{L}$ with* $\mathbf{1}$ *a right eigenvector, and all non-zero eigenvalues are real and positive. Furthermore, zero is a simple eigenvalue of $\mathscr{L}$ if and only if $\mathscr{G}$ has a directed spanning tree. Then, the non-zero eigenvalues of $\mathscr{L}$ can be given as $0 < \lambda_2 < \lambda_3 < \cdots < \lambda_N$. (2) For an undirected graph $\mathscr{G}$, the smallest non-zero eigenvalue λ_2 of $\mathscr{L}$ satisfies $\lambda_2 = \min_{x\neq 0, \mathbf{1}^T x = 0} \frac{x^T \mathscr{L} x}{x^T x}$.*

Lemma 2.2 *[145, 200] Suppose that graph $\mathscr{G}$ is strongly connected. Then there exists a positive vector $\zeta = [\zeta_1,\cdots,\zeta_N]^T \in \mathbf{R}^N$, (i.e., $\zeta_i > 0, i = 1,\cdots,N$) such that $\zeta^T \mathscr{L} = \mathbf{0}_N^T$ and $\mathbf{1}_N^T \zeta = 1$. In addition, $\Xi\mathscr{L} + \mathscr{L}^T \Xi$ for $\Xi = diag\{\zeta_1,\cdots,\zeta_N\}$ is positive semidefinite, with zero being its simple eigenvalue.*

Definition 2.3 *[145, 200] For a strongly connected network $\mathcal{G}$ with Laplacian matrix $\mathcal{L}$, its generalized algebraic connectivity is defined by*

$$\alpha(\mathcal{L}) = \min_{x^T\zeta=0, x\neq 0} \left(\frac{x^T \hat{\mathcal{L}} x}{x^T \Xi x}\right)$$

where $\hat{\mathcal{L}} = \frac{1}{2}(\Xi\mathcal{L} + \mathcal{L}^T\Xi)$, and Ξ and ζ are as given in Lemma 2.2.

In order to compute $\alpha(\mathcal{L})$, Algorithm 2.1 can be obtained as follows, by using similar matrix manipulations as that of Lemma 8 in [200].

Algorithm 2.1 *The general algebraic connectivity of a strongly connected network can be computed as follows:*

$$\begin{array}{rl} \min & \varsigma \\ \textit{subject to} & \varsigma Q^T \hat{\mathcal{L}} Q - Q^T \Xi Q \geq 0 \end{array}$$

where $\alpha(\mathcal{L}) = \frac{1}{\varsigma}$, $Q = \begin{bmatrix} I_{N-1} \\ -\frac{\hat{\zeta}^T}{\zeta_N} \end{bmatrix} \in \mathbf{R}^{N\times(N-1)}$ *and* $\hat{\zeta} = [\zeta_1, \cdots, \zeta_{N-1}]^T$.

Remark 2.1 *The above minimization problem (instead of maximization problem as in [200]) can be easily solved by using the solver function mincx in the LMI Toolbox.*

2.5 H_∞ Performance Index

A popular performance measure of a stable linear time-invariant system is the H_∞ of its *transfer function*. It is defined as follows.

Definition 2.4 *[215] Consider a linear time-invariant continuous-time system*

$$\begin{aligned} \dot{x}(t) &= Ax(t) + B_1 w(t) \\ z(t) &= Cx(t) + D_1 w(t) \end{aligned} \tag{2.1}$$

where $x(t) \in \mathbf{R}^n$ is the state, $w(t) \in \mathbf{R}^s$ is an exogenous disturbance in $L_2[0,\infty]$, that is,

$$\|\omega(t)\|_2^2 = \int_0^\infty \omega^T(t)\omega(t)dt < \infty$$

and $z(t) \in \mathbf{R}^r$is the regulated output, respectively. A,B_1,C,D_1 are known constant matrices of appropriate dimensions.

Let $\gamma > 0$ be a given constant, then the system (2.1) is said to be with an H_∞ performance index no larger than γ, if the following conditions hold

(1) System (2.1) is asymptotically stable
(2) Subject to initial conditions $x(0) = 0$, the transfer function matrix $T_{wz}(s)$ satisfies,

$$\|T_{wz}(s)\|_\infty := \sup_{\|w\|_2 \leq 1} \frac{\|z\|_2}{\|w\|_2} \leq \gamma \tag{2.2}$$

which is equivalent to

$$\int_0^\infty z^T(t)z(t)dt \leq \gamma^2 \int_0^\infty w^T(t)w(t)dt, \quad \forall w(t) \in L_2[0,\infty) \tag{2.3}$$

It is easy to see that the inequality (2.3) describes the ability to restrain the effect of the disturbance on the system performance. Moreover, the smaller the value of γ is, the better the system performance is.

2.6 Some Other Definitions and Lemmas

Some other definitions and lemmas that will be used in the monograph are presented as follows.

Definition 2.5 *The Kronecker product of matrices $A \in \mathbf{R}^{m\times n}$ and $B \in \mathbf{R}^{p\times q}$ is defined as*

$$A \otimes B = \begin{bmatrix} a_{11}B & \cdots & a_{1n}B \\ \vdots & \ddots & \vdots \\ a_{m1}B & \cdots & a_{mn}B \end{bmatrix}$$

which has the following properties:

$$\begin{aligned}
k(A\otimes B) &= (kA)\otimes B = A\otimes(kB) \\
A\otimes(B+C) &= A\otimes B + A\otimes C \\
(A+B)\otimes C &= A\otimes C + B\otimes C \\
(A\otimes B)(C\otimes D) &= (AC\otimes BD) \\
(A\otimes B)^T &= A^T\otimes B^T \\
(A\otimes B)^{-1} &= A^{-1}\otimes B^{-1}, \textit{if } A, B \textit{ are invertible} \\
(A\otimes B)\otimes C &= A\otimes(B\otimes C)
\end{aligned}$$

where k is a constant scalar, and A, B, C and D are matrices with appropriate dimensions.

Lemma 2.3 *(Barbalat Lemma [90]) If $\psi(t) : \mathbf{R} \to \mathbf{R}$ is a uniformly continuous function for $t \geq 0$ and if the limit of the integral*

$$\lim_{t\to\infty} \int_0^t \psi(\tau)d\tau$$

exists and is finite, then

$$\lim_{t\to\infty} \psi(t) = 0.$$

Lemma 2.4 *(Finite-Time Stability [9]) Suppose there is a positive definite Lyapunov function $V(x,t)$ defined on $\mathbf{U}\times\mathbf{R}^+$ where $\mathbf{U}$ is the neighborhood of the origin, and there are positive real constants $c>0$ and $0<a<1$, such that $\dot{V}(x,t)+cV^a(x,t)\le 0$ on $\mathbf{U}$. Then, $V(x,t)$ is locally finite-time convergent with a settling time $T=\frac{V^{1-a}(x_0,t_0)}{c(1-a)}$ for any given initial condition $x(t_0)\in\mathbf{U}$.*

Lemma 2.5 *[144] For $\forall\varepsilon>0$ and $\xi\in\mathbf{R}$, the following inequality holds*

$$0\le|\xi|-\xi\tanh(\tfrac{\xi}{\varepsilon})\le\kappa\varepsilon$$

where κ is a constant that satisfies $\kappa=e^{-(\kappa+1)}$, that is, $\kappa=0.2785$.

Lemma 2.6 *[97] Let function $V(t)\ge 0$ be a continuous function defined $\forall t\ge 0$ and $\dot{V}(t)\le-\gamma V(t)+\varepsilon$, where $\gamma>0$ and ε are constants, then,*

$$V(t) \;\le\; (V(0)-\tfrac{\varepsilon}{\gamma})e^{-\gamma t}+\tfrac{\varepsilon}{\gamma}.$$

Lemma 2.7 *(Young's inequality) [70, 78] For $\forall(x,y)\in\mathbf{R}^2$, the following inequality holds*

$$xy\le\tfrac{\varepsilon^p}{p}|x|^p+\tfrac{1}{q\varepsilon^q}|y|^q$$

where $\varepsilon>0$, $p>1$, $q>1$, and $(p-1)(q-1)=1$.

Lemma 2.8 *[137] For $\forall a,b\in\mathbf{R}^n$ and any symmetric positive-definite matrix $\Phi\in\mathbf{R}^{n\times n}$, $2a^Tb\le a^T\Phi^{-1}a+b^T\Phi b$.*

Lemma 2.9 *[12, 141]: Given matrices Y, M and N with the appropriate dimensions, the following statements are equivalent:*

(i)

$$Y+M\Delta N+N^T\Delta^TM^T<0$$

holds for all Δ satisfying $\Delta^T\Delta\le\theta I$.

(ii)

$$Y+M\Delta N+N^T\Delta^TM^T+\theta I<0$$

holds for all Δ satisfying $\Delta^T\Delta\le\theta I$ and some $\theta>0$.

(iii) there exists a constant $\varepsilon>0$ such that

$$Y+\varepsilon MM^T+\frac{\theta}{\varepsilon}N^TN<0$$

The following lemma will be useful in the derivation of our main result.

Lemma 2.10 *Let E, F and $\tilde{\Re}$ be real matrices of appropriate dimensions with $\tilde{\Re} = diag\{\delta_1, \cdots, \delta_r\}$, $|\delta_l| \leq \bar{\delta}, l = 1 \cdots r$. Then, for any real matrix $\Lambda = diag\{\rho_1, \cdots \rho_r\} > 0$, the following inequality holds:*

$$F\tilde{\Re}E + E^T\tilde{\Re}^T F^T \leq F\Lambda F^T + \bar{\delta}^2 E^T \Lambda^{-1} E. \tag{2.4}$$

Proof 2.1 *Let $\Psi = \Lambda^{\frac{1}{2}} F^T - \Lambda^{-\frac{1}{2}} \tilde{\Re} E$, then*

$$\Psi^T\Psi = F\Lambda F^T - F\tilde{\Re}E - E^T\tilde{\Re}^T F^T + E^T\tilde{\Re}^T\Lambda^{-1}\tilde{\Re}E$$

where

$$\begin{aligned} E^T\tilde{\Re}^T\Lambda^{-1}\tilde{\Re}E &= E^T diag\{\rho_1^{-1}\delta_1^2, \rho_2^{-1}\delta_2^2, \cdots, \rho_n^{-1}\delta_r^2\}E \\ &\leq \bar{\delta}^2 E^T \Lambda^{-1} E \end{aligned}$$

by using the fact $|\delta_l| \leq \bar{\delta}, l = 1 \cdots r$.

Then, from the fact $\Psi^T\Psi \geq 0$, we can obtain (2.4). Therefore, the proof of this lemma is complete. □

Remark 2.2 *In [190, 191], the authors pointed out that the type of interval-bounded variations has a numerical problem because the number of the LMIs involved in the design conditions grows exponentially with the number of uncertain parameters. Although many effective methods have been proposed to deal with the numerical problem caused by interval-bounded uncertainty such as structured vertex separator method [190], randomized algorithm [29], the sparse structure method [13] and transfer function sensitivity approach [50]. The number of LMI constraints involved in the design conditions by using the first three methods is still large, while the fourth method involves the closed-loop system's transfer function and cannot be applied to multi-agent systems with quantization information. The method proposed in Lemma 2.10 has significantly lower computational complexity compared to these four existing methods.*

3

String Stability of Vehicle Platoons with External Disturbances

The task of this chapter is to introduce a distributed *finite time* adaptive integral *sliding mode control* (ISMC) approach for a platoon of vehicles consisting of a leader and multiple followers subjected to bounded unknown disturbances. No knowledge is assumed on the bounds of the disturbances in advance except for the existence of such bounds. A *modified constant time headway* (MCTH) policy is introduced to remove the assumption of zero initial spacing and velocity errors. It is shown that the proposed control protocols can ensure *strong string stability* of the whole *vehicle platoon*. In addition, an effective method is also proposed to reduce the *chattering phenomenon* caused by the indicator function.

3.1 Introduction

The field of *string stability* research has been active for decades due to its broad applications in various areas, such as autonomous formation flying [7, 155, 151], *intelligent vehicle highway systems* [43, 45, 46, 186], etc. In these applications, each vehicle in the platoon should be controlled to maintain a *desired safe spacing* from its leading vehicle and *avoid collision* [27, 142, 143, 164]. As shown in Chapter 1, the CS and VS policies are generally applied to maintain the desired spacing. It is known that the *CTH policy* as the simplest and most common *VS policy* [2, 164] does not require a lot of data from other vehicles, and it can ensure *string stability* using on-board information only and thus improving *string stability* [164]. In addition, since using variable spacing, this policy is more tolerant to the *external disturbances* than the CS policy [92, 154]. Therefore, we are interested in investigating the *string stability* with the *CTH policy* instead of the *CS policy* in this chapter.

On the other hand, during the past several years, many interesting and important approaches have been developed to guarantee *string stability*, such as *linear matrix inequality* (LMI) approach [45, 46], backstepping approach [138], adaptive sliding mode approach [93, 116], and distributed receding horizon control approach [34], etc. Because the sliding mode control method has the robustness and insensitivity to withstand certain types of *external disturbances* and model uncertainties [84], it has been implemented successfully by many studies to solve the problems of *string stability* [93, 116, 164]. It is worth mentioning that, recently, a new result based on

CS policy and adaptive coupled sliding mode control method appeared in [93]. The drawback of the approach in [93], however, is that only a special case is considered, i.e., ignoring the effect of initial spacing and velocity errors to the vehicle platoon system. Consequently, the practical relevance of this approach is limited, since *non-zero initial* spacing and velocity errors are of utmost importance in practice due to the fact that *non-zero initial* spacing and velocity errors may cause large transient engine and brake torques [164]. One may therefore legitimately ask, what happens if the initial spacing and velocity errors are uncertain or non-zero, and how to design a controller to eliminate the impact of this type of uncertainty? In addition, as the vehicle platoon is an interconnected coupled system, disturbances acting on one vehicle may affect other vehicles and even amplify spacing errors along the string, namely *string instability* [120, 124]. Thus, it is interesting and important to synthesize control laws to provide attenuation on both exogenous disturbances and *non-zero initial* spacing and velocity errors.

Motivated by the aforementioned observations, this chapter proposes a distributed adaptive *integral sliding mode control* (ISMC) strategy for vehicle platoon systems to guarantee the *finite time stability* of each vehicle and the *strong string stability* of the whole vehicle platoon systems using the ISMC method developed in [84]. Compared to the asymptotic stability control approach, the *finite time stability* control method has faster convergence rate and better disturbance rejection to systems uncertainty and *external disturbances* [102, 193]. The main contributions of this chapter are summarized as follows.

(a) An adaptive ISMC strategy based on *MCTH policy* is proposed to guarantee the *finite time stability* of each vehicle and the *strong string stability* of the whole vehicle platoon. It is worth mentioning that, in contrast to requiring zero initial spacing and zero initial velocity errors simultaneously in [38, 43, 45, 46, 93, 169, 212] where *CS policy* is used, initial spacing and velocity errors are not required to be zero in this chapter.

(b) A *MCTH policy* is proposed to solve the case of *non-zero initial* spacing and velocity errors. The main idea of this policy is to transform a *non-zero initial* spacing and velocity error problem to a zero initial spacing and velocity error problem.

(c) Unlike the control strategies based on *CS policy* proposed in [38, 43, 45, 46, 93, 169, 212], the preceding vehicle's acceleration information is not required here and thus reduces the communication load.

(d) The upper and lower bounds of the disturbance are adaptively updated online. In addition, the drawback of *chattering phenomenon* in the *sliding mode control* method is well known, thus some effective methods are proposed by introducing continuous approximations to reduce the *chattering phenomenon* in Remark 3.5.

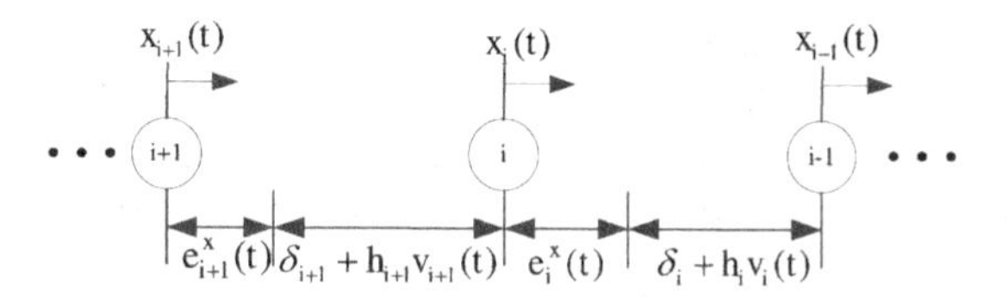

FIGURE 3.1
Configuration of a vehicle platoon system based on *CTH policy*.

3.2 Model Description and Problem Formulation

3.2.1 Vehicle Dynamics

Consider a string of $N+1$ vehicles with vehicles following one another along a straight line, with the leader dynamics being described as:

$$\dot{x}_0(t) = v_0(t) \tag{3.1}$$

where $x_0(t)$ and $v_0(t)$ are the position and the velocity of the leader vehicle, respectively. The velocity $v_0(t)$ is a given function of time, that is, the leader vehicle is not adjusting its trajectory based on feedback from the platoon. Let $x_i(t)$ and $v_i(t), i \in \mathscr{V}_N$ denote respectively the position and velocity of the ith follower vehicle in the platoon in an inertial frame. Then, the behavior of the ith follower vehicle is mathematically described by a simple linear model without considering parasitic time delays and lags, as follows

$$\begin{aligned} \dot{x}_i(t) &= v_i(t) \\ \dot{v}_i(t) &= \frac{1}{m_i}u_i(t) + w_i(t), \text{ for } i \in \mathscr{V}_N \end{aligned} \tag{3.2}$$

where $u_i(t)$ denotes the control input, m_i is the mass of the ith vehicle assumed to be constant, and $w_i(t)$ denotes the *external disturbance* of the vehicle. Since small disturbances acting on one vehicle may cause large spacing errors between vehicles down the platoon [120], the derivation of a controller to attenuate the effect of the disturbances along the vehicle string is an important problem. Compared with existing work in [27, 186] where the *external disturbances* of all followers are assumed to be zero, the evolution of the followers in the present model is subject to some bounded *external disturbances* $w_i(t)$, i.e.,

$$\underline{w}_i \leq w_i(t) \leq \overline{w}_i \tag{3.3}$$

where $\underline{w}_i$ and $\overline{w}_i$ are unknown constants. This disturbance $w_i(t)$ can be used to model atmospheric disturbances, etc. Thus, this model is more general than the one studied in [27, 186]. It is worth mentioning that $w_i(t)$ can be seen as acceleration disturbances. Here, we use $a_i(t)$ to denote the acceleration of the ith follower.

Figure 3.1 describes a vehicle platoon system. The safety spacing between the ith and the $(i-1)$th vehicles is defined as $h_i v_i(t)$ with an additional standstill spacing

δ_i. That is, as the velocity of the ith vehicle increases, the safety spacing between the ith and $(i-1)$th vehicles should also increase to facilitate collision avoidance [164]. The value of h_i also plays an important role in deciding the time at which the brake has to be applied. From the configuration, the spacing error $e_i(t)$ can be written as:

$$e_i(t) = d_i(t) - \delta_i - h_i v_i(t), \text{ for } i \in \mathcal{V}_N \tag{3.4}$$

where

$$d_i(t) = x_{i-1}(t) - x_i(t),$$

which denotes the actual distance between two consecutive vehicles, h_i is the time headway and δ_i is the standstill spacing. In addition, for convenience, we denote the velocity errors by

$$e_i^v(t) = v_{i-1}(t) - v_i(t), \text{ for } i \in \mathcal{V}_N.$$

Remark 3.1 *For $h_i > 0$, the spacing error policy (3.4) is called a CTH policy [27, 142, 143], while for $h_i = 0$, the spacing error policy (3.4) reduces to a CS policy, which has been extensively investigated in [38, 43, 45, 46, 93, 169, 212]. Compared with the CS policy ($h_i = 0$), the CTH policy is known to improve string stability and safety [143].*

3.2.2 Control Objective

Design a distributed adaptive ISMC law for a vehicle platoon to maintain a desired neighboring-vehicle spacing and meet the following requirements:

1. While both upper and lower bounds of the *external disturbances* $w_i(t)$ are unknown, the adaptive law is proposed to estimate the upper and lower bounds of disturbances online.

2. Individual vehicle stability: *Finite time stability* of each vehicle system is guaranteed.

3. *Strong string stability* defined in Lemma 2.2 of Chapter 2 can be guaranteed.

4. The final control objective is to achieve

$$d_i(t) \rightarrow \delta_i + h_i v_i(t), v_i(t) \rightarrow v_0(t) \text{ for for } i \in \mathcal{V}_N$$

3.3 Design of Distributed Adaptive Integral Sliding Mode Control

3.3.1 Zero Initial Spacing Error Case

In this subsection, based on *CTH policy*, an adaptive integral sliding mode (ISM) control law is proposed for the case of zero initial spacing errors but not necessarily requiring zero initial velocity errors. In contrast, zero initial spacing and zero initial velocity errors are required simultaneously to achieve string stability under the *CS policy* in [38, 43, 45, 46, 93, 169, 212]. As mentioned earlier, the main objective here is to construct adaptive sliding mode controllers $u_i(t)$ for vehicle systems (3.1)-(3.2) to make $e_i(t)$ converge to zero in a finite time and to guarantee *strong string stability*. Then, to develop an ISM controller, the integral sliding surface is defined by

$$s_i(t) = e_i(t) + \int_0^t \lambda e_i(\tau) d\tau \tag{3.5}$$

where λ is a positive design parameter. By using the ISM approach, the preceding vehicle's acceleration information will not be needed. However, it should be pointed out that the convergence of the sliding surface $s_i(t)$ can make $e_i(t)$ zero, but cannot guarantee *strong string stability*. Therefore, we define a new sliding surface as follows

$$S_i(t) = \begin{cases} qs_i(t) - s_{i+1}(t), \text{ for } i \in \mathcal{V}_N / \{N\} \\ qs_i(t), \; i = N \end{cases} \tag{3.6}$$

where $q \neq 0$ is a constant parameter, which is different from the one in [93] where q must satisfy $q > 0$. Then, we can obtain the following relationship between $S_i(t)$ and $s_i(t)$:

$$S(t) = Qs(t) \tag{3.7}$$

where

$$\begin{aligned} Q &= \begin{bmatrix} q & -1 & \cdots & 0 & 0 \\ 0 & q & -1 & \cdots & 0 \\ & & \ddots & & \\ 0 & 0 & \cdots & q & -1 \\ 0 & 0 & \cdots & 0 & q \end{bmatrix} \\ s(t) &= \begin{bmatrix} s_1(t) & s_2(t) & \cdots & s_N(t) \end{bmatrix}^T \\ S(t) &= \begin{bmatrix} S_1(t) & S_2(t) & \cdots & S_N(t) \end{bmatrix}^T. \end{aligned}$$

Since $q \neq 0$, we can obtain the inverse of Q as

$$Q^{-1} = \begin{bmatrix} \frac{1}{q} & \frac{1}{q^2} & \cdots & \frac{1}{q^{N-1}} & \frac{1}{q^N} \\ 0 & \frac{1}{q} & \frac{1}{q^2} & \cdots & \frac{1}{q^{N-1}} \\ & & \ddots & & \\ 0 & 0 & \cdots & \frac{1}{q} & \frac{1}{q^2} \\ 0 & 0 & \cdots & 0 & \frac{1}{q} \end{bmatrix}.$$

Consequently, it follows the equivalence of $S_i(t)$ and $s_i(t)$ for any fixed platoon size N.

Remark 3.2 *It should be pointed out that the transformation (3.7) is of course true for any fixed platoon size N. However, the element $\frac{1}{q^N}$ in Q^{-1} approaches 0 as N grows towards infinity. In particular choosing a smaller value of q even increases the rate with which $\frac{1}{q^N}$ approaches 0. Fortunately, the number of vehicles in every platoon will be finite in practice. For a long platoon, a relatively large value of q (the extreme case $N = \infty$, let $q = 1$) is required to guarantee the validity mapping between $s_i(t)$ and $S_i(t)$.*

Taking the time derivative of $S_i(t)$ along the definitions of (3.4), (3.5) and (3.6) yields

$$\begin{aligned} \dot{S}_i(t) &= q\dot{s}_i(t) - \dot{s}_{i+1}(t) \\ &= q(\dot{e}_i(t) + \lambda e_i(t)) - (\dot{e}_{i+1}(t) + \lambda e_{i+1}(t)) \\ &= -\tfrac{qh_i}{m_i}(u_i(t) + m_i w_i(t)) + q(v_{i-1}(t) - v_i(t)) + A_i(t) \end{aligned} \tag{3.8}$$

where

$$A_i(t) = q\lambda e_i(t) - \dot{e}_{i+1}(t) - \lambda e_{i+1}(t), \text{ for } i \in \mathcal{V}_N / \{N\}.$$

When $i = N$, using the definition of (3.6), one can obtain $S_N(t) = qs_N(t)$, therefore, one can further obtain that

$$\begin{aligned} \dot{S}_N(t) &= q\dot{s}_N(t) \\ &= q(\dot{e}_N(t) + \lambda e_N(t)) \\ &= -\tfrac{qh_N}{M_N}(u_N(t) + M_N w_N(t)) \\ &\quad + q(v_{N-1}(t) - v_N(t)) + A_N(t) \end{aligned} \tag{3.9}$$

where

$$A_N(t) = q\lambda e_N(t).$$

To facilitate the development of the main result, we denote

$$\begin{aligned} \tilde{\overline{w}}_i(t) &= \hat{\overline{w}}_i(t) - \overline{w}_i, \ \tilde{\underline{w}}_i(t) = \hat{\underline{w}}_i(t) - \underline{w}_i \\ |\tilde{\overline{w}}_i(t)| &\le \tilde{\overline{w}}_{maxi}, \ |\tilde{\underline{w}}_i(t)| \le \tilde{\underline{w}}_{maxi} \end{aligned} \tag{3.10}$$

where $\hat{\overline{w}}_i(t)$ and $\hat{\underline{w}}_i(t)$ are the estimates of the upper bound $\overline{w}_i$ and the lower bound $\underline{w}_i$ of the disturbance $w_i(t)$ in (3.3), respectively.

Then, the following theorem, which guarantees the *finite time stability* of each vehicle and the *strong string stability* of the whole vehicle platoon systems, can be obtained.

Theorem 3.1 *Consider the vehicle platoon system (3.2) with leader (3.1) and CTH policy (3.4) under the assumption of zero initial spacing errors. For a sufficiently large positive constant V_{max}, if the initial condition satisfies*

$$\|\tilde{S}(0)\|^2 \le \tfrac{2V_{max}}{\vartheta} \tag{3.11}$$

where

$$\begin{aligned}
\tilde{S}(t) &= \begin{bmatrix} S^T(t) & \tilde{\bar{w}}^T(t) & \tilde{\underline{w}}^T(t) \end{bmatrix}^T \\
S(t) &= \begin{bmatrix} S_1(t) & \cdots & S_N(t) \end{bmatrix}^T \\
\tilde{\bar{w}}(t) &= \begin{bmatrix} \sqrt{1-\mu_1}\tilde{\bar{w}}_1(t) & \cdots & \sqrt{1-\mu_N}\tilde{\bar{w}}_N(t) \end{bmatrix}^T \\
\tilde{\underline{w}}(t) &= \begin{bmatrix} \sqrt{\mu_1}\tilde{\underline{w}}_1(t) & \cdots & \sqrt{\mu_N}\tilde{\underline{w}}_N(t) \end{bmatrix}^T \\
\vartheta &= max\{1, \min_{1\le i\le N}\{\eta_i\}\}
\end{aligned}$$

with η_i being a positive constant, and μ_i being denoted below, then the finite time stability of individual vehicle can be achieved by using the following adaptive ISM control law

$$\begin{aligned}
u_i(t) &= \tfrac{m_i}{h_i}(v_{i-1}(t)-v_i(t)) - m_i[(1-\mu_i)\hat{\bar{w}}_i(t) \\
&\quad +\mu_i\hat{\underline{w}}_i(t)] + \tfrac{m_i}{qh_i}A_i(t) + \tfrac{km_i}{qh_i}sgn(S_i(t))
\end{aligned} \tag{3.12}$$

where q is a real non-zero constant, and

$$A_i(t) = \begin{cases} q\lambda e_i(t) - \dot{e}_{i+1}(t) - \lambda e_{i+1}(t), \; for\; i \in \mathcal{V}_N/\{N\} \\ q\lambda e_i(t), \; for\; i = N \end{cases} \tag{3.13}$$

while λ and k are positive constants with k satisfying

$$k > |q| \max_{1\le i\le N}\{h_i\tilde{\bar{w}}_{maxi}, h_i\tilde{\underline{w}}_{maxi}\}. \tag{3.14}$$

The adaptive laws for $\hat{\bar{w}}$ and $\hat{\underline{w}}$ are given by

$$\dot{\hat{\bar{w}}}_i(t) = -\eta_i q h_i S_i(t), \; \dot{\hat{\underline{w}}}_i(t) = -\eta_i q h_i S_i(t) \tag{3.15}$$

and μ_i is defined as

$$\begin{cases} q>0, \; \mu_i = \begin{cases} 1, S_i(t) > 0 \\ 0, S_i(t) \le 0 \end{cases} \\ q<0, \; \mu_i = \begin{cases} 1, S_i(t) \le 0 \\ 0, S_i(t) > 0. \end{cases} \end{cases} \tag{3.16}$$

The parameters $\eta_i > 0, i = 1,2,\cdots,N$ are the adaptive law gains to be designed. Moreover, the strong string stability of the whole vehicle platoon system can be guaranteed (i.e., $\|G_i(s)\| = \|\frac{E_{i+1}(s)}{E_i(s)}\| \le 1$) after some finite time if $0 < |q| \le 1$ holds.

Proof 3.1 *The proof is composed of three steps: vehicle stability, reachability of the sliding surface in a finite time and strong string stability. First of all, the Lyapunov stability method is introduced to show the stability of the vehicle system. Consider a positive definite Lyapunov function candidate in the following form:*

$$V(t) = \tfrac{1}{2}\sum_{i=1}^{N}[S_i^2(t) + \tfrac{1-\mu_i}{\eta_i}\tilde{\bar{w}}_i^2(t) + \tfrac{\mu_i}{\eta_i}\tilde{\underline{w}}_i^2(t)] \tag{3.17}$$

where $\tilde{\overline{w}}_i(t)$ and $\tilde{\underline{w}}_i(t)$ are defined as (3.10). Since $\overline{w}_i$ and $\underline{w}_i$ are unknown constants, it follows that $\dot{\tilde{\overline{w}}}_i(t) = \dot{\hat{\overline{w}}}_i(t), \dot{\tilde{\underline{w}}}_i(t) = \dot{\hat{\underline{w}}}_i(t)$.

Taking the time derivative of (3.17), we have

$$\dot{V}(t) = \sum_{i=1}^{N} [S_i(t)\dot{S}_i(t) + \frac{1-\mu_i}{\eta_i}\tilde{\overline{w}}_i(t)\dot{\hat{\overline{w}}}_i(t) + \frac{\mu_i}{\eta_i}\tilde{\underline{w}}_i(t)\dot{\hat{\underline{w}}}_i(t)].$$

Substituting (3.8) and (3.9) into the above equation, we have

$$\begin{aligned}\dot{V}(t) = & \sum_{i=1}^{N} \{S_i(t)[-\frac{qh_i}{m_i}(u_i(t) + m_i w_i(t)) \\ & +q(v_{i-1}(t) - v_i(t)) + A_i(t)] \\ & +\frac{1-\mu_i}{\eta_i}\tilde{\overline{w}}_i(t)\dot{\hat{\overline{w}}}_i(t) + \frac{\mu_i}{\eta_i}\tilde{\underline{w}}_i(t)\dot{\hat{\underline{w}}}_i(t)\}\end{aligned} \tag{3.18}$$

where $A_i(t)$ is defined as (3.13).

Using (3.16), one has

$$-qh_i w_i(t)S_i(t) \leq -qh_i[(1-\mu_i)\overline{w}_i + \mu_i \underline{w}_i]S_i(t). \tag{3.19}$$

Introducing (3.19) and the control law (3.12) into (3.18), one gets

$$\begin{aligned}\dot{V}(t) = & \sum_{i=1}^{N} \{S_i(t)[-q(v_{i-1}(t) - v_i(t)) + qh_i((1-\mu_i)\hat{\overline{w}}_i(t) \\ & +\mu_i\hat{\underline{w}}_i(t)) - A_i(t) - ksgn(S_i(t)) - qh_i w_i(t) \\ & +q(v_{i-1}(t) - v_i(t)) + A_i(t)] + \frac{1-\mu_i}{\eta_i}\tilde{\overline{w}}_i(t)\dot{\hat{\overline{w}}}_i(t) \\ & +\frac{\mu_i}{\eta_i}\tilde{\underline{w}}_i(t)\dot{\hat{\underline{w}}}_i(t)\} \\ = & \sum_{i=1}^{N} \{qh_i[(1-\mu_i)\hat{\overline{w}}_i(t) + \mu_i\hat{\underline{w}}_i(t)]S_i(t) - k|S_i(t)| \\ & +\frac{1-\mu_i}{\eta_i}\tilde{\overline{w}}_i(t)\dot{\hat{\overline{w}}}_i(t) + \frac{\mu_i}{\eta_i}\tilde{\underline{w}}_i(t)\dot{\hat{\underline{w}}}_i(t) - qh_i w_i(t)S_i(t)\} \\ \leq & \sum_{i=1}^{N} \{qh_i[(1-\mu_i)\tilde{\overline{w}}_i(t) + \mu_i\tilde{\underline{w}}_i(t)]S_i(t) \\ & +\frac{1-\mu_i}{\eta_i}\tilde{\overline{w}}_i(t)\dot{\hat{\overline{w}}}_i(t) + \frac{\mu_i}{\eta_i}\tilde{\underline{w}}_i(t)\dot{\hat{\underline{w}}}_i(t) - k|S_i(t)|\}\end{aligned}$$

With the adaptation laws in (3.15), the above equation can be rewritten as

$$\dot{V}(t) \leq -k\sum_{i=1}^{N}|S_i(t)| = -\psi(t) \leq 0 \tag{3.20}$$

where $\psi(t) = k\sum_{i=1}^{N}(|S_i(t)|) \geq 0$. Integrating (3.20) from zero to t yields $V(0) - V(t) \geq \int_0^t \psi(\tau)d\tau$. Using the fact $\dot{V}(t) \leq 0$ in (3.20), $V(0) - V(t) \geq 0$ is positive and bounded if $V(0)$ is bounded as in (3.11), hence $\tilde{\overline{w}}_i(t)$ and $\tilde{\underline{w}}_i(t)$ are also bounded as in (3.10). In addition, it further implies that $\lim_{t\to\infty}\int_0^t \psi(\tau)d\tau$ exists and is bounded. Then, according to the Lemma 2.3 in Chapter 2, it can be concluded that

$$\lim_{t\to\infty}\psi(t) = \lim_{t\to\infty} k\sum_{i=1}^{N}(|S_i(t)|) = 0. \tag{3.21}$$

Since k is positive, (3.21) implies $\lim_{t\to\infty} S_i(t) = 0$. Then, according to (3.5) and (3.7), $s_i(t)$ and the spacing error $e_i(t)$ converges to zero.

Now consider the candidate Lyapunov function $\bar{V}(t) = \frac{1}{2}\sum_{i=1}^{N} S_i^2(t)$, whose time derivative becomes

$$\dot{\bar{V}}(t) = \sum_{i=1}^{N}\{qh_i[(1-\mu_i)\tilde{\overline{w}}_i(t) + \mu_i\tilde{\underline{w}}_i(t)]S_i(t) - k|S_i(t)|\}. \qquad (3.22)$$

If (3.14) holds, let $k - |q|\max_{1\le i\le N}\{h_i\tilde{\overline{w}}_{maxi}, h_i\tilde{\underline{w}}_{maxi}\} \ge \bar{k} > 0$, where $\tilde{\overline{w}}_{maxi}$ and $\tilde{\underline{w}}_{maxi}$ are defined as (3.10), then, (3.22) can be rewritten as $\dot{\bar{V}}(t) \le -\bar{k}\sum_{i=1}^{N}|S_i(t)| \le -\bar{k}\sqrt{2\bar{V}(t)}$, which is a standard reachability condition, and is sufficient to guarantee that a sliding motion is maintained for all time [71]. After some manipulation, we get that $\sqrt{2\bar{V}(t)} \le \sqrt{2\bar{V}(0)} - \bar{k}t$. Therefore, by Lemma 2.4 in Chapter 2, we have $\bar{V}(t) \equiv 0$ and equivalently $\|S_i(t)\| \equiv 0$ $(i = 1, 2, \cdots, N)$ when $T_1 \ge \frac{\sqrt{2\bar{V}(0)}}{\bar{k}}$. Thus, we can conclude that $S_i(t)$ converges to zero in a finite time.

In the sequel, we will prove the strong string stability of the whole vehicle platoon system. Since $S_i(t) = qs_i(t) - s_{i+1}$ converges to zero in a finite time, we can get the relationship

$$q(e_i(t) + \textstyle\int_0^t \lambda e_i(\tau)d\tau) = e_{i+1}(t) + \textstyle\int_0^t \lambda e_{i+1}(\tau)d\tau. \qquad (3.23)$$

Since $e_i(0) = 0$ and $\int_{-\infty}^{0} e_i(t)dt = 0$, taking Laplace transform of (3.23), we can get

$$q(E_i(s) + \tfrac{\lambda}{s}E_i(s)) = E_{i+1}(s) + \tfrac{\lambda}{s}E_{i+1}(s).$$

Then, $G_i(s) = \frac{E_{i+1}(s)}{E_i(s)} = q$. Since the strong string stability requires that the transient spacing errors are not enlarging from one vehicle to another vehicle downstream along the platoon, q should satisfy $0 < |q| \le 1$ to guarantee $\|G_i(s)\| \le 1$. Then, within the presented controller framework, strong string stability can be guaranteed. This completes the proof. □

Remark 3.3 *Recalling (3.15), notice that the estimates $\hat{\overline{w}}_i(t)$ and $\hat{\underline{w}}_i(t)$ may not necessarily converge to the true values $\overline{w}_i$ and $\underline{w}_i$, respectively, since the estimates $\hat{\overline{w}}_i(t)$ and $\hat{\underline{w}}_i(t)$ will stop converging when $S_i(t)$ reaches the surface $S_i(t) = 0$ in a finite time.*

Remark 3.4 *To overcome the effect caused by external disturbances, control law $u_i(t)$ in (3.12) is developed. Moreover, different from the protocols in [91, 93], the estimations of the upper and lower bounds of external disturbances in the protocol (3.12) are adaptively updated via (3.15).*

Remark 3.5 *It is known that the signum function $sgn(S_i(t))$ may trigger chattering in practical implementation. To avoid chattering, a sigmoid-like function $\frac{S_i(t)}{\|S_i(t)\|+\sigma}$ [75, 211] can be used, where σ is a small positive constant. However, a large/small σ*

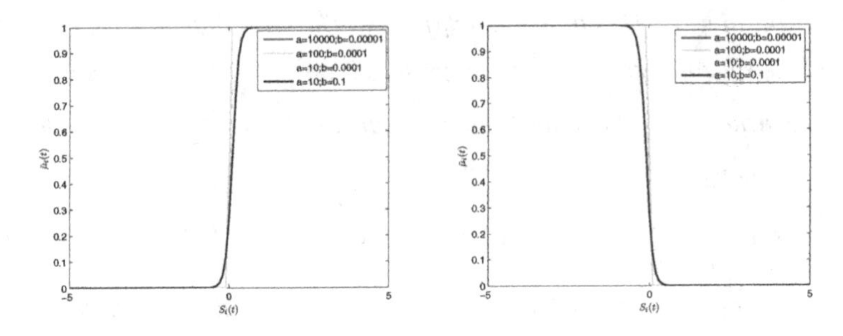

FIGURE 3.2
Graph of $\bar{\mu}_i(t)$ for different values of a and b (Left: $q > 0$; Right: $q < 0$).

can more/less effectively alleviate the chattering phenomenon, but leads to less/more accurate control results. Therefore, it would involve a trade-off between the value of σ and the control accuracy according the practical applications. In addition, it must be noted, however, that μ_i in (3.16) is an indicator function, which is also a chattering function. To solve the chattering problem, a smooth continuous function is introduced as follows

$$\bar{\mu}_i(t) = \begin{cases} \frac{1}{1+e^{-a(S_i(t)-b)}}, & if\, q > 0 \\ \frac{1}{1+e^{a(S_i(t)+b)}}, & if\, q < 0 \end{cases} \tag{3.24}$$

where $a \gg 1$ and $0 < b \ll 1$. Figure 3.2 is a graph of $\bar{\mu}_i(t)$ for different values of a and b. Observe that from $\bar{\mu}_i(t)$ by choosing bigger value for a and smaller value for b, (3.16) is most likely to be satisfied. Therefore, $\bar{\mu}_i(t)$ can be seen as a smooth approximation of μ_i by choosing bigger value for a and smaller value for b. Then, $u_i(t)$ can be rewritten as

$$\begin{aligned} u_i(t) &= \tfrac{m_i}{h_i}(v_{i-1}(t) - v_i(t)) - m_i[(1-\bar{\mu}_i(t))\hat{\overline{w}}_i(t) \\ &\quad + \bar{\mu}_i(t)\hat{\underline{w}}_i(t)] + \tfrac{m_i}{qh_i}A_i(t) + \tfrac{km_i}{qh_i}\tfrac{S_i(t)}{\|S_i(t)\|+\sigma} \end{aligned} \tag{3.25}$$

where $\bar{\mu}_i(t)$ is as (3.24) and $A_i(t)$ is as defined in (3.13).

In order to show the contrasts between the existing *CS policy* in [38, 43, 45, 46, 93, 169, 212] and the *CTH policy* in (3.4), similar to [93], a differential sliding surface which appeared different from (3.5) is defined as

$$\begin{aligned} s_i(t) &= \dot{e}_i(t) + \lambda e_i(t) \\ e_i(t) &= x_{i-1}(t) - x_i(t) - \delta_i \end{aligned} \tag{3.26}$$

where λ is a positive design parameter. Then, the following corollary can be obtained by following similar steps as in Theorem 1 under the condition $q > 0$.

Corollary 3.1 *Consider the vehicle platoon system (3.2) with leader (3.1) and CS policy (3.26) under the assumption of zero initial spacing and zero initial velocity* *errors. For a sufficiently large positive constant V_{max}, if the initial condition satisfies (3.11), the stability of individual vehicle can be achieved by using the following adaptive ISM control input*

$$\begin{aligned} u_i(t) &= \tfrac{m_i}{q+1}A_i(t) - m_i[(1-\bar{\mu}_i(t))\hat{\bar{w}}_i(t) + \bar{\mu}_i(t)\hat{\underline{w}}_i(t)] \\ &\quad + \tfrac{km_i}{q+1}\tfrac{S_i(t)}{\|S_i(t)\|+\sigma}, \, for\ i = 1,2,\cdots,N-1; \\ u_N(t) &= -m_N[(1-\bar{\mu}_N(t))\hat{\bar{w}}_N(t) + \bar{\mu}_N(t)\hat{\underline{w}}_N(t)] \\ &\quad + \tfrac{m_N}{q}A_N(t) + \tfrac{km_N}{q}\tfrac{S_N(t)}{\|S_N(t)\|+\sigma} \end{aligned} \tag{3.27}$$

with

$$\begin{aligned} A_i(t) &= q\ddot{x}_{i-1}(t) + \ddot{x}_{i+1}(t) + \lambda(q\dot{e}_i(t) - \dot{e}_{i+1}(t)),\, for\ i \in \mathscr{V}_N/\{N\} \\ A_N(t) &= q\ddot{x}_{N-1}(t) + q\lambda\dot{e}_N(t) \end{aligned}$$

where λ, k and q are positive constants with k satisfying $k > (q+1)\max\limits_{1\le i\le N}\{\tilde{\bar{w}}_{maxi}, \tilde{\underline{w}}_{maxi}\}$. The adaptive laws for $\hat{\bar{w}}$ and $\hat{\underline{w}}$ are given by

$$\begin{aligned} \dot{\hat{\bar{w}}}_i(t) &= -\eta_i(q+1)S_i(t),\ \dot{\hat{\underline{w}}}_i(t) = -\eta_i(q+1)S_i(t) \\ &\qquad for\ i = 1,2,\cdots,N-1 \\ \dot{\hat{\bar{w}}}_i(t) &= -\eta_i qS_i(t),\ \dot{\hat{\underline{w}}}_i(t) = -\eta_i qS_i(t),\, for\ i = N \end{aligned} \tag{3.28}$$

and $\bar{\mu}_i(t) = \frac{1}{1+e^{-a(S_i(t)-b)}}$. The parameters $\eta_i > 0, i = 1,2,\cdots,N$ are the adaptive law gains to be designed. Moreover, the strong string stability of the whole vehicle platoon system can be guaranteed (i.e., $\|G_i(s)\| = \|\frac{E_{i+1}(s)}{E_i(s)}\| \le 1$) after some finite time if $0 < q \le 1$ holds.

Remark 3.6 *The information flow topology known as bidirectional topology [212] is used in this chapter. In Corollary 3.1, every control vehicle needs the position, velocity and acceleration signals of both its preceding vehicle and following vehicle for control. However, in contrast to (3.27) in Corollary 3.1, the preceding vehicle's acceleration information (i.e., $\ddot{x}_{i-1}(t)$) is not required in (3.12) in Theorem 3.1, which decreases the communication load. In addition, Theorem 3.1 is more tolerant of the external disturbances than Corollary 3.1 since Theorem 3.1, based on CTH policy, has variable spacing [92, 154]. The above facts will be verified by using simulations.*

Remark 3.7 *It should be pointed out that zero initial spacing and zero initial velocity errors are required simultaneously to achieve strong string stability in Corollary 3.1, which is also required by other works using CS policy, such as [38, 43, 45, 46, 93, 169, 212], while only zero initial spacing error is needed in Theorem 3.1. However, the assumption of zero initial spacing errors is still restrictive as in most cases the initial spacing error is not zero, which may cause large transient response and may lead to loss of strong string stability. Therefore, a MCTH policy is proposed in the next subsection.*

3.3.2 Non-zero Initial Spacing Error Case

In this subsection, a new distributed adaptive sliding mode protocol is proposed to overcome the problem of large transient caused by non-zero initial spacing errors. To deal with *non-zero initial* spacing errors, a modified CTH policy is defined as

$$\bar{e}_i(t) = e_i(t) - \gamma_i(t) \tag{3.29}$$

with

$$\gamma_i(t) = [e_i(0) + (\zeta_i e_i(0) + \dot{e}_i(0))t]e^{-\zeta_i t}$$

where $e_i(0) = e_i(t)|_{t=0}$, $\dot{e}_i(0) = \dot{e}_i(t)|_{t=0}$ and ζ_i are strictly positive constants. Now, we can obtain that

$$\bar{e}_i(t)|_{t=0} = 0, \ \dot{\bar{e}}_i(t)|_{t=0} = 0 \tag{3.30}$$

which shows that the modified spacing errors are initially zero for arbitrary initial spacing errors. The main idea is to transform the original *non-zero initial* spacing error problem to a zero initial spacing error problem. It is worth noting that $\bar{e}_i(t)$ converges to $e_i(t)$ with a rate of convergence determined by ζ_i since $\gamma_i(t)$ converges to zero. In order to make the convergence rate of $\bar{e}_i(t)$ to $e_i(t)$ faster, ζ_i should be large. Then, similar to (3.5), a new integral sliding surface is defined by

$$\bar{s}_i(t) = \bar{e}_i(t) + \int_0^t \lambda \bar{e}_i(\tau)d\tau. \tag{3.31}$$

At the same time, in order to achieve the *strong string stability* of the whole vehicle platoon system, an equivalent sliding surface is given out as follows:

$$S_i(t) = \begin{cases} q\bar{s}_i(t) - \bar{s}_{i+1}(t), \text{ for } i \in \mathscr{V}_N/\{N\} \\ q\bar{s}_i(t), \ i = N. \end{cases} \tag{3.32}$$

Taking the time derivative of $S_i(t)$ along the definitions of (3.29), (3.31) and (3.32) yields

$$\begin{aligned} \dot{S}_i(t) = \ & q[v_{i-1}(t) - v_i(t) - h_i(\tfrac{1}{m_i}u_i(t) + w_i(t)) \\ & -\dot{\gamma}_i(t)] + \bar{A}_i(t) \end{aligned} \tag{3.33}$$

where

$$\begin{aligned} \dot{\gamma}_i(t) &= -\zeta_i[e_i(0) + (\zeta_i e_i(0) + \dot{e}_i(0))t]e^{-\zeta_i t} \\ &\quad + (\zeta_i e_i(0) + \dot{e}_i(0))e^{-\zeta_i t} \\ \bar{A}_i(t) &= \begin{cases} q\lambda\bar{e}_i(t) - \dot{\bar{e}}_{i+1}(t) - \lambda\bar{e}_{i+1}(t), \text{ for } i \in \mathscr{V}_N/\{N\} \\ q\lambda\bar{e}_i(t), \text{ for } i = N. \end{cases} \end{aligned} \tag{3.34}$$

Then, the following theorem, which guarantees the stability of each vehicle and the *strong string stability* of the whole vehicle platoon system, can be obtained.

Theorem 3.2 *Consider the vehicle platoon system (3.2) with leader (3.1) and modified CTH policy (3.29) with non-zero initial spacing errors. For a sufficiently large positive constant V_{max}, if the initial condition satisfies (3.11), the stability of individual vehicle can be achieved by using the following adaptive ISM control input*

$$\begin{aligned} u_i(t) &= \tfrac{m_i}{h_i}(v_{i-1}(t) - v_i(t) - \dot{\gamma}_i(t)) - m_i[(1-\mu_i) \\ &\quad \hat{\overline{w}}_i(t) + \mu_i \hat{\underline{w}}_i(t)] + \tfrac{m_i}{qh_i}\bar{A}_i(t) + \tfrac{km_i}{qh_i} sgn(S_i(t)) \end{aligned} \quad (3.35)$$

with $S_i(t)$ as in (3.32), and $\dot{\gamma}_i(t)$ and $\bar{A}_i(t)$ as defined in (3.34). The adaptive laws for $\hat{\overline{w}}_i(t)$ and $\hat{\underline{w}}$ are given by (3.15). Moreover, the strong string stability of the whole vehicle platoon system can be guaranteed (i.e., $\|G_i(s)\| = \|\frac{E_{i+1}(s)}{E_i(s)}\| \leq 1$) after some finite time if $0 < |q| \leq 1$ holds.

Proof 3.2 *It can be completed by following steps as in the proof of Theorem 3.1 by considering the same Lyapunov function candidate as in (3.17). Therefore, the details of the proof are omitted.* □

Remark 3.8 *It must be pointed out that $\bar{s}_i(t)$ converges to $s_i(t)$ in (3.5) with a rate of convergence determined by ζ_i since $\bar{e}_i(t)$ converges to $e_i(t)$ with same rate as pointed out below (3.30). In order to make the convergence rate of $\bar{s}_i(t)$ to $s_i(t)$ in (3.5) faster, ζ_i should be large. Similar to Theorem 3.1, the strong string stability of the whole vehicle platoon system can be guaranteed if $\|\frac{\bar{E}_{i+1}(s)}{\bar{E}_i(s)}\| \leq 1$ holds under the condition of $0 < |q| \leq 1$. Simultaneously, $\|\frac{E_{i+1}(s)}{E_i(s)}\| \leq 1$ will also be satisfied since $\bar{s}_i(t)$ converge to $s_i(t)$ with a rate of convergence determined by ζ_i.*

Remark 3.9 *In order to reduce the chattering effect as in Remark 3.5, $u_i(t)$ in (3.35) can be rewritten as*

$$\begin{aligned} u_i(t) &= \tfrac{m_i}{h_i}(v_{i-1}(t) - v_i(t) - \dot{\gamma}_i(t)) \\ &\quad - m_i[(1-\bar{\mu}_i(t))\hat{\overline{w}}_i(t) + \bar{\mu}_i(t)\hat{\underline{w}}_i(t)] \\ &\quad + \tfrac{m_i}{qh_i}\bar{A}_i(t) + \tfrac{km_i}{qh_i}\tfrac{S_i(t)}{\|S_i(t)\|+\sigma} \end{aligned} \quad (3.36)$$

where $\bar{\mu}_i(t)$ is defined as (3.24), $\dot{\gamma}_i(t)$ and $\bar{A}_i(t)$ are defined as (3.33).

3.4 Numerical Examples

In this section, the proposed control protocols are applied to a platoon of six follower vehicles and a leader. Comparisons are made between the new methods (3.25) and (3.36) and the approaches in [93]. In addition, the comparisons are also made between the two adaptive sliding mode control strategies (3.25) and (3.36) for the case that the initial spacing errors are not zero. The initial position of the leader is set as

$x_0(0) = 20m$, $v_0(0) = 1m/s$ and the desired trajectory is given by

$$v_0(t) = \begin{cases} 1\ m/s, & 0 \le t < 2s, \\ 0.5t\ m/s, & 2s \le t < 6s, \\ 3\ m/s, & otherwise. \end{cases}$$

The required string velocity is $3m/s$ to be achieved and a standstill distance $\delta_i = 0.5m$ with a time headway $h_i = 1s$ for all $i \in \mathscr{V}_6$. Then, the final desired distance and velocity are $3.5m$ and $3m/s$, respectively. The disturbances $w_i(t) = 1.5\sin(3t)e^{\frac{-(t-5-0.2i)^2}{4}}, i = 1,2,\cdots,6$ enter into the follower vehicles at the beginning ($t = 0s$). Furthermore, without loss of generality, we consider the case of homogeneous vehicle, i.e., $m_i = m$ for $i \in \mathscr{V}_6$. The mass m can be chosen any value according to practical applications. The accelerations $a_i(t) = \frac{1}{m_i}u_i(t)$ will be given out in the following simulation. In numerical simulations, all controller's parameters are chosen as $k = 3$, $q = 0.9$, $\lambda = 0.2$, $\eta_i = 0.01$, $\zeta_i = 10$, $\sigma = 0.3$, $a = 10$ and $b = 0.0001$. Finally, in this section, we always choose the *initial conditions* of the estimates of the upper and lower bounds of the disturbances as $\hat{\overline{w}}(0) = 1.5[1,1,1,1,1,1]$ and $\hat{\underline{w}}(0) = -1.5[1,1,1,1,1,1]$.

First of all, the proposed method in Theorem 3.1 by using CTH policy (3.4) is considered. The initial position and velocity are chosen as $x(0) = [16.5, 13.5, 11, 8,\ 5.5, 3]$ and $v(0) = [3, 2.5, 2, 2.5, 2, 2]$, respectively, which means zero initial spacing errors $e_i(0)$ but *non-zero initial* velocity errors $e_i^v(0)$ as $e^v(0) = [-2, 0.5, 0.5, -0.5, 0.5, 0]$ from the definition of CTH policy (3.4). By using CTH policy, the final desired distance and velocity are $3.5m$ and $3m/s$, respectively. The simulation results are shown in Figure 3.3. From the simulation results in Figure 3.3, the proposed adaptive control protocol (3.25) works well. The spacing errors in Figure 3.3(a) converge to zero in a finite time ($t \le 10s$) and the amplitudes of the spacing errors decrease through the string of vehicles, which show the stability of each vehicle and the *strong string stability* of the whole vehicle platoon (i.e., $\|e_6(t)\| \le \|e_5(t)\| \le \cdots \le \|e_1(t)\|$). Figure 3.3(b) shows the collision avoidance, not only in steady-state condition, but also during the initial transient. It can be seen from Figure 3.3(c) that the distances between all consecutive vehicles converge to the constant distance $3.5m$ and the velocities of the vehicles follow that of the leader as shown in Figure 3.3(d). The acceleration $a_i(t)$ and sliding surface $S_i(t)$ are shown in Figures 3.3(e) and (f), respectively. Figure 3.3(f) shows $S_i(t)$ reaches the surface $S_i(t) = 0$ in a finite time and the *chattering phenomenon* caused by signum and indicator functions is almost eliminated. The results of the estimations of the upper and lower bounds of the disturbances are presented in Figures 3.3(g) and (h), which show the estimates converge to constants. The above simulation results illustrate the effectiveness of the proposed protocol (3.25).

To have fair comparison with Theorem 3.1, for the existing method in Corollary 3.1 with *CS policy* (3.26), we choose the initial position and velocity as $x(0) = [19.5, 19, 18.5, 18, 17.5, 17]$ and $v(0) = [3, 2.5, 2, 2.5, 2, 2]$, respectively, which mean zero initial spacing errors $e_i(0)$ but non-zero initial velocity errors $e_i^v(0)$ as $e^v(0) = [-2, 0.5, 0.5, -0.5, 0.5, 0]$ from the definition of CS policy (3.26). Then, the simulation results are shown in Figure 3.4. It can be observed from Figure 3.4(b) that some neighboring vehicles collide with each other not only during the initial transient, but also when the *external disturbances* take place since there exist cross and overlapped positions. Therefore, the effect of *non-zero initial* velocity errors and *external disturbances* can result in collisions among the vehicles despite zero initial spacing errors. From Figure 3.4(a), we can find the condition for *strong string stability* is not satisfied during the initial transient and the *external disturbances*, which causes collisions among the vehicles. In order to avoid collision due to the disturbances, the safety distance δ_i can be set larger. However, it is difficult to know how large is enough because the magnitude of the disturbances is not known in advance. Consequently, CTH policy (variable-spacing policy) is superior to the *CS policy*. Moreover, it can be seen from Figure 3.4, the distances and velocities are slower to converge than those in Figure 3.3. Taking all these results together, one can conclude that the *non-zero initial* velocity errors and *external disturbances* result in a loss of string stability and may cause collisions among the vehicles by using the *CS policy* protocol (3.27) in Corollary 3.1, which shows the advantage of the protocol (3.25).

In the following, in order to demonstrate the necessity to consider the case of *non-zero initial* spacing errors in the control design procedure, the initial position and velocity for all vehicles are set as $x(0) = [19.5, 18.5, 18, 17, 16.5, 16]$ and $v(0) = [3, 2.5, 2, 2.5, 2, 2]$, respectively. Then, using the definition of CTH policy (3.4), we can obtain *non-zero initial* spacing error $e(0) = [-3, -2, -2, -2, -2, -2]$ and *non-zero initial* velocity error $e^v(0) = [2, 1.5, 1, 1.5, 1, 1]$. The spacing errors, the positions, distances, velocities, accelerations, sliding mode and the estimations using the proposed adaptive control protocols (3.25) and (3.36) are shown in Figure 3.5 and Figure 3.6, respecively. Note that, in Figure 3.5(a), *non-zero initial* spacing and velocity errors result in *string instability*, which brings about collisions between vehicles in Figure 3.5(b). On the other hand, the controller (3.36) designed in Theorem 3.2 is applied to the *non-zero initial* spacing errors. The results show that the distant converges to the desired distance $3.5m$ very fast and accurately in Figure 3.6(c) and the velocities of the vehicles follow that of the leader vehicle in Figure 3.6(d). Figure 3.6(a) shows the *strong string stability* of the vehicle platoons, while Figure 3.6(b) shows no collision among vehicles. Therefore, by comparing between the simulation results obtained by using the protocol (3.25) and the simulation results obtained by using the protocol (3.36), one can conclude that it is necessary to pay attention to the impact of the initial spacing errors on the *strong string stability* problem.

3.5 Conclusion

This chapter presents a new distributed adaptive ISM control protocol to achieve the *finite time stability* of each vehicle and the *strong string stability* of the whole vehicle platoon system in the presence of *external disturbances*. The adaptive control

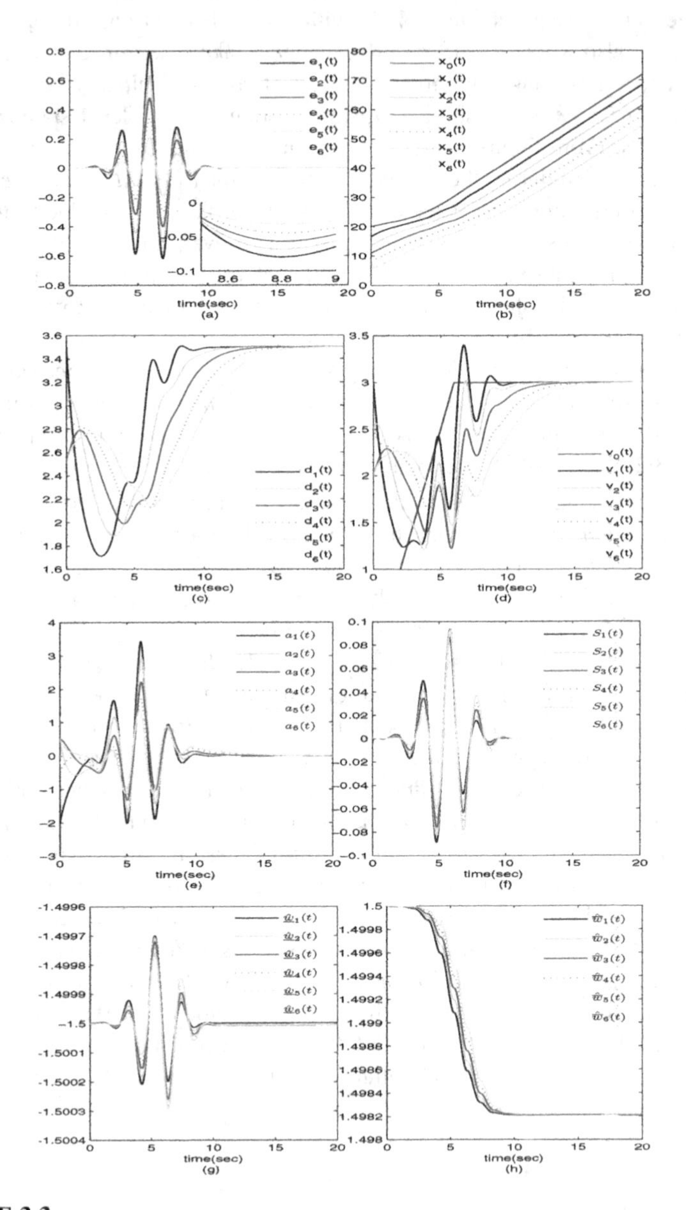

FIGURE 3.3

(a) Spacing error $e_i(t)$; (b) position $x_i(t)$; (c) distance $d_i(t)$; (d) velocity $v_i(t)$; (e) acceleration $a_i(t)$; (f) sliding surface $S_i(t)$; (g) lower bound estimation $\underline{\hat{w}}_i(t)$; (h) upper bound estimation $\hat{\overline{w}}_i(t)$ by using (3.25) under $e(0) = 0$ and $e^v(0) \neq 0$.

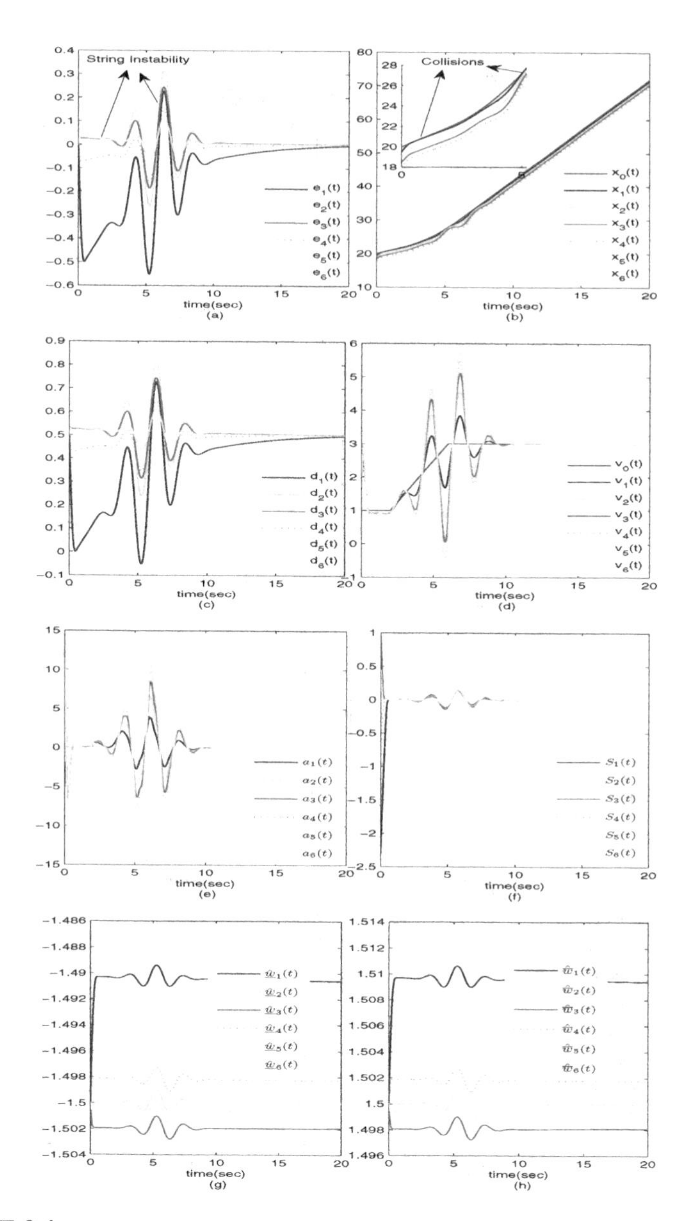

FIGURE 3.4
(a) Spacing error $e_i(t)$; (b) position $x_i(t)$; (c) distance $d_i(t)$; (d) velocity $v_i(t)$; (e) acceleration $a_i(t)$; (f) sliding surface $S_i(t)$; (g) lower bound estimation $\hat{\underline{w}}_i(t)$; (h) upper bound estimation $\hat{\overline{w}}_i(t)$ by using (3.27) under $e(0) = 0$ and $e^v(0) \neq 0$.

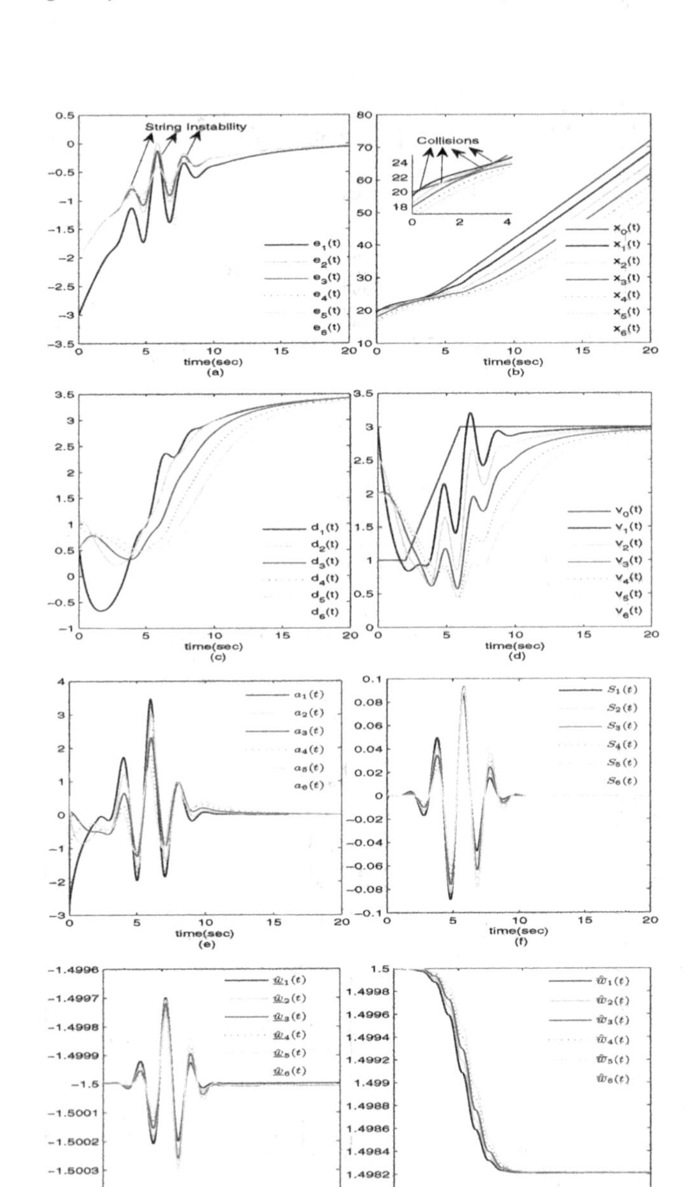

FIGURE 3.5
(a) Spacing error $e_i(t)$; (b) position $x_i(t)$; (c) distance $d_i(t)$; (d) velocity $v_i(t)$; (e) acceleration $a_i(t)$; (f) sliding surface $S_i(t)$; (g) lower bound estimation $\underline{\hat{w}}_i(t)$; (h) upper bound estimation $\hat{\overline{w}}_i(t)$ by using (3.25) under $e(0) \neq 0$ and $e^v(0) \neq 0$.

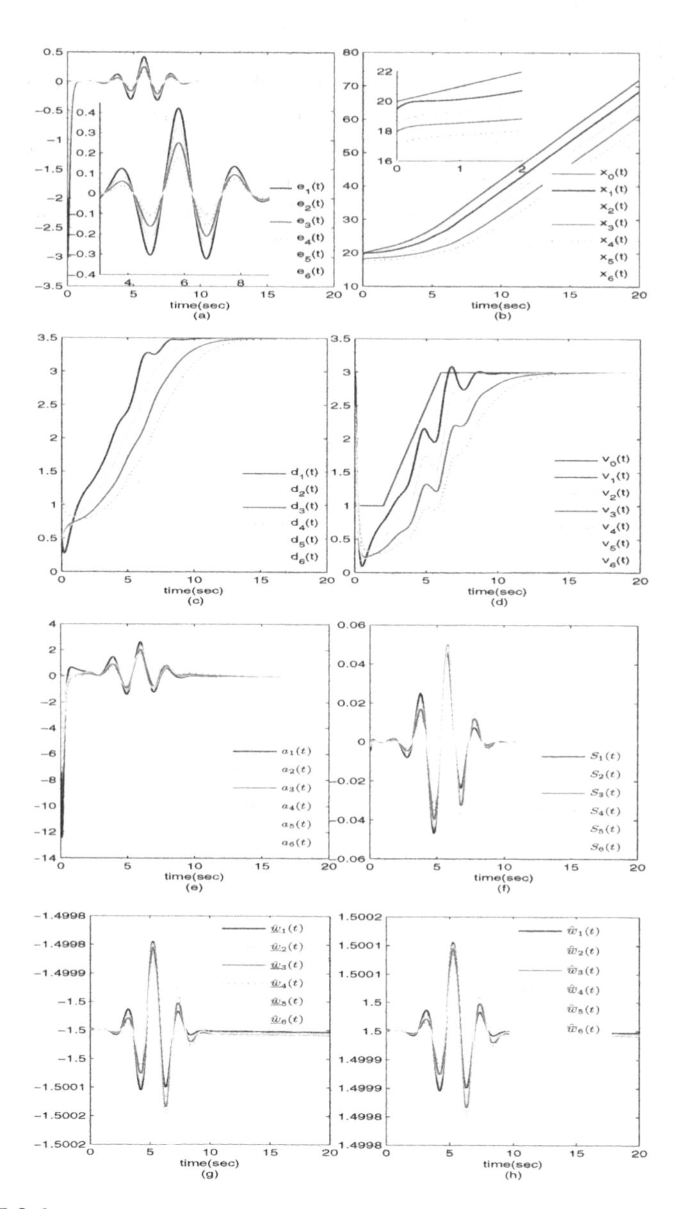

FIGURE 3.6

(a) Spacing error $e_i(t)$; (b) position $x_i(t)$; (c) distance $d_i(t)$; (d) velocity $v_i(t)$; (e) acceleration $a_i(t)$; (f) sliding surface $S_i(t)$; (g) lower bound estimation $\underline{\hat{w}}_i(t)$; (h) upper bound estimation $\hat{\overline{w}}_i(t)$ by using (3.36) under $e(0) \neq 0$ and $e^v(0) \neq 0$.

scheme does not require the exact knowledge of maximum and minimum values of the *external disturbances*, which is desirable for practical applications. In order to avoid collisions in the inter-connected vehicles, control strategies based on CTH policy which is known to improve *strong string stability* are developed for both cases of zero and *non-zero initial* spacing errors. In addition, the *chattering phenomenon* caused by ISM control method is reduced by introducing continuous approximations. Future work will be directed toward extension of the presented control strategies to the cases where the vehicles have actuator delays, plant/control mismatch uncertainties, *actuator saturation* and nonlinear dynamic constraints.

4

String Stability of Vehicle Platoons with Nonlinear Acceleration Uncertainties

This chapter investigates the distributed adaptive control problems for nonlinear vehicle-following systems subject to nonlinear acceleration uncertainties involving vehicle acceleration disturbances, wind gust, and parameter uncertainties. It is worth mentioning that the acceleration of the leader in most existing works is always assumed to be zero or constant, the evolution of the leader in this chapter may be subject to some unknown bounded input. Distributed adaptive control strategies based on ISMC technique are proposed to maintain a rigid formation for a string of vehicle platoon in 1D. Compared with the *MCTH policy* in Chapter 3, a new *MCTH policy* is developed to not only remove the assumption of zero initial spacing and velocity errors but also effectively decrease the neighboring-vehicle spacing (i.e., increase the traffic density), making them nearly equal to those by using the constant-spacing (CS) policy. *Adaptive compensation* terms without requiring *a prior* knowledge of upper bounds of the uncertainties are constructed to compensate for the time-varying effects caused by nonlinear acceleration uncertainties.

4.1 Introduction

In Chapter 3, a *MCTH policy* is introduced to remove the assumption of zero initial spacing and velocity errors. Nevertheless, the steady-state neighboring-vehicle spacings of the *MCTH policy* in Chapter 3 are very large at high speed, and thus the traffic density is low. Recently, in order to reduce the neighboring-vehicle spacing, a *MCTH policy* by introducing a virtual truck was proposed in [2], but it will increase the risk of collision. Thus, in this chapter, some efforts will be made in the direction of reducing the neighboring-vehicle spacing and simultaneously removing the assumption of zero initial spacing and velocity errors, which is important and challenging in both theory and practice. On the other hand, as the vehicle platoon is an interconnected coupled system, disturbances acting on one vehicle may affect other vehicles and even amplify spacing errors along the string, namely *string instability* [120, 124, 201]. Generally speaking, disturbances and parameter uncertainties in the system are two main underlying causes of the *string instability* problem [45]. The primary disturbances existing in a vehicle platoon include the vehicle acceleration disturbances caused by environmental circumstances such as gust, friction on grounds and rolling resistance, while the uncertainties in a vehicle platoon consist of the aerodynamic drag, the mass of passenger vehicles [45] and intermediate uncertainties induced by networks [120]. Thus, how to reduce the effect of disturbances and parameter uncertainties on *string stability* has attracted considerable interest (see [38, 45, 46, 55, 93, 120, 201], and the references therein). Note that in the above-mentioned works (except for [120]) on *string stability*, the research is all based on a common assumption that the acceleration of the leader is always assumed to be zero or constant, which has clear shortcomings in practice due to the complexity of the environment. Furthermore, in practice, it is almost impossible to get an exact mathematical model of a dynamic system due to sensor measurement noise or environmental disturbances, and so on [62, 63, 65]. Then, it is natural to build our analysis and design on full account of nonlinear vehicle dynamics directly, which can potentially lead to a more accurate *platoon control* model, although highly nonlinear platoon dynamics indeed make the controller design more complex and challenging.

Motivated by the above points, based on traditional CTH (TCTH) and MCTH policies, novel distributed control strategies combining integral sliding mode control (ISMC) with adaptive control technique are proposed to solve the *string stability* of vehicle-following systems subject to unknown bounded nonlinear acceleration uncertainties in both the leader and the followers. Based on MCTH in Chapter 3, a new *MCTH policy* is proposed to increase the traffic density. Besides the new *MCTH policy* and based on it, effective *platoon control* strategies are proposed to guarantee all signals bounded (i.e., bounded stability) of individual vehicle and *string stability* of the platoon. In addition, the bound on the spacing errors can be rendered arbitrarily small by adjusting design parameters. The main contributions of this chapter lie in the following.

(1) Distributed adaptive ISMC strategies: Based on TCTH and MCTH policies, new distributed adaptive control strategies combining ISMC method with adaptive control technique are proposed to guarantee bounded stability of the spacing errors and the *strong string stability* of the whole vehicle platoon.

(2) New *MCTH policy*: This policy is effective to guarantee smooth transient response for non-zero initial spacing and velocity deviations and simultaneously increase the traffic density, which is nearly equal to that of *CS policy*. Unfortunately, increasing the traffic density is at the cost of increasing the communication required since the information of the leader needs to be broadcast to all followers via wireless communication.

(3) Nonlinear acceleration uncertainties: Unlike most existing work which assumes that the acceleration of the leader is zero or constant, both the leader and the followers in this chapter may be subject to some unknown bounded nonlinear uncertainties. Moreover, adaptive compensation terms without requiring *a prior* knowledge of upper bound of the uncertainties are constructed to effectively reduce the effect of the unknown bounded nonlinear acceleration uncertainties.

4.2 Vehicle Platoon and Problem Formulation

4.2.1 Vehicle Platoon

Consider N follower vehicles and a leader that are organized into a platoon running in a straight line. Let $x_i(t)$, $v_i(t)$ and $a_i(t)$ $(i = 0, 1, \cdots, N)$ denote respectively the position, velocity and acceleration of the ith vehicle in the platoon with $i = 0$ standing for the leader vehicle and the others being followers. For brevity, we let $\mathscr{V}_N/\{N\}$ with $\mathscr{V}_N = \{1, 2, \cdots, N\}$ denote the relative complement of $\{N\}$ in $\mathscr{V}_N$, and $\mathscr{V}_N \cup \{0\}$ denote the union of $\mathscr{V}_N$ and $\{0\}$. A nonlinear vehicle model is utilized to describe the dynamic behavior of the vehicles as realistically as possible. Let the ith follower vehicle in the platoon be represented by the following nonlinear differential equations

$$\begin{aligned} \dot{x}_i(t) &= v_i(t) \\ \dot{v}_i(t) &= u_i(t) + f_i(x_i(t), v_i(t), t), i \in \mathscr{V}_N \end{aligned} \tag{4.1}$$

and the leader dynamic be described as

$$\begin{aligned} \dot{x}_0(t) &= v_0(t) \\ \dot{v}_0(t) &= a_0(t) + f_0(x_0(t), v_0(t), t) \end{aligned} \tag{4.2}$$

where $u_i(t)$ denotes the control input, and $f_i(x_i(t), v_i(t), t)$ for vehicle i is an unknown bounded nonlinear time-varying uncertainty involving vehicle acceleration disturbances, wind gust, parameters uncertainties and intermediate uncertainties induced by networks. Since $f_i(x_i(t), v_i(t), t)$ is introduced in the leader and follower acceleration dynamics, the model is thus more general than the one studied in [27, 26, 55, 72, 125]. The acceleration $a_0(t)$ is a known function of time. According to the practical case, it is reasonable to assume $\|v_0(t)\| \le \bar{v}_0$ since the desired acceleration $a_0(t)$ and the nonlinear time-varying uncertainty $f_0(x_0(t), v_0(t), t)$ acting on the leader vehicle are bounded.

Throughout this chapter, the following assumption is made for the development of control strategies.

Assumption 4.1 *Nonlinearities $f_i(x_i(t), v_i(t), t)$ $(i \in \mathscr{V}_N \cup \{0\})$ satisfy the following bounded condition*

$$\begin{aligned} &\|f_{i-1}(x_{i-1}(t), v_{i-1}(t), t) - f_i(x_i(t), v_i(t), t)\| \\ &\le \mu_i \|x_{i-1}(t) - x_i(t)\| + \eta_i \|v_{i-1}(t) - v_i(t)\| + \theta_i \end{aligned} \tag{4.3}$$

where $\mu_i > 0$, $\eta_i > 0$ and $\theta_i \ge 0$ are unknown non-negative constants. In addition, $f_0(x_0(t), v_0(t), t)$ is bounded as $\|f_0(x_0(t), v_0(t), t)\| \le \bar{f}$.

Remark 4.1 *In Assumption 4.1, the nonlinear uncertain term $f_i(x_i(t), v_i(t), t)$ involves the position $x_i(t)$ and velocity $v_i(t)$ simultaneously, and can be used to denote vehicle acceleration disturbances. This is a more general and relaxant condition of nonlinear coupling function compared with that studied in [84, 85]. It should be* *pointed out that the condition (4.3) reduces to the one in [131] when $\theta_i = 0$. It is well known that the acceleration disturbance caused by environmental circumstances such as gust, friction on grounds and rolling resistance is often encountered in a practical system. As the vehicle platoon is an interconnected coupled system, disturbances acting on one vehicle may affect other vehicles and even amplifying spacing errors along the platoon [120, 124]. Therefore, it is necessary to deal with the effect of the acceleration disturbances on vehicle platoon. However, the existence of nonlinear acceleration disturbance renders the controller design more complex and challenging.*

4.2.2 Problem Formulation

The goal of the *platoon control* is to maintain *desired safe spacing* between any two neighboring vehicles and reach the velocity consensus among vehicles. Then, our control objective is formally stated as follows.

(1) Design distributed adaptive ISMC strategies to ensure bounded stability of each vehicle system based on TCTH and MCTH policies;

(2) Provide a *MCTH policy* to remove the assumption of zero initial spacing and velocity errors and increase the traffic density simultaneously;

(3) Maintain the neighboring-vehicle distance equal to a desired safe distance, and make all the followers' velocity follow the leader's velocity;

(4) Ensure *strong string stability* of the vehicle platoon;

(5) Construct *adaptive compensation* terms to compensate for the time-varying effects caused by nonlinear acceleration uncertainties.

4.3 Distributed Adaptive Integral Sliding Mode Control Strategy

In this section, two distributed adaptive integral sliding mode control (ISMC) strategies based on TCTH policy and *MCTH policy*, respectively, are proposed to attain the *strong string stability* of the nonlinear vehicle control system (4.1) with a leader (4.2).

4.3.1 Control Strategy 1: TCTH Control Law

The aim of the *platoon control* is that of imposing on the followers the velocity of the leader as well as a given spacing policy between two consecutive vehicles. Consider

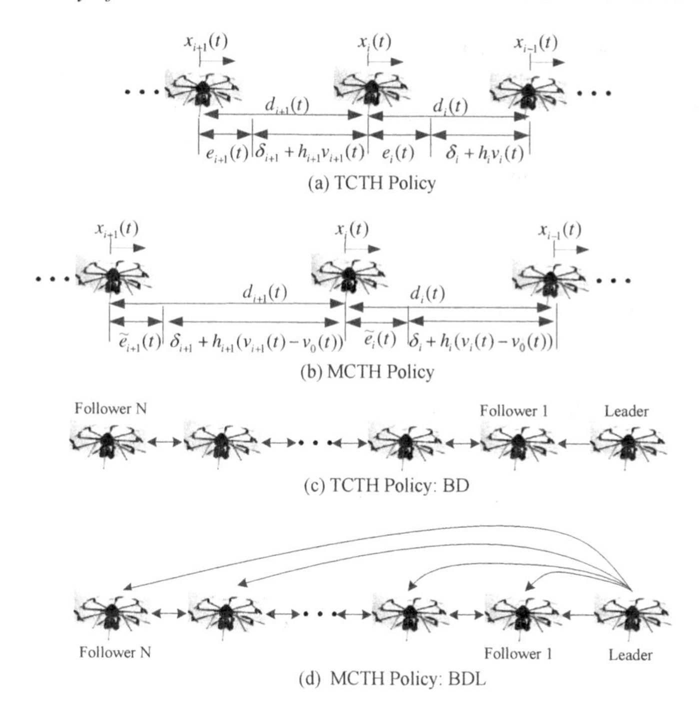

FIGURE 4.1
(a) TCTH Policy; (b) MCTH Policy; (c) TCTH Policy: Bidirectional Topology (BD); (d) MCTH Policy: Bidirectional-Leader Topology (BDL).

the TCTH policy used in [2, 55, 186] as follows

$$\begin{aligned} e_i^x(t) &= d_i(t) - \delta_i - h_i v_i(t) \\ &= d_i(t) - \delta_i - h_i e_i^v(t) + h_i v_0(t) \\ e_i^v(t) &= v_i(t) - v_0(t) \end{aligned} \tag{4.4}$$

where $\delta_i > 0$ and $h_i > 0$ are the required ith standstill distance and *constant-time headway*, respectively. $d_i(t)$ is the relative distance between the preceding vehicle and the ego vehicle and is defined as follows

$$d_i(t) = x_{i-1}(t) - x_i(t). \tag{4.5}$$

The parameters relevant to any two adjacent vehicles in a platoon by using TCTH policy is shown in Figure 4.1(a). The network topology of TCTH policy is bidirectional topology as shown in Figure 4.1(c).

Similar to Chapter 3, for the nonlinear vehicle control system (4.1) with a leader (4.2), we design an integral sliding surface and the coupled sliding mode as:

$$\begin{aligned} s_i(t) &= e_i^x(t) + \int_0^t \lambda e_i^x(\tau) d\tau \\ S_i(t) &= \begin{cases} q s_i(t) - s_{i+1}(t), i \in \mathscr{V}_N / \{N\} \\ q s_i, i = N \end{cases} \end{aligned} \tag{4.6}$$

where λ is a positive design parameter. Unlike the one in Chapter 3, q in this chapter is a positive constant. Since $q > 0$ is a constant, from the relationship in (3.7) of Chapter 3, it concludes that when $S_i(t)$ becomes zero, $s_i(t)$ also becomes zero at the same time, and vice versa. The coupled sliding mode is used to achieve *strong string stability* defined in Definition 2.2 of Chapter 2 as

$$|e_N^x(t)| \leq |e_{N-1}^x(t)| \leq \cdots \leq |e_1^x(t)|, \tag{4.7}$$

i.e., the error propagation transfer function $G_i(s) := \frac{E_{i+1}^x(s)}{E_i^x(s)}$ satisfies $|G_i(s)| \leq 1$ for all $i \in \mathscr{V}_N$, where $E_i^x(s)$ denotes the *Laplace transform* of $e_i^x(t)$.

Next, to keep the system on the integral sliding surface, we take the time derivative of $S_i(t)$ along the TCTH policy (4.4) as

$$\begin{aligned} \dot{S}_i(t) &= q\dot{s}_i(t) - \dot{s}_{i+1}(t) \\ &= q[\dot{e}_i^x(t) + \lambda e_i^x(t)] - [\dot{e}_{i+1}^x(t) + \lambda e_{i+1}^x(t)] \\ &= q[v_{i-1}(t) - v_i(t) - h_i[u_i(t) + \lambda e_i^x(t) \\ &\quad + f_i(x_i(t), v_i(t), t)] - [\dot{e}_{i+1}^x(t) + \lambda e_{i+1}^x(t)] \\ &= q\{v_{i-1}(t) - v_i(t) - h_i[u_i(t) \\ &\quad + f_i(x_i(t), v_i(t), t)]\} + A_i(t) \end{aligned} \tag{4.8}$$

where

$$A_i(t) = q\lambda e_i^x(t) - \dot{e}_{i+1}^x(t) - \lambda e_{i+1}^x(t)$$

for $i = 1, 2, \cdots, N-1$. When $i = N$, one can obtain $S_N(t) = qs_N(t)$, therefore, one can further obtain that

$$\begin{aligned} \dot{S}_N(t) &= q[\dot{e}_N(t) + \lambda e_N(t)] \\ &= q[v_{N-1}(t) - v_N(t) - h_N(u_N(t)) \\ &\quad + f_N(x_N(t), v_N(t), t)] + A_N(t) \end{aligned} \tag{4.9}$$

where $A_N(t) = q\lambda e_N(t)$. Then, the following distributed adaptive control law is designed for vehicle i:

$$\begin{aligned} u_i(t) &= \tfrac{1}{h_i}[v_{i-1}(t) - v_i(t)] + [\rho_i \textstyle\int_0^t |S_i(\tau)| d\tau \\ &\quad + \gamma_i(t)] \tanh(\tfrac{S_i(t)}{\varepsilon_i}) + \tfrac{1}{qh_i} A_i(t) + \tfrac{k}{2qh_i} S_i(t) \end{aligned} \tag{4.10}$$

where

$$A_i(t) = \begin{cases} q\lambda e_i^x(t) - \dot{e}_{i+1}^x(t) - \lambda e_{i+1}^x(t), \text{for } i \in \mathscr{V}_N / \{N\} \\ q\lambda e_i^x(t), \text{for } i = N. \end{cases} \tag{4.11}$$

Moreover, ρ_i and k are positive design constants, ε_i is a small positive parameter, and $\gamma_i(t)$ is updated by the following adaptive law:

$$\dot{\gamma}_i(t) = \iota_i q h_i S_i(t) \tanh(\tfrac{S_i(t)}{\varepsilon_i}) - \iota_i q h_i \sigma_i(t) \gamma_i(t). \tag{4.12}$$

Here, ι_i is a positive design constant, and $\sigma_i(t)$ is any positive uniform continuous and bounded function which satisfies

$$\lim_{t\to\infty}\int_{t_0}^{t}\sigma_i(\tau)d\tau \quad \leq \quad \bar{\sigma}_i < +\infty.$$

with $\bar{\sigma}_i$ a finite bound.

Remark 4.2 *The control law in (4.10) is composed of four components. The first component* $\frac{1}{h_i}[v_{i-1}(t) - v_i(t)]$ *is constructed to achieve whole platoon synchronization. The second component* $[\rho_i \int_0^t |S_i(\tau)|d\tau + \gamma_i(t)]\tanh(\frac{S_i(t)}{\varepsilon_i})$ *is developed to compensate for the time-varying effects caused by nonlinear acceleration uncertainties and external disturbances. The third component* $\frac{1}{qh_i}A_i(t)$ *contains the coupling of the vehicle to both its preceding and following vehicles, which plays an important role in guaranteeing the strong string stability, while the last component* $\frac{k}{2qh_i}S_i(t)$ *relates to the convergence rate in the stability analysis.*

Then, the following theorem, which guarantees the bounded stability of each vehicle and the *strong string stability* of the whole vehicle-following system, can be obtained.

Theorem 4.1 *Consider the vehicle-following platoon with N followers (4.1) and a leader (4.2) under the assumption that the initial spacing and velocity errors are zero. Under Assumption 4.1, the distributed adaptive ISMC laws (4.10)-(4.12) guarantee that the spacing errors* $e_i^x(t)$ *in TCHT (4.4) converge to an arbitrarily small neighborhood of zero by choosing the design parameters appropriately, while strong string stability of the vehicle platoon also can be guaranteed for* $i = 1, 2, \cdots, N$ *when* q *satisfies* $0 < q \leq 1$.

Proof 4.1 *Employing Assumption 4.1 and the assumption* $\|v_0(t)\| \leq \bar{v}_0$, *one can obtain that*

$$\begin{aligned}
\| & f_i(x_i(t), v_i(t), t)\| \\
= & \ \|\sum_{j=1}^{i}[f_j(x_j(t), v_j(t), t) - f_{j-1}(x_{j-1}(t), v_{j-1}(t), t)] \\
& + f_0(x_0(t), v_0(t), t)\| \\
\leq & \ \sum_{j=1}^{i}\|f_j(x_j(t), v_j(t), t) - f_{j-1}(x_{j-1}(t), v_{j-1}(t), t)\| \\
& + \|f_0(x_0(t), v_0(t), t)\| \\
\leq & \ \sum_{j=1}^{i}[\mu_j\|e_j^x\| + \mu_j\|\delta_j + h_j v_j(t)\| + \theta_j \\
& + \eta_j\|v_{j-1}(t) - v_j(t)\|] + \bar{f} \\
= & \ \sum_{j=1}^{i}[\mu_j\|e_j^x\| + \mu_j\|\delta_j + h_j e_j^v(t) + h_j v_0(t)\| \\
& + \eta_j\|e_{j-1}^v(t) - e_j^v(t)\| + \theta_j] + \bar{f} \\
\leq & \ \Omega_i^*(t) + \gamma_i^*
\end{aligned} \tag{4.13}$$

where

$$\begin{aligned}\Omega_i^*(t) &= \sum_{j=1}^{i} [\mu_j \|e_j^x(t)\| + \mu_j h_j \|e_j^v(t)\| \\ &\quad + \eta_j \|e_{j-1}^v(t) - e_j^v(t)\|] \\ \gamma_i^* &\geq \sum_{j=1}^{i} [\mu_j \delta_j + \mu_j h_j \bar{v}_0 + \theta_j] + \bar{f}.\end{aligned} \tag{4.14}$$

Then, we construct the following Lyapunov function candidate for vehicle i

$$V_i(t) = \tfrac{1}{2} S_i^2(t) + \tfrac{1}{2\iota_i} (\gamma_i^* - \gamma_i(t))^2$$

where γ_i^* is defined as in (4.14).

Taking the time derivative of $V_i(t)$ for $t > 0$ yields

$$\dot{V}_i(t) = S_i(t)\dot{S}_i(t) - \tfrac{1}{\iota_i}(\gamma_i^* - \gamma_i(t))\dot{\gamma}_i(t). \tag{4.15}$$

Based on (4.13), we have

$$-qh_i f_i(x_i(t), v_i(t), t) S_i(t) \leq qh_i(\Omega_i^*(t) + \gamma_i^*)|S_i(t)|. \tag{4.16}$$

Then, it follows from (4.8), (4.9) and (4.16), we have

$$\begin{aligned}\dot{S}_i(t)S_i(t) &= q[v_{i-1}(t) - v_i(t) - h_i(u_i(t) \\ &\quad + f_i(x_i(t), v_i(t), t))]S_i(t) + A_i(t)S_i(t) \\ &\leq [q(v_{i-1}(t) - v_i(t)) - qh_i u_i(t) \\ &\quad + A_i(t)]S_i(t) + qh_i(\Omega_i^*(t) + \gamma_i^*)|S_i(t)|\end{aligned} \tag{4.17}$$

where $A_i(t)$ is as (4.11).

Note that all $|e_i^x(t)||S_i(t)|$, $|e_i^v(t)||S_i(t)|$, $\rho_i \int_0^t |S_i(\tau)|d\tau|S_i(t)|$ and $|e_{i-1}^v(t) - e_i^v(t)|$ $|S_i(t)|$ are monotone functions of $|S_i(t)|$, which depends on $e_i^x(t)$ and $e_i^v(t)$ according to the definition of (4.4) and (4.6). (4.6) can be changed via the Laplace transform as $E_i^x(s) = \frac{s}{s+\lambda} s_i(s)$, where $E_i^x(s)$ and $s_i(s)$ are the Laplace transform of $e_i^x(t)$ and $s_i(t)$, respectively. This implies that if $S_i(t)$ is bounded, all terms will also be bounded by using the facts that $s_i(t)$ is equivalent to $S_i(t)$ for all $i = 1, 2, \cdots, N$ and $v_0(t)$ is bounded by $\bar{v}_0$. Then, $\Omega_i^*(t)$ in (4.14) will also be bounded. In addition, it should be mentioned that $\rho_i \int_0^t |S_i(\tau)|d\tau$ is always increasing monotonically with time. Thus, similar to [84], if we choose adjustable parameter ρ_i large enough, we can guarantee that the speed increase of $\rho_i \int_0^t |S_i(\tau)|d\tau$ is faster than $\Omega_i^*(t)$. Thus,

$$\rho_i \int_0^t |S_i(\tau)|d\tau \geq \Omega_i^*(t) \tag{4.18}$$

can be achieved. Noted that $\rho_i \int_0^0 |S_i(\tau)|d\tau = \Omega_i^*(0) = 0$ under the assumption that initial spacing and velocity errors are zero.

Substituting (4.10) and (4.18) into (4.17) yields

$$\begin{aligned}\dot{S}_i(t)S_i(t) &\leq -qh_i\gamma_i(t)\tanh\left(\tfrac{S_i(t)}{\varepsilon_i}\right)S_i(t) - \tfrac{k}{2}S_i^2(t) \\ &\quad + qh_i[\gamma_i^*|S_i(t)| + 0.2785\varepsilon_i\Omega_i^*],\end{aligned} \tag{4.19}$$

where the following fact is used

$$\begin{aligned} q \quad & h_i\Omega_i^*|S_i(t)| - qh_i\rho_i \int_0^t |S_i(\tau)|d\tau S_i(t)\tanh(\tfrac{S_i(t)}{\varepsilon_i}) \\ \leq \ & qh_i\Omega_i^*|S_i(t)| - qh_i\Omega_i^* S_i(t)\tanh(\tfrac{S_i(t)}{\varepsilon_i}) \\ \leq \ & 0.2785qh_i\varepsilon_i\Omega_i^*. \end{aligned}$$

By using Lemma 2.5 in Chapter 2, one obtains

$$qh_i[\gamma_i^*|S_i(t)| - \gamma_i^* S_i(t)\tanh(\tfrac{S_i(t)}{\varepsilon_i})] \quad \leq \ 0.2785qh_i\varepsilon_i\gamma_i^*. \tag{4.20}$$

Since $\sigma_i(t) > 0$, combining (4.12), (4.19) and (4.20) with the following fact

$$\begin{aligned} q \quad & h_i[\gamma_i^* - \gamma_i(t)]\sigma_i(t)\gamma_i(t) \\ = \ & qh_i\sigma_i(t)\{-\tfrac{1}{2}\gamma_i^2(t) - \tfrac{1}{2}[\gamma_i^* - \gamma_i(t)]^2 + \tfrac{1}{2}\gamma_i^{*2}\} \\ \leq \ & qh_i\sigma_i(t)\{-\tfrac{1}{2}[\gamma_i^* - \gamma_i(t)]^2 + \tfrac{1}{2}\gamma_i^{*2}\}, \end{aligned}$$

(4.15) can be rewritten as

$$\begin{aligned} \dot{V}_i(t) \ \leq \ & -\tfrac{k}{2}S_i^2(t) - \tfrac{\iota_i qh_i\sigma_i(t)}{2\iota_i}[\gamma_i^* - \gamma_i(t)]^2 \\ & + qh_i[\tfrac{\sigma_i(t)}{2}\gamma_i^{*2} + 0.2785\varepsilon_i(\gamma_i^* + \Omega_i^*)]. \end{aligned} \tag{4.21}$$

Now, we construct a global Lyapunov function $V(t)$ as follows:

$$V(t) \ = \ \sum_{i=1}^{N} V_i(t).$$

In light of (4.21), it then follows that

$$\dot{V}(t) \ \leq \ -\zeta_1 V(t) + \zeta_2 \tag{4.22}$$

where the positive parameters ζ_1 and ζ_2 are given as follows

$$\begin{aligned} \zeta_1 \ &= \ \min\{k, \min_{1\leq i\leq N} \iota_i qh_i\sigma_i(t)\} \\ \zeta_2 \ &= \ \sum_{i=1}^{N} qh_i[\tfrac{\sigma_i(t)}{2}\gamma_i^{*2} + 0.2785\varepsilon_i(\gamma_i^* + \Omega_i^*)]. \end{aligned}$$

Now, multiplying both sides of (4.22) by $e^{\zeta_1 t}$ and then integrating both sides with respect to t over the range $(0,t)$ gives

$$\begin{aligned} V(t) \ & \leq \ [V(0) - \tfrac{\zeta_2}{\zeta_1}]e^{-\zeta_1 t} + \tfrac{\zeta_2}{\zeta_1} \\ & \leq \ V(0) + \tfrac{\zeta_2}{\zeta_1} \end{aligned} \tag{4.23}$$

which implies that

$$\begin{aligned} \tfrac{1}{2}\sum_{i=1}^{N} S_i^2(t) \ & \leq \ V(0) + \tfrac{\zeta_2}{\zeta_1} \\ \sum_{i=1}^{N} \tfrac{1}{2\iota_i}[\gamma_i^* - \gamma_i(t)]^2 \ & \leq \ V(0) + \tfrac{\zeta_2}{\zeta_1}. \end{aligned} \tag{4.24}$$

It follows from (4.24) that

$$\begin{aligned} \|S(t)\| &\leq \sqrt{2(V(0)+\tfrac{\zeta_2}{\zeta_1})} \\ \|\tilde{\gamma}_1(t)\| &\leq \sqrt{2\bar{\iota}(V(0)+\tfrac{\zeta_2}{\zeta_1})}, \bar{\iota} = \min_{1\leq i\leq N} \iota_i \end{aligned} \tag{4.25}$$

where

$$\begin{aligned} S(t) &= \begin{bmatrix} S_1(t) & S_2(t) & \cdots & S_N(t) \end{bmatrix}^T \\ \tilde{\gamma}(t) &= \begin{bmatrix} \gamma_1^* - \gamma_1(t) & \gamma_2^* - \gamma_2(t) & \cdots & \gamma_N^* - \gamma_N(t) \end{bmatrix}^T. \end{aligned}$$

Consequently, the sliding mode $S_i(t)$ and the adaptive updated law $\gamma_i(t)$ are bounded from (4.25). Moreover, following the fact (4.23), $S(t)$ and $\gamma(t)$ will converge exponentially to $\sqrt{\frac{\zeta_2}{\zeta_1}}$. Therefore, it is clear that reducing ζ_2, meanwhile increasing ζ_1 will lead to smaller bounds of $S_i(t)$, i.e., $S_i(t)$ can converge to an arbitrary small neighborhood of zero by choosing the design parameters appropriately. Then, according to (4.6), $s_i(t)$ and the spacing error $e_i^x(t)$ can also practically converge to zero.

Next, the strong string stability of the whole vehicle control system is established by using a similar method as in [93]. Since $S_i(t) = qs_i(t) - s_{i+1}$ practically converges to zero by choosing the design parameters appropriately, we have

$$q(e_i^x(t) + \textstyle\int_0^t \lambda e_i^x(\tau)d\tau) \approx e_{i+1}^x(t) + \textstyle\int_0^t \lambda e_{i+1}^x(\tau)d\tau. \tag{4.26}$$

Since $e_i^x(0) = 0$ and $\int_{-\infty}^{0} e_i^x(t)dt = 0$, taking Laplace transform of (4.26), we can get

$$q(E_i^x(s) + \tfrac{\lambda}{s}E_i^x(s)) \approx E_{i+1}^x(s) + \tfrac{\lambda}{s}E_{i+1}^x(s).$$

Then, $G_i(s) \triangleq \frac{E_{i+1}^x(s)}{E_i^x(s)} \approx q$. Since strong string stability requires that neighboring-vehicle spacing errors do not amplify as they propagate along the platoon, q should satisfy $0 < q \leq 1$ to guarantee $\|G_i(s)\| \leq 1$. Thus, the proof is completed. □

Remark 4.3 *As shown in many existing results [75, 108, 211], the discontinuity of the sign function in the control law causes the chattering. Therefore, in order to avoid the chattering, the discontinuous sign function is replaced by the continuous* tanh *function in (4.10). Furthermore, a centralized control problem is considered in [45, 46] where the information of all the vehicles in the string is used to obtain the optimal controller, while a distributed control method suitable for platoon control implementation is used in Theorem 4.1.*

Remark 4.4 *Although the TCTH policy has the advantage of improving string stability without requiring a lot of data from other vehicles and it can ensure string stability using on-board information only [2, 164], one obvious drawback of this policy is low traffic density. In practice, with the same velocity, the higher the traffic density is, the bigger the traffic flow is. Consequently, in order to overcome the above two problems, a MCTH policy will be provided in the next subsection.*

4.3.2 Control Strategy 2: MCTH Control Law

In order to remove the assumption of zero initial spacing errors in Theorem 4.1 and simultaneously increase the traffic density, we develop a new *MCTH policy* as follows:

$$\begin{aligned} e_i^x(t) &= \tilde{e}_i^x(t) - \Xi_i(t) \\ \tilde{e}_i^x(t) &= x_{i-1}(t) - x_i(t) - \delta_i - h_i e_i^v(t) \\ e_i^v(t) &= v_i(t) - v_0(t) \\ \Xi_i(t) &= [\tilde{e}_i^x(0) + (\zeta_i \tilde{e}_i^x(0) + \dot{\tilde{e}}_i^x(0))t]e^{-\alpha_i t} \end{aligned} \tag{4.27}$$

where $\tilde{e}_i^x(0) = \tilde{e}_i^x(t)|_{t=0}$, $\dot{\tilde{e}}_i^x(0) = \dot{\tilde{e}}_i^x(t)|_{t=0}$ and α_i are strictly positive constants. In addition, we can obtain that

$$e_i^x(t)|_{t=0} = 0, \ \dot{e}_i^x(t)|_{t=0} = 0$$

which shows that the modified spacing errors are initially zero for arbitrary initial spacing errors. Obviously, $\Xi_i(t)$ is bounded and denoted as $\|\Xi_i(t)\| \leq \Xi_M$. Then, use the same sliding mode surfaces $s_i(t)$ and $S_i(t)$ in (4.6), to obtain a new control strategy.

Remark 4.5 *The parameters of MCTH policy and its network topology are shown in Figure 4.1(b) and Figure 4.1(c), respectively. The steady state vehicle spacing of the MCTH policy $\tilde{e}_i^x(t)$ in (4.27) will be the same as the one of the CS policy considered in [38, 43, 45, 46, 93, 169, 212]. In contrast to the CS policy, because of introducing the time headway h_i, the MCTH policy $\tilde{e}_i^x(t)$ has better dynamic performance than that of the CS policy. Moreover, in contrast to the TCTH policy in (4.4), the MCTH policy $\tilde{e}_i^x(t)$ in (4.27) is an efficient way toward satisfactorily improved traffic flow and greatly increased highway capacity by reducing the vehicles' interspacing at high speed. Since the information of leader vehicle needs to be broadcast to all followers in (4.27), it will reduce the risk of collisions as opposed to the one proposed in [2] where a virtual truck is introduced. In addition, the method proposed in [2] involving the frequency-domain transfer function cannot be applied to nonlinear vehicle-following systems. However, all the above advantages for (4.27) can be obtained at the cost of communication load.*

Remark 4.6 *Study of the information flow and interaction among multiple vehicles in a platoon shows that the TCTH policy (4.4) and the MCTH policy (4.27) have different communication topologies, i.e., bidirectional topology for (4.4) and bidirectional-leader topology for (4.27), which are shown in Figures 4.1(b) and (c), respectively.*

Then, the derivative of $S_i(t)$ with respect to t is given by

$$\begin{aligned} \dot{S}_i(t) = & \ q\{v_{i-1}(t) - v_i(t) - \dot{\Xi}_i(t) - h_i[u_i(t) \\ & + f_i(x_i(t), v_i(t), t)]\} + A_i(t) \end{aligned} \tag{4.28}$$

where

$$\begin{aligned} A_i(t) &= \begin{cases} qh_i a_0(t) + q\lambda e_i^x(t) - \dot{e}_{i+1}^x(t) - \lambda e_{i+1}^x(t) \\ \qquad \text{for } i \in \mathscr{V}_N/\{N\} \\ qh_i a_0(t) + q\lambda e_i^x(t), \text{for } i = N. \end{cases} \\ \dot{\Xi}_i(t) &= -\alpha_i[\tilde{e}_i^x(0) + (\alpha_i \tilde{e}_i^x(0) + \dot{\tilde{e}}_i^x(0))t]e^{-\alpha_i t} \\ &\quad + [\alpha_i \tilde{e}_i^x(0) + \dot{\tilde{e}}_i^x(0)]e^{-\alpha_i t}. \end{aligned} \tag{4.29}$$

In order to obtain the bounded stability of all individual vehicles and guarantee the *strong string stability* of the whole vehicle control system, a new distributed adaptive ISMC law based on the MCTH policy (4.27) is given by

$$\begin{aligned} u_i(t) &= \tfrac{1}{h_i}[v_{i-1}(t) - v_i(t) - \dot{\Xi}(t)] + [\rho_i \textstyle\int_0^t |S_i(\tau)| d\tau \\ &\quad + \gamma_i(t)] \tanh(\tfrac{S_i(t)}{\varepsilon_i}) + \tfrac{1}{qh_i} A_i(t) + \tfrac{k}{2qh_i} S_i(t) \end{aligned} \tag{4.30}$$

where $A_i(t)$ is defined as in (4.29). In addition, the adaptive update law $\gamma_i(t)$ is the same as (4.12).

The following theorem based on the MCTH policy (4.27) is proposed by following similar steps as in Theorem 1.

Theorem 4.2 *Consider the vehicle-following system with N followers (4.1) and a leader (4.2). Under Assumption 4.1, the distributed adaptive ISMC laws (4.12) and (4.30) guarantee that the spacing errors $e_i^x(t)$ in (4.27) converge to an arbitrarily small neighborhood of zero by choosing the design parameters appropriately, while strong string stability of the vehicle platoon also can be guaranteed for $i = 1, 2, \cdots, N$ when q satisfies $0 < q \leq 1$.*

Proof 4.2 *It can be completed by following similar steps as in the proof of Theorem 4.1 and by using the following assertion:*

$$\begin{aligned} &\| f_i(x_i(t), v_i(t), t) - f_0(x_0(t), v_0(t), t)\| \\ = &\ \|\sum_{j=1}^{i} [f_j(x_j(t), v_j(t), t) - f_{j-1}(x_{j-1}(t), v_{j-1}(t), t)]\| \\ \leq &\ \sum_{j=1}^{i} [\mu_j \|x_{j-1}(t) - x_j(t) - \delta_j - h_j e_j^v(t) - \Xi_j(t) + \delta_j \\ &\ + h_j e_j^v(t) + \Xi_j(t)\| + \eta_j \|v_{j-1}(t) - v_j(t)\| + \theta_j] \\ = &\ \sum_{j=1}^{i} [\mu_j \|e_j^x(t)\| + \mu_j \|\delta_j + h_j e_j^v(t) + \Xi_j(t)\| \\ &\ + \eta_j \|e_{j-1}^v(t) - e_j^v(t)\| + \theta_j] \\ \leq &\ \sum_{j=1}^{i} [\mu_j \|e_j^x(t)\| + \mu_j h_j \|e_j^v(t)\| + \eta_j \|e_{j-1}^v(t) - e_j^v(t)\|] \\ &\ + \sum_{j=1}^{i} (\mu_j \delta_j + \mu_j \Xi_M + \theta_j) \\ \leq &\ \Omega_i^* + \gamma_i^* \end{aligned} \tag{4.31}$$

where

$$\begin{aligned}\Omega_i^* &= \sum_{j=1}^{i}[\mu_j\|e_j^x(t)\|+\mu_j h_j\|e_j^v(t)\|+\eta_j\|e_{j-1}^v(t)-e_j^v(t)\|]\\ \gamma_i^* &\geq \sum_{j=1}^{i}(\mu_j\delta_j+\mu_j\Xi_M+\theta_j).\end{aligned}$$ □

Remark 4.7 *It should be emphasized that the steady-state traffic density of the MCTH policy (4.27) will be the same as the CS policy considered in [38, 43, 45, 46, 93, 169, 212] and will be much higher than the TCTH policy (4.4). However, the MCTH policy (4.27) will improve the control performance at the cost of communication load since the leader vehicle's information needs to be broadcast through a wireless network.*

Remark 4.8 *In practical applications, selecting suitable controller parameters in (4.10) and (4.30) is crucial to realize good control performance. However, how to choose a set of optimal values for these design parameters is still an open problem for the control community. All of the design parameters are chosen by trial and error method. Nevertheless, we would like to provide some general guidelines as follows.*

(a) The positive parameter k in the control law (4.10) and (4.30) is very important, and it is one of the parameters relating to the convergence rate in the stability analysis. A larger k leads to a faster convergence rate. However, in practice, a compromise is needed between the convergence rate and control input. Since too big a k will require a very high control input, which may result in control saturation, thus the parameter k cannot be selected to be too large.

(b) The design parameter q satisfies $0 < q \leq 1$, which couples the sliding surfaces $s_i(t)$ and $s_{i+1}(t)$. It plays an important role in guaranteeing strong string stability. The larger the value of q is, the stronger the string stability can be achieved. However, the control energy will also become larger.

(c) An integral term $\int_0^t \lambda e_i^x(\tau)d\tau$ is introduced in (4.6) to decrease the stable/steady-state errors (similar to an integral controller), where λ is a positive design parameter. It should be pointed out that too large a λ may result in the degradation of the whole platoon's transient performance when there exist large initial spacing errors.

(d) If the design parameters ε_i and $\sigma_i(t)$ are set smaller but $\rho_i > 0$ and $\iota_i > 0$ are chosen larger, then the effect of the adaptive compensation for the nonlinear acceleration uncertainties and external disturbances could be better. However, too small ε_i and $\sigma_i(t)$ and/or too large ρ_i and ι_i may result in control shock or chattering phenomenon. Furthermore, for how to choose the values of these adaptive parameters, one can refer to the many existing results on adaptive techniques such as [84, 85, 108].

(e) In order to make the convergence rate of $e_i^x(t)$ to $\tilde{e}_i^x(t)$ in (4.27) faster, α_i needs to be set a little large.

Therefore, in practice, these design parameters should be adjusted carefully according to the required control performance and the available control resource.

4.4 Simulation Study and Performance Results

In this section, the feasibility of the proposed method and the control performances are illustrated by the following two examples. All experiments were performed on MATLAB version 2007b and a notebook computer with 1.8GHz Intel (R) Core (TM) i5-3337U CPU and 8.00GB RAM using Windows 10.

4.4.1 Example 1 (Numerical Example)

In this subsection, a numerical example and comparisons between TCTH control law (4.10) and MCTH control law (4.30) are provided to confirm the effectiveness of the proposed control method.

Suppose there is a vehicle platoon of six follower vehicles and a leader. The initial position of the leader is set as $x_0(0) = 20m$, $v_0(0) = 1m/s$ and the acceleration of the leader is given by

$$a_0(t) = \begin{cases} 0.5 \quad m/s^2, & 2s \leq t < 5s \\ -1 \quad m/s^2, & 15s \leq t < 20s \\ 0 \quad m/s^2, & otherwise. \end{cases}$$

The standstill distance is $\delta_i = 0.5m$ with a time headway $h_i = 1s$ for all $i = 1, 2, \cdots, 6$. The nonlinear acceleration disturbances are

$$f_i(x_i(t), v_i(t), t) = (\frac{x_i^2(t)}{1+x_i^2(t)} + 0.1v_i^2(t) + 0.5v_i(t))e^{-(t-3+0.5i)^2} + 0.1\sin(t) \quad \text{for } i = 0, 1, \cdots, 6.$$

The controller parameters are $k = 100$, $q = 0.8$, $\lambda = 2$, $\rho_i = 20$, $\alpha_i = 20$, $\varepsilon_i = 0.1$, $\sigma_i(t) = 10e^{-5t}$, and the gain $\iota_i = 0.1$. In addition, for simulation purposes, the *initial conditions* of the follower vehicles are assumed as follows.

$$\begin{aligned} x(0) &= [18.5, 17, 15.5, 14, 12.5, 11] \\ v(0) &= [1, 1, 1, 1, 1, 1] \\ \gamma_i(0) &= 1, s_i(0) = 0, i = 1, 2, \cdots, 6. \end{aligned}$$

When using TCTH policy (4.4), the initial values of $e^x(0)$ and $e^v(0)$ are zero. The simulation results by using TCTH policy (4.4) are shown in Figure 4.2. The simulation results of Figures 4.2(b) and (c) show that the neighboring-vehicle distances are proportional to vehicle speed, and can become very large at high velocity. The collision is avoided not only in steady-state condition, but also during the initial transient since there do not exist cross and overlapped positions in Figure 4.2(b). The bounded stability of the spacing errors is clearly shown in Figure 4.2(a) and the amplitude of the spacing error decreases through the string of vehicles (i.e., $\|e_6(t)\| \leq \cdots, \leq \|e_1(t)\|$), which means that the *strong string stability* of the whole vehicle platoon can be guaranteed despite the existence of nonlinear acceleration uncertainties. Figure 4.2(d) shows that the velocity of the follower vehicles converges towards the velocity of the leader. The input control $u_i(t)$ and sliding surface $S_i(t)$ are shown in Figures 4.2(e) and (f), respectively. As we can see, the *chattering phenomenon* is almost eliminated fully. Figure 4.2(g) shows that the adaptive gains $\gamma_i(t)$ are bounded.

To highlight the advantages of MCTH control law (4.30) over TCTH control law (4.10), the simulation results are shown in Figure 4.3 for comparison. From the definition of $\tilde{e}_i^x(t)$ in (4.27), the initial spacing errors $\tilde{e}_i^x(0)$ are non-zero as $\tilde{e}_i^x(0) = 1m$. From Figure 4.3(c), with the same speed $13.5m/s$, the neighboring-vehicle distances using MCTH policy ($0.5m$) are much smaller than those in Figure 4.2(c) ($14m$). The neighboring-vehicle distances by using MCTH policy are not dependent on the velocity and are nearly equal to $0.5m$, which is the same as those using *CS policy*. However, before reaching a steady state, the MCTH policy has better dynamic performance than that of the *CS policy* since h_i is introduced in MCTH policy. The *strong string stability* of the vehicle platoon subject to nonlinear acceleration uncertainties is verified in Figures 4.3(a) and (h). In addition, comparing Figure 4.3(d) with Figure 4.2(d), one can conclude that using the leader's velocity information leads to faster convergence. Considering these facts, MCTH control law (4.30) provides higher performance than that of TCTH control law (4.10). These results show the advantages of the MCTH policy (4.27).

4.4.2 Example 2 (Practical Example)

To show the effectiveness of the proposed strategy in a real system, we study a platoon of HSTs moving on the railway line. The dynamic motion equations for each train can be established as follows [101]:

$$\begin{aligned} \dot{x}_i(t) &= v_i(t), \text{for } i \in \mathscr{V}_N \cup \{0\} \, (N=6) \\ m_i \dot{v}_i(t) &= F_i(t) - m_i(c_{i0} + c_{i1} v_i(t) + c_{i2} v_i^2(t)) \end{aligned}$$

where $x_i(t)$ and $v_i(t)$ are the position and speed of train i, respectively, m_i is the ith train's mass, and $F_i(t)$ is the control force of the ith train. The above dynamic motion equations can be rewritten as

$$\begin{aligned} \dot{x}_i(t) &= v_i(t) \\ \dot{v}_i(t) &= u_i(t) + f_i(x_i(t), v_i(t), t), i \in \mathscr{V}_N \cup \{0\} \, (N=6) \end{aligned}$$

where

$$u_i(t) = \frac{F_i(t)}{m_i}, \; f_i(x_i(t), v_i(t), t) = -(c_{i0} + c_{i1} v_i(t) + c_{i2} v_i^2(t)).$$

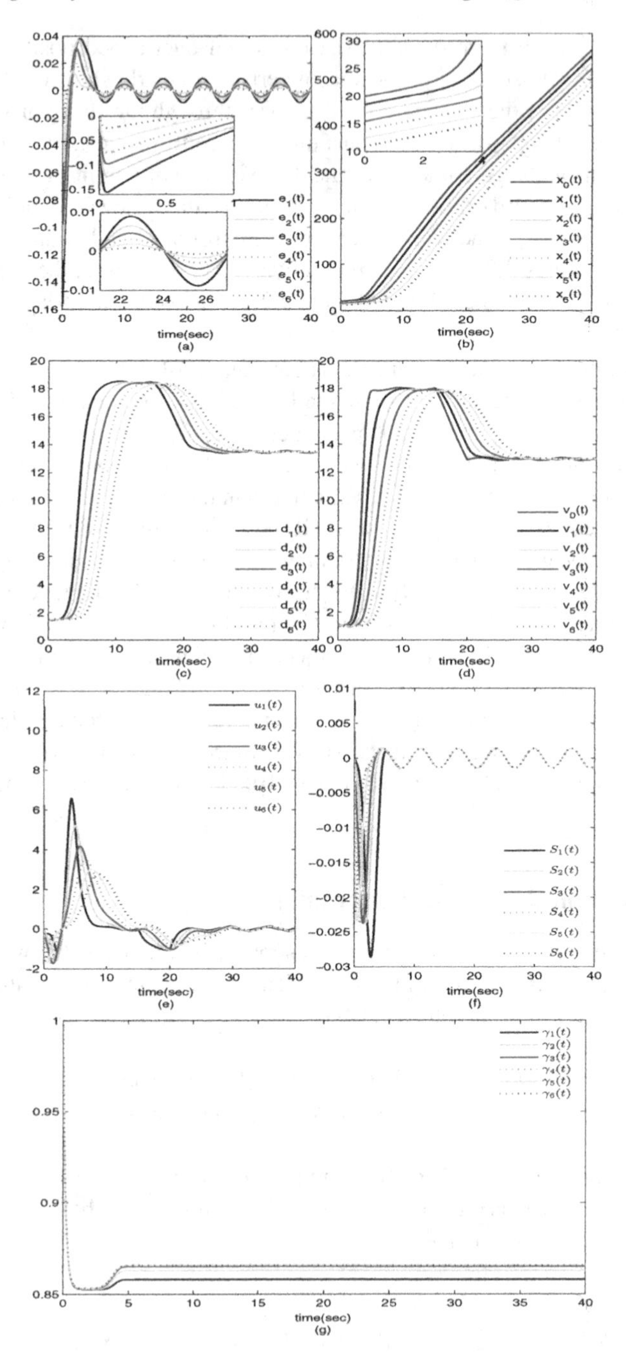

FIGURE 4.2
(a) Spacing error $e_i(t)$; (b) position $x_i(t)$; (c) distance $d_i(t)$; (d) velocity $v_i(t)$; (e) control input $u_i(t)$; (f) sliding surface $S_i(t)$; (g) estimation $\gamma_i(t)$ by using TCTH control law (4.10).

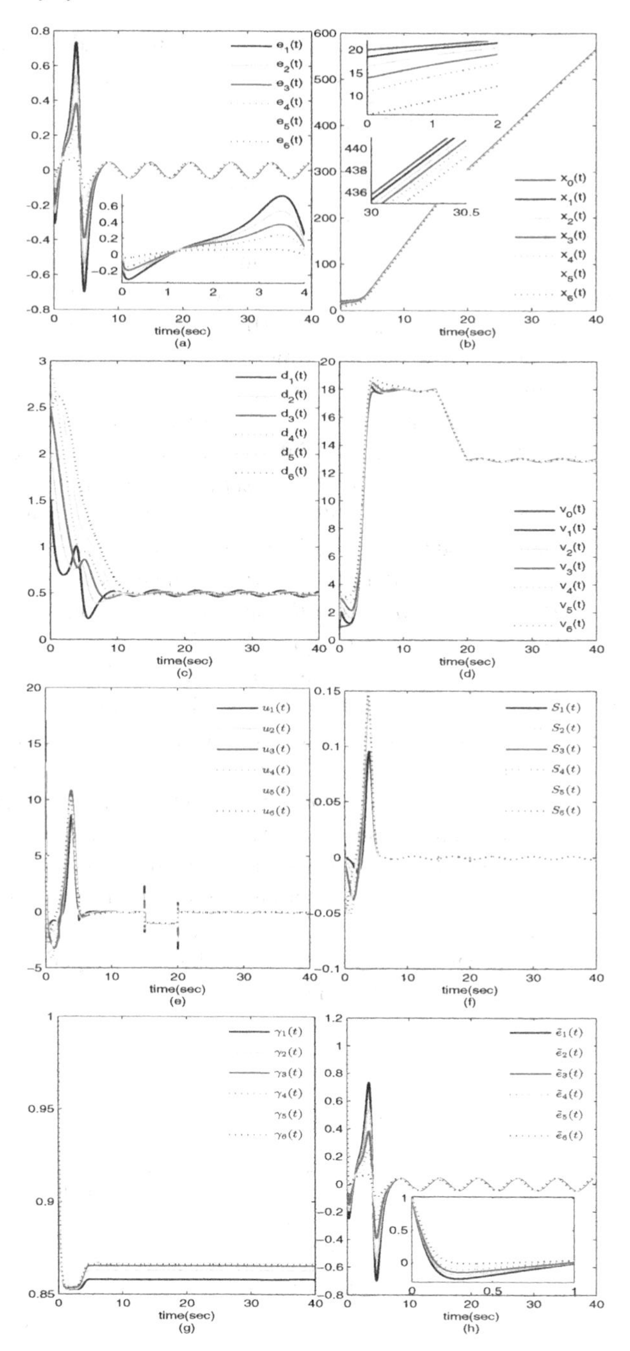

FIGURE 4.3
(a) Spacing error $e_i(t)$; (b) position $x_i(t)$; (c) distance $d_i(t)$; (d) velocity $v_i(t)$; (e) control input $u_i(t)$; (f) sliding surface $S_i(t)$; (g) estimation $\gamma_i(t)$; (h) spacing error $\tilde{e}_i(t)$ by using MCTH control law (4.30).

We assume that the leader high speed train's behavior is independent of its followers, and $u_0(t)$ (that is, $a_0(t)$ in (4.2)) keeps changing during the entire motion process as

$$u_0(t) = \begin{cases} 0.5t & km/h^2, \quad 2h \leq t < 4h \\ -1 & km/h^2, \quad 10h \leq t < 15h \\ 0 & km/h^2, \quad otherwise. \end{cases}$$

We also assume that the *high speed trains* are homogenous, i.e., each train has the same system parameters. The system parameters of each high speed train are given in Table 4.1 [101, 194], while the initial positions and speeds of each high speed train are presented as in Table 4.2 [101]. Assume that the minimal safety headway distances between two neighboring trains are 20 km, i.e., $\delta_i = 20$. Then, we choose the control parameters as $k = 10$, $q = 0.8$, $\lambda = 5$, and other parameters ρ_i, α_i, ε_i, $\sigma_i(t)$, and ι_i are the same as those in Example 1. According to the definition of $\tilde{e}_i^x(t)$ in (4.27), the initial spacing errors $\tilde{e}_i^x(0)$ are non-zero as $\tilde{e}^x(0) = [20, 34, 82.5, 103, 64.5, 213]km$. At the same time, the initial velocity errors $e^v(0)$ are non-zero as $e^v(0) = [-10, -30, -80, -100, -60, -210]$. In order to highlight the effectiveness and superiority of the proposed control strategy, only the responses of spacing errors $e_i^x(t)$ and the neighboring-vehicle distances $d_i(t)$ are shown here. When the effect of the *non-zero initial* spacing errors is not considered as in [38, 43, 45, 46, 93, 169, 212], $e_i^x(t)$ is equal to $\tilde{e}_i^x(t)$, i.e., $\Xi_i(t) = 0$. For this case, the simulation results are shown in Figures 4.4(a) and (b). When $\Xi_i(t) \neq 0$, the simulation results are shown in Figures 4.5(a) and (b). From Figure 4.4(b) and Figure 4.5(b), we can see that the distances between two neighboring trains converge to 20 km, which means that the velocity of the whole platoon can converge to that of the leader. Furthermore, it is clear that the *strong string stability* cannot be guaranteed in Figure 4.4(a), which may result in collision between neighboring trains since one distance between neighboring trains is smaller than the minimal safety headway distance in Figure 4.4(b). In comparison, *strong string stability* can be guaranteed in Figure 4.5(a), and the minimal safety headway distances also can be ensured despite the effect of *non-zero initial spacing errors*, which shows the superiority and the necessity of our design method (4.30) with the consideration of the effect of *non-zero initial spacing errors*.

TABLE 4.1
Parameters of high speed train for $i \in \mathcal{V}_N \cup \{0\} (N = 6)$

Symbol	Value	Unit
m_i	80×10^4	kg
c_{i0}	0.01176	N/kg
c_{i1}	0.00077616	Ns/mkg
c_{i2}	1.6×10^{-5}	Ns^2/m^2kg
δ_i	20	km
h_i	1	h

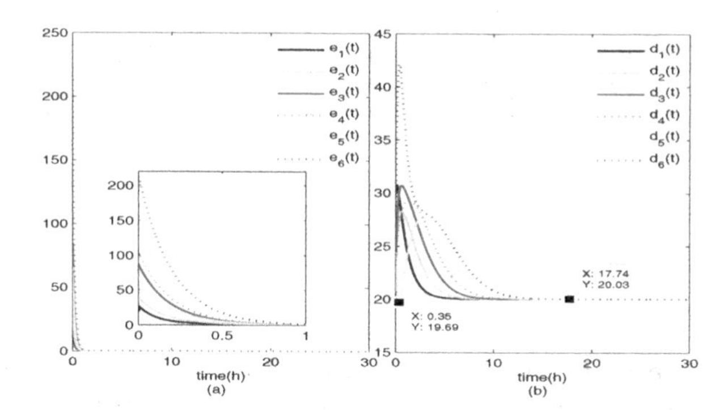

FIGURE 4.4
(a) Spacing error $e_i(t)$; (b) distance $d_i(t)$ by using MCTH control law (4.30) with $\Xi_i(t) = 0$.

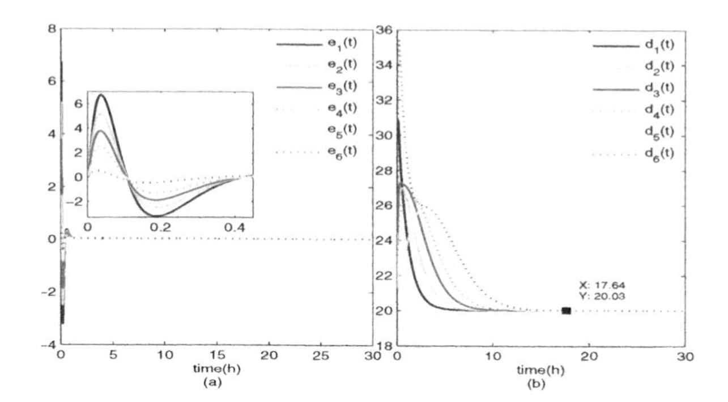

FIGURE 4.5
(a) Spacing error $e_i(t)$; (b) distance $d_i(t)$ by using MCTH control law (4.30) with $\Xi_i(t) \neq 0$.

TABLE 4.2
The parameters $x_i(0)$ and $v_i(0)$ for $i \in \mathscr{V}_N \cup \{0\} (N = 6)$

i	0	1	2	3	4	5	6
$x_i(0)(km)$	148	118	94	71.5	48.5	24	1
$v_i(0)(km/h)$	220	210	190	140	120	160	10

4.5 Conclusion

This chapter proposes two distributed adaptive ISMC strategies for *strong string stability* of nonlinear vehicle-following systems subject to nonlinear acceleration uncertainties in both the leader and the followers. The control strategies guarantee the uniformly and ultimately bounded stability of all spacing errors and *strong string stability* of the whole vehicle platoon. Moreover, a new MCTH policy is introduced to remove the assumption of zero *initial conditions* and simultaneously increase the traffic density, which is nearly equal to that of *CS policy*. Finally, *adaptive compensation* terms without requiring the prior knowledge of maximum values of the uncertainties are constructed to compensate for the time-varying effects caused by nonlinear acceleration uncertainties. Future research may include the effects of other nonlinearities in the actuator including saturation, hysteresis and quantized in later chapters.

5

CNN-Based Adaptive Control for Vehicle Platoon with Input Saturation

This chapter investigated the effect of input saturation on *strong string stability*. Combining Chebyshev neural networks (CNNs) with sliding mode technique, a new neural network-based adaptive control scheme is provided. In order to eliminate the effect of input saturation, a simple and straightforward strategy by adjusting only a single parameter is developed. By using CNN technique, the assumption as in the existing results that the nonlinearities of consecutive vehicles are assumed to satisfy the matching condition is removed.

5.1 Introduction

Intelligent vehicle highway systems and intelligent transportation systems have recently attracted considerable attention among researchers for addressing awfully crowded urban traffic [150]. The vehicle-following platoon architecture, which requires vehicles within each lane to be organized into platoons and capable of maintaining a small neighboring-vehicle spacing, may help to improve lane capacity and reduce energy consumption [27]. It is desirable, from the point of preventing collisions, that the real distances between adjacent vehicles maintain a *desired safe spacing* [2, 27, 142] and neighboring-vehicle *spacing errors* do not amplify as they propagate along the platoon [2], i.e., guaranteeing *string stability* [140]. Control methods for ensuring platoon string stability exist in [2, 34, 93, 142, 212], however, under the unrealistic assumption that initial spacing/velocity errors are zero. Since the non-zero initial spacing/velocity errors may result in *string instability*, Chapters 3 and 4 remove this assumption and further avoid the low traffic density of the traditional constant time headway (TCTH) policy. However, the nonlinear vehicle dynamics considered in Chapter 4 have two assumptions that the nonlinearities of the consecutive vehicles (including the leader) should satisfy the *Lipschitz* constraints and should exactly match each other. These may restrict the application domain of the proposed approach in Chapter 4 in real applications. On the other hand, although the unreliable vehicular networking with packet loss, transmission error and communication time delay has recently been investigated based on graph theory [8, 27, 82, 153], input saturation unavoidable in almost all real applications due to the limited capability of any physical actuator and/or safety constraints is not considered. If its adverse effect is neglected, input saturation will usually result in performance degradation and even instability [20, 76, 175]. For this reason, many effectiveness methods have been proposed to handle input saturation, such as *adaptive compensation* [76, 175] and neural network control [20] methods, etc. However, the control problem for vehicle-following platoon with input saturation becomes much more complex and remains an open problem. How to deal with this problem has been a task of major practical interest as well as theoretical significance.

Unmodeled nonlinear dynamics often exist in vehicle-following systems because of the measurement noise, modeling errors, and environment disturbances such as wind gusts and rough road surface. *Platoon control* with unmodeled nonlinear dynamics is challenging since they can severely degrade the performance of the platoon and even result in *string instability* of the whole vehicle platoon. Therefore, how to mitigate the effects of unmodeled nonlinear dynamics is very interesting. It is known that neural networks (NNs) have the capability to approximate any continuous functions over a compact set to arbitrary accuracy [217]. In particular, *Chebyshev neural network* (CNN) is a functional link network whose input is generated by using a subset of Chebyshev polynomials (CPs), and it has been shown that CNN has powerful approximation capabilities [135]. Although radial basis function neural network (RBFNN) has been successfully employed to solve the control problem for vehicle-following platoon in [116, 138], we are interested in investigating the *string stability* issue with CNN (instead of RBFNN) due to the fact that only one parameter (i.e., the order of the CP basis) is required to determine the CP basis [217], which will significantly simplify our control design.

Motivated by the above discussion, the problem of distributed adaptive NN control for vehicle-following platoon with unmodeled nonlinear dynamics, unknown *external disturbances* and input saturation is investigated. The proposed control schemes can guarantee the *strong string stability* of the whole vehicle-following platoon and the uniform ultimate boundedness of all signals. The *spacing errors* will converge to a small neighborhood of the origin. The main contributions of the proposed schemes are highlighted as follows.

(1) As shown in Remark 5.3, unlike the existing methods in [20, 76, 175], by exploring a simple and straightforward method of adjusting only one parameter, the effect of input saturation can be attenuated.

(2) Unlike the method in Chapter 4, vehicles are governed by a more general model where the nonlinearities of the consecutive vehicles are not required to satisfy matching conditions. In addition, two control schemes respectively based on TCTH policy and MCTH policy introduced in Chapter 4 are proposed.

(3) Sliding mode technique [209] combined with adaptive NN is developed for the control of vehicle-following platoon with unmodeled nonlinear dynamics, unknown *external disturbances* and input saturation. The implementation of the basis functions of CNN depends only on the leader's velocity and acceleration. By adopting the CNN technique, the requirement of matching conditions for

the nonlinearities of the consecutive vehicles as in Chapter 4 is removed. It is worth mentioning that a common assumption in [20, 217] that the reference signal and its derivatives are assumed to be bounded will not be required.

5.2 Vehicle-Following Platoon Model and Preliminaries

5.2.1 Vehicle-Following Platoon Description

Suppose a vehicle platoon with N follower vehicles and one leader vehicle (labeled as 0) runs in a straight line as shown in Figure 5.1. Let $(x_i(t), v_i(t)) \in \mathbf{R}^2$ denote the position and velocity of vehicle $i \in \mathscr{V}_N$, and $(x_0(t), v_0(t), a_0(t)) \in \mathbf{R}^3$ denote the position, velocity and acceleration of the leader, respectively. Each follower vehicle i in the platoon is characterized by the following second-order nonlinear time-variant dynamics

$$\begin{aligned} \dot{x}_i(t) &= v_i(t), i \in \mathscr{V}_N \\ \dot{v}_i(t) &= \mathrm{sat}(u_i(t)) + f_i(x_i(t), v_i(t), t) + w_i(t) \end{aligned} \tag{5.1}$$

where $u_i(t)$ denotes the control input, $f_i(x_i(t), v_i(t), t)$ is the unknown nonlinear effect, which is assumed to be smooth continuous bounded on a compact set Ω, and models vehicle acceleration disturbances, wind gust, parameters uncertainties and intermediate uncertainties induced by networks. This unknown nonlinear function $f_i(x_i(t), v_i(t), t)$ can be considered as a bounded unmodeled dynamic, which is more general than the one considered in [58]. The disturbance input $w_i(t)$ is assumed bounded by $\|w_i(t)\| \leq \bar{w}$. The control input $u_i(t)$ is subject to saturation type nonlinearity described by

$$\mathrm{sat}(u_i(t)) = \begin{cases} u_{Mi}\mathrm{sgn}(u_i(t)), & \text{if} |u_i(t)| \geq u_{Mi} \\ u_i(t), & \text{if} |u_i(t)| \leq u_{Mi} \end{cases} \tag{5.2}$$

where u_{Mi} is a known bound of $\mathrm{sat}(u_i(t))$. Obviously, the relationship between the applied control $\mathrm{sat}(u_i(t))$ and the input control $u_i(t)$ has a sharp corner when $|u_i(t)| = u_{Mi}$. Similar to [175, 20], $\mathrm{sat}(u_i(t))$ is approximated by a smooth function defined as

$$g(u_i(t)) = u_{Mi}\tanh(u_i(t)/u_{Mi}). \tag{5.3}$$

The approximation of the saturation function and the approximation error $\Delta u_i(t) = \mathrm{sat}(u_i(t)) - g(u_i(t))$ are shown in Figure 5.2(a). From Figure 5.2(a), one can find the approximation error $\Delta u_i(t)$ is a bounded function in time and its bound can be obtained as

$$\begin{aligned} |\Delta u_i(t)| &= |\mathrm{sat}(u_i(t)) - g(u_i(t))| \\ &\leq u_{Mi}(1 - \tanh(1)) := \bar{D}_i. \end{aligned}$$

Note that, according to the mean-value theorem, $g(u_i(t))$ can be rewritten as

$$g(u_i(t)) = g(u_0) + g_{u_i}(t)(u_i(t) - u_0) \tag{5.4}$$

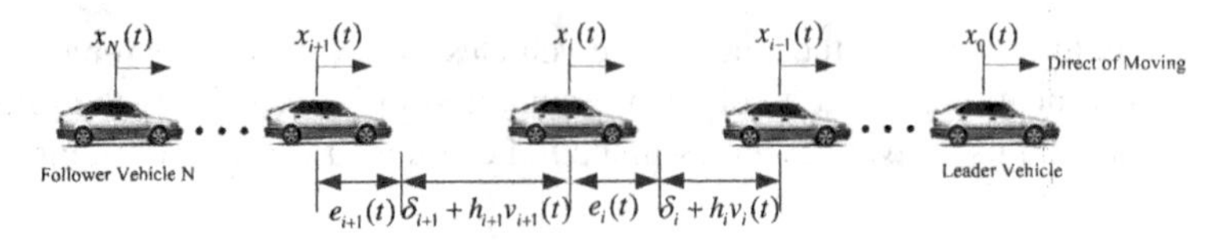

FIGURE 5.1
Configuration of vehicle-following platoon under TCTH policy.

where

$$\begin{aligned} g_{u_i}(t) &= \frac{\partial g(u_i(t))}{\partial u_i(t)}\Big|_{u_i(t)=u_{\mu i}} \\ &= \frac{4}{(e^{u_i(t)/u_{Mi}}+e^{-u_i(t)/u_{Mi}})^2}\Big|_{u_i(t)=u_{\mu i}} \le 1 \end{aligned}$$

with

$$u_{\mu i} = \mu u_i(t) + (1-\mu)u_0, \ (0<\mu<1).$$

The above relation (5.4) has been used extensively in the literature, such as [20, 175]. Since $g(0)=0$, by choosing $u_0=0$, (5.4) can be further rewritten as

$$g(u_i(t)) = g_{u_i}(t)u_i(t). \tag{5.5}$$

Then, the follower dynamics (5.1) can be transformed as follows

$$\begin{aligned} \dot{x}_i(t) &= v_i(t), i \in \mathscr{V}_N \\ \dot{v}_i(t) &= g_{u_i}(t)u_i(t) + f_i(x_i(t), v_i(t), t) + \Delta u_i(t) + w_i(t) \end{aligned} \tag{5.6}$$

where $\Delta u_i(t) = \text{sat}(u_i(t)) - g(u_i(t))$. In addition, from Figure 5.2(b), one can find $0 < \frac{\partial g(u_i(t))}{\partial u_i(t)} \le 1$ for any $u_i(t)$, therefore, there exists a sufficiently small parameter $g_m > 0$ such that $0 < g_m \le g_{u_i}(t) \le 1$ as in [175].

Remark 5.1 *In fact, the parameter $g_{u_i}(t) \in (0,1]$ in (5.5) can be viewed as an indicator for the degree of saturation at the actuator. When $g_{u_i}(t)$ approaches zero, there is almost no feedback from input $u_i(t)$, while $g_{u_i}(t)=1$ means that $u_i(t)$ does not saturate, and $\Delta u_i(t)=0$. When $u_i(t) \to \infty$, $g_{u_i}(t)$ will approach zero, which will cause deep saturation, and there is no miracle control method for this case. Therefore, the parameter g_m can be viewed as a true reflection of the extent of saturation.*

Suppose that the dynamics of the leader is governed by

$$\dot{x}_0(t) = v_0(t); \ \dot{v}_0(t) = a_0(t) \tag{5.7}$$

where the acceleration $a_0(t)$ is a known function of time. According to practical experience, it is reasonable to assume that $v_0(t)$ and the acceleration $a_0(t)$ acting on the leader are bounded but the position $x_0(t)$ may be unbounded. For example, the velocity and acceleration of a bus on the road should be limited, where the velocity constraint is due to safety considerations and acceleration constraint is due to hardware limitations and comfortableness of drivers and passengers [83]. It is worth mentioning that all exiting results about tracking control problems by using

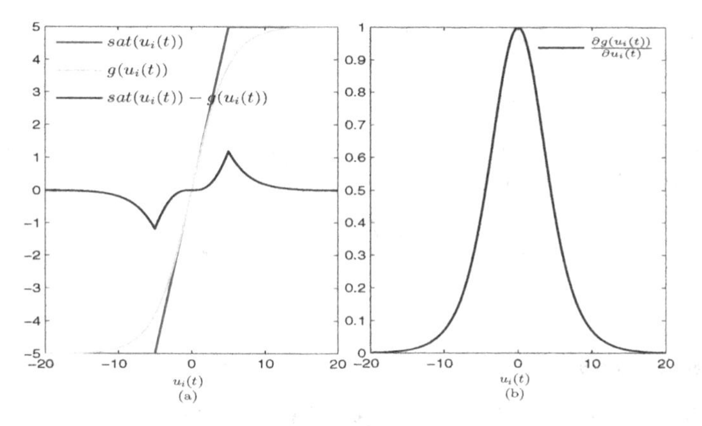

FIGURE 5.2
(a) Saturation function $\mathrm{sat}(u_i(t))$; approximation function $g(u_i(t))$; approximation error $\mathrm{sat}(u_i(t)) - g(u_i(t))$; (b) $\frac{\partial g(u_i(t))}{\partial u_i(t)}$.

neural network approach require that the reference signals (i.e., the leader's signals $(x_0(t), v_0(t), a_0(t))$ here) are assumed to be bounded (see, e.g., [20, 83, 217] and the references therein). Obviously, this assumption is not reasonable to the problem considered in this chapter since the position $x_0(t)$ is unbounded here.

Remark 5.2 *In the vehicle follower dynamics (5.1), each vehicle's nonlinear dynamic function $f_i(x_i, v_i, t)$ is supposed to be unknown, hence it can meet the requirement of a lot of practical engineering. However, most of the existing nonlinear control schemes are limited to many strict assumptions, for example, in [182], the nonlinearities $f_i(x_i, v_i, t)$ of the agents (including the leader) should exactly match each other.*

5.2.2 Chebyshev Neural Network

The neural network (NN) structure employed in this chapter is a single layer *Chebyshev neural network* (CNN). The CNN has been shown to be capable of universally approximating nonlinear systems with any degree of nonlinearity to any degree of accuracy due to their inherent approximation capabilities [217]. CNN is a functional link network based on Chebyshev polynomials (CP). Thus, according to the universal approximation property of CNN, a nonlinear function $f(Z) : \mathbf{R}^m \to \mathbf{R}$ can be approximated by CNN as

$$f(Z) \quad = \quad W^* \xi(Z) + \varepsilon(Z) \tag{5.8}$$

where $Z \in \mathbf{R}^m$ is the input vector of CNN, $W^* \in \mathbf{R}^{1\times(mN_1+1)}$ is the optimal weight vector of the CNN with N_1 denoting the order of the CP and $\xi(Z) \in \mathbf{R}^{(mN_1+1)\times 1}$ being the CP basis function, and $\varepsilon(Z)$ is the function reconstruction error. For a given vector $Z \in \mathbf{R}^m$, $\xi(Z)$ is given by

$$\xi(Z) = \begin{bmatrix} 1 & T_1(z_1) & \cdots & T_{N_1}(z_1) & \cdots & T_1(z_m) & \cdots & T_{N_1}(z_m) \end{bmatrix}^T$$

where $T_k(z_j)$ $(k = 1, \cdots, N_1; j = 1, \cdots, m)$ can be generated by the following two-term recursive formula [217]

$$\begin{aligned} T_{k+1}(z_j) &= 2zT_k(z_j) - T_{k-1}(z_j) \\ T_0(z_j) &= 1, T_1(z_j) = z_j, z_j \in \mathbf{R}. \end{aligned}$$

5.3 Distributed Adaptive NN Control Design and Stability Analysis

In this section, the sliding mode control and CNN techniques are used to construct the distributed adaptive NN control schemes based on TCTH and MCTH policies to achieve the bounded stability of individual vehicle and *strong string stability* of the whole vehicle platoon. In addition, by adopting CNN technique, the requirement of matching conditions for the nonlinearities of the consecutive vehicles as in [58] is removed.

5.3.1 Control Scheme I: TCTH Policy

The aim of the *platoon control* is to track the speed of the leader while maintaining a desired safety spacing between consecutive vehicles. One of major spacing policies for vehicle-following platoons is *constant time headway* (CTH) policy, which is shown in Figure 5.1. For the *CTH policy*, the desired neighboring-vehicle spacing varies with vehicle velocity, which accords with driver behaviors to some extent but limits the achievable traffic capacity [212]. The TCTH policy is given out by (4.4) in Chapter 4. $d_i(t) = x_{i-1}(t) - x_i(t)$ is also used to denote the spacing between consecutive vehicles. The control objective here is to design a distributed adaptive neural network sliding mode algorithm for (5.1) with a leader (5.7) such that $v_i(t) \to v_0(t)$ and $d_i(t) \to \delta_i + h_i v_i(t)$. Simultaneously, for the vehicle-following *platoon control* problem, *strong string stability* defined in (4.7) of Chapter 4 is considered. Then, as in exiting results, the case of zero initial spacing/velocity errors is considered first in this subsection, while the case of *non-zero initial* spacing/velocity errors is considered in next subsection. Since the initial spacing errors $e_i^x(0)$ are assumed to be zero, define the same integral sliding surface and coupled sliding mode as in (4.6) of Chapter 4 to achieve the above objectives.

It is to be noted that if all vehicles' velocities $v_i(t)$ $(i \in \mathscr{V}_N)$ approach the leader's velocity $v_0(t)$, then the nonlinear function $f_i(x_i(t), v_i(t), t)$ converges to the *desired* nonlinear function $F_i(x_0(t), v_0(t), t)$ defined by

$$F_i(x_0(t), v_0(t), t) = f_i(\textstyle\int_0^t v_0(\tau)d\tau - \sum\limits_{j=1}^{i} \delta_j - \sum\limits_{j=1}^{i} h_j v_0(t), v_0(t), t).$$

Using the approximation property of the CNN as (5.8), the unknown function $F_i(x_0(t), v_0(t), t)$ $(i \in \mathscr{V}_N)$ can be approximated over a compact set Ω_0 by

$$F_i(x_0(t), v_0(t), t) = W_i^* \xi_i(v_0(t), a_0(t)) + \varepsilon_i(v_0(t), a_0(t)) \quad (5.9)$$

where $W_i^* \in \mathbf{R}^{1\times(2N_1+1)}$ with N_1 being the order of the CP is the optimal weight matrix of CNN, $\varepsilon_i(v_0(t), a_0(t)) \in \mathbf{R}$ is the CNN approximation error, and $\xi_i(v_0(t), a_0(t)) \in \mathbf{R}^{(2N_1+1)\times 1}$ is called the CP basis function. To simplify writing, $\xi_i(v_0(t), a_0(t))$ and $\varepsilon_i(v_0(t), a_0(t))$ are abbreviated to $\xi_i(t)$ and $\varepsilon_i(t)$, respectively, in some subsequent formulae. In addition, there exist optimal constant weights W_i^* such that the absolute value of $\varepsilon_i(t)$ is assumed to be less than a small positive constant $\bar{\varepsilon}_i$ (i.e., $|\varepsilon_i(t)| \leq \bar{\varepsilon}_i$). Moreover, W_i^* is bounded by $tr\{W_i^{*T} W_i^*\} \leq \bar{W}_i$. However, it is difficult to determine the bounds $\bar{W}_i$ and $\bar{\varepsilon}_i$; therefore, we further assume that $\bar{W}_i$ and $\bar{\varepsilon}_i$ are unknown positive parameters.

To facilitate the development of the main result, we denote

$$\Phi_i(t) = f_i(x_i(t), v_i(t), t) - F_i(x_0(t), v_0(t), t). \quad (5.10)$$

Since $f_i(x_i(t), v_i(t), t)$ is assumed to be smooth continuous bounded on the compact set Ω, the variable $\Phi_i(t)$ has a maximum $\bar{\Phi}_i$ on the compact set Ω. Then, recalling $|\Delta u_i(t)| \leq \bar{D}_i$, $|w_i(t)| \leq \bar{w}_i$ and $|\varepsilon_i(t)| \leq \bar{\varepsilon}_i$, one can obtain that

$$-qh_i[\Phi_i(t) + \Delta u_i(t) + w_i(t) + \varepsilon_i(t)]S_i(t) \leq qh_i\gamma_i^*|S_i(t)| \quad (5.11)$$

where $\gamma_i^* = \bar{\Phi}_i + \bar{D}_i + \bar{w}_i + \bar{\varepsilon}_i$.

Next, define $\tilde{W}_i(t) = W_i^* - W_i(t)$ and $\tilde{\gamma}_i(t) = \gamma_i^* - \gamma_i(t)$ $(i \in \mathscr{V}_N)$, where $W_i(t)$ is the estimation of W_i^* denoted in (5.9), and $\gamma_i(t)$ is used to estimate γ_i^* denoted in (5.11). Then, the control law for the ith vehicle in the formation is defined as

$$\begin{aligned} u_i(t) = {} & \frac{1}{h_i g_m}\frac{(v_{i-1}(t)-v_i(t))^2 S_i(t)}{|(v_{i-1}(t)-v_i(t))S_i(t)|+\sigma_{1i}} + \frac{K}{2qh_i g_m}S_i(t) \\ & + \frac{1}{qh_i g_m}\frac{A_i^2(t)S_i(t)}{|A_i(t)S_i(t)|+\sigma_{1i}} + \frac{1}{g_m}\gamma_i(t)\tanh\left(\frac{S_i(t)}{\varepsilon_i}\right) \\ & + \frac{1}{g_m}\frac{|W_i(t)\xi_i(t)|^2 S_i(t)}{|W_i(t)\xi_i(t)S_i(t)|+\sigma_{1i}} \end{aligned} \quad (5.12)$$

where

$$A_i(t) = \begin{cases} q\lambda e_i^x(t) - \dot{e}_{i+1}^x(t) - \lambda e_{i+1}^x(t) \\ \quad \text{for } i \in \mathscr{V}_N \backslash N \\ q\lambda e_i^x(t), \text{for } i = N. \end{cases} \quad (5.13)$$

The adaptive laws for $\gamma_i(t)$ and $W_i(t)$ are given by

$$\begin{aligned} \dot{\gamma}_i(t) &= \iota_{1i} q h_i [S_i(t)\tanh(\tfrac{S_i(t)}{\varepsilon_i}) - \sigma_{2i}\gamma_i(t)] \\ \dot{W}_i(t) &= -\iota_{2i} q h_i \xi_i^T(t) S_i(t) - \iota_{2i}\sigma_{3i} W_i(t) \end{aligned} \quad (5.14)$$

with $\gamma_i(0) \geq 0$. Here, K, ι_{1i} and ι_{2i} are any positive constants, while σ_{1i} σ_{2i}, σ_{3i} and ε_i are small positive constants.

Then, the following theorem, which guarantees the bounded stability of the individual vehicle and the *strong string stability* of the whole vehicle-following platoon, can be derived.

Theorem 5.1 *Consider the vehicle-following platoon (5.1) with a leader (5.7) subject to input saturation constraint (5.2) under the assumption that the initial spacing errors $e_i^x(0)(i \in \mathscr{V}_N)$ are zero. For a sufficiently large positive constant V_{max}, if the initial condition satisfies*

$$\sum_{i=1}^{N} tr\{\tilde{W}_i^T(0)\tilde{W}_i(0)\} + \|\tilde{S}(0)\|^2 \leq \frac{2V_{max}}{\kappa} \tag{5.15}$$

where

$$\begin{aligned}
\tilde{S}(t) &= \begin{bmatrix} S^T(t) & \tilde{\gamma}^T(t) \end{bmatrix}^T \\
S(t) &= \begin{bmatrix} S_1(t) & \cdots & S_N(t) \end{bmatrix}^T \\
\tilde{\gamma}(t) &= \begin{bmatrix} \tilde{\gamma}_1(t) & \cdots & \tilde{\gamma}_N(t) \end{bmatrix}^T \\
\kappa &= max\{1, \min_{1\leq i\leq N}\{\iota_{1i}\}, \min_{1\leq i\leq N}\{\iota_{2i}\}\},
\end{aligned}$$

and $\tilde{W}_i(t)$ and $\tilde{\gamma}_i(t)$ are as defined just above (5.12), then, the distributed adaptive NN control law (5.12)-(5.14) guarantees that the spacing errors $e_i^x(t)$ in (4.4) of Chapter 4 converge to a small neighborhood of the origin by appropriately choosing design parameters, while strong string stability of the vehicle-following platoon also can be guaranteed for $i \in \mathscr{V}_N$ when q satisfies $0 < q \leq 1$.

Proof 5.1 *Construct the following Lyapunov function candidate for vehicle i:*

$$V_i(t) = \frac{1}{2}S_i^2(t) + \frac{1}{2\iota_{1i}}\tilde{\gamma}_i(t)^2 + \frac{1}{2\iota_{2i}}tr\{\tilde{W}_i^T(t)\tilde{W}_i(t)\}$$

where $\tilde{\gamma}_i(t)$ and $\tilde{W}_i(t)$ are the parameter estimation errors.

Taking the time derivative of $V_i(t)$ for $t > 0$ results in

$$\dot{V}_i(t) = S_i(t)\dot{S}_i(t) - \frac{1}{\iota_{1i}}\tilde{\gamma}_i(t)\dot{\gamma}_i(t) - \frac{1}{\iota_{2i}}tr\{\tilde{W}_i^T(t)\dot{W}_i(t)\}. \tag{5.16}$$

Recalling the definitions of the integral sliding mode and coupled sliding mode in (4.6) of Chapter 4 and invoking $s_{N+1}(t) = 0$, one can obtain that

$$\begin{aligned}
\dot{S}_i(t) &= q[\dot{e}_i^x(t) + \lambda e_i^x(t)] - [\dot{e}_{i+1}^x(t) + \lambda e_{i+1}^x(t)] \\
&= q\{v_{i-1}(t) - v_i(t) - h_i[\text{sat}(u_i(t)) + w_i(t) \\
&\quad + f_i(x_i(t), v_i(t), t)] + \lambda e_i^x(t)\} \\
&\quad - [\dot{e}_{i+1}^x(t) + \lambda e_{i+1}^x(t)] \\
&= q\{v_{i-1}(t) - v_i(t) - h_i[g_{u_i}(t)u_i(t) \\
&\quad + \Delta u_i(t) + \Phi_i(t) + W_i^*\xi_i(t) + \varepsilon_i(t) \\
&\quad + w_i(t)]\} + A_i(t), for\ i \in \mathscr{V}_N \setminus N \qquad (5.17) \\
\dot{S}_i(t) &= q\{v_{i-1}(t) - v_i(t) - h_i[\text{sat}(u_i(t)) \\
&\quad + f_i(x_i(t), v_i(t), t)]\} + q\lambda e_i(t) \\
&= q\{v_{i-1}(t) - v_i(t) - h_i[g_{u_i}(t)u_i(t) \\
&\quad + \Delta u_i(t) + \Phi_i(t) + W_i^*\xi_i(t) + \varepsilon_i(t) \\
&\quad + w_i(t)]\} + A_i(t), for\ i = N,
\end{aligned}$$

where $\Phi_i(t)$ and $A_i(t)$ are defined as in (5.10) and (5.13), respectively.

The parameter γ_i^* is estimated by adaptive law $\gamma_i(t)$ in (5.14), and the initial value of $\gamma_i(t)$ has to be selected as $\gamma_i(0) \geq 0$ to guarantee that $\gamma_i(t) \geq 0$ for all $t \in [0,\infty)$. Then, using the control law (5.12), we have

$$
\begin{aligned}
& - qh_i g_{u_i}(t)u_i(t)S_i(t) \\
= & -\frac{qg_{u_i}(t)}{g_m}\frac{(v_{i-1}(t)-v_i(t))^2S_i^2(t)}{|(v_{i-1}(t)-v_i(t))S_i(t)|+\sigma_{1i}} - \frac{g_{u_i}(t)K}{2g_m}S_i^2(t) \\
& -\frac{qh_ig_{u_i}(t)}{g_m}\frac{|W_i(t)\xi_i(t)|^2S_i^2(t)}{|W_i(t)\xi_i(t)S_i(t)|+\sigma_{1i}} - \frac{g_{u_i}(t)}{g_m}\frac{A_i^2(t)S_i^2(t)}{|A_i(t)S_i(t)|+\sigma_{1i}} \\
& -\frac{qh_ig_{u_i}(t)}{g_m}\gamma_i(t)\tanh(\frac{S_i(t)}{\varepsilon_i})S_i(t) \\
\leq & -\frac{q(v_{i-1}(t)-v_i(t))^2S_i^2(t)}{|(v_{i-1}(t)-v_i(t))S_i(t)|+\sigma_{1i}} - qh_i\gamma_i(t)\tanh(\frac{S_i(t)}{\varepsilon_i})S_i(t) \\
& -\frac{qh_i|W_i(t)\xi_i(t)|^2S_i^2(t)}{|W_i(t)\xi_i(t)S_i(t)|+\sigma_{1i}} - \frac{A_i^2(t)S_i^2(t)}{|A_i(t)S_i(t)|+\sigma_{1i}} - \frac{K}{2}S_i^2(t)
\end{aligned}
\tag{5.18}
$$

where we have used the facts that $0 < g_m \leq g_{u_i}(t) < 1$ and $\gamma_i(t) \geq 0$.

Applying the following inequalities

$$
\begin{aligned}
& - \frac{q(v_{i-1}(t)-v_i(t))^2S_i^2(t)}{|(v_{i-1}(t)-v_i(t))S_i(t)|+\sigma_{1i}} \\
\leq & -q|v_{i-1}(t)-v_i(t)||S_i(t)| + q\sigma_{1i} \\
& - \frac{A_i^2(t)S_i^2(t)}{|A_i(t)S_i(t)|+\sigma_{1i}} \\
\leq & -|A_i(t)S_i(t)| + \sigma_{1i} \\
& - \frac{qh_i|W_i(t)\xi_i(t)|^2S_i^2(t)}{|W_i(t)\xi_i(t)S_i(t)|+\sigma_{1i}} \\
\leq & -qh_i|W_i(t)\xi_i(t)||S_i(t)| + qh_i\sigma_{1i} \\
\leq & \; qh_iW_i(t)\xi_i(t)S_i(t) + qh_i\sigma_{1i}
\end{aligned}
$$

into (5.18), one can obtain that

$$
\begin{aligned}
& - qh_ig_{u_i}(t)u_i(t)S_i(t) \\
\leq & -q|v_{i-1}(t)-v_i(t)||S_i(t)| - |A_i(t)S_i(t)| \\
& + qh_iW_i(t)\xi_i(t)S_i(t) - \frac{K}{2}S_i^2(t) \\
& - qh_i\gamma_i(t)\tanh(\frac{S_i(t)}{\varepsilon_i})S_i(t) + \Sigma_i
\end{aligned}
\tag{5.19}
$$

where $\Sigma_i = (q+1+qh_i)\sigma_{1i}$.

By using the property $0 \leq |\eta| - \eta\tanh(\frac{\eta}{\varepsilon}) \leq 0.2785\varepsilon$ for $\forall\varepsilon > 0$ and $\eta \in \mathbf{R}$ [58], one obtains

$$
\gamma_i^*|S_i(t)| \quad \leq \quad \gamma_i^*S_i(t)\tanh(\frac{S_i(t)}{\varepsilon_i}) + 0.2785\gamma_i^*\varepsilon_i. \tag{5.20}
$$

It then follows from (5.11), (5.17), (5.19) and (5.20) that

$$\begin{aligned}
&\dot{S}_i \quad (t)S_i(t) \\
\leq & -qh_i\gamma_i(t)\tanh(\tfrac{S_i(t)}{\varepsilon_i})S_i(t) - \tfrac{K}{2}S_i^2(t) \\
& -qh_i\tilde{W}_i(t)\xi_i(t)S_i(t) + qh_i\gamma_i^* S_i(t)\tanh(\tfrac{S_i(t)}{\varepsilon_i}) \\
& +0.2785qh_i\gamma_i^*\varepsilon_i + \Sigma_i \\
= & \ qh_i\tilde{\gamma}_i(t)\tanh(\tfrac{S_i(t)}{\varepsilon_i})S_i(t) - qh_i\sigma_{2i}\tilde{\gamma}_i(t)\gamma_i(t) \\
& +qh_i\sigma_{2i}\tilde{\gamma}_i(t)\gamma_i(t) - \tfrac{K}{2}S_i^2(t) - qh_i\tilde{W}_i(t)\xi_i(t)S_i(t) \\
& -\sigma_{3i}tr\left\{\tilde{W}_i^T(t)W_i(t)\right\} + \sigma_{3i}tr\left\{\tilde{W}_i^T(t)W_i(t)\right\} \\
& +0.2785qh_i\gamma_i^*\varepsilon_i + \Sigma_i
\end{aligned} \tag{5.21}$$

where the following relationships have been used

$$\begin{aligned}
(v_{i-1}(t) - v_i(t))S_i(t) - |v_{i-1}(t) - v_i(t)||S_i(t)| &\leq 0 \\
A_i(t)S_i(t) - |A_i(t)S_i(t)| &\leq 0.
\end{aligned}$$

Noted that $W_i(t) \in \mathbf{R}^{1\times(2N_1+1)}, \tilde{W}_i(t) \in \mathbf{R}^{1\times(2N_1+1)}$, $\xi_i(t) \in \mathbf{R}^{(2N_1+1)\times 1}$ *and* $S_i(t) \in \mathbf{R}$, *we can obtain that*

$$\begin{aligned}
& - qh_i\tilde{W}_i(t)\xi_i(t)S_i(t) - \sigma_{3i}tr\left\{\tilde{W}_i^T(t)W_i(t)\right\} \\
& = -qh_i tr\left\{\xi_i^T(t)\tilde{W}_i^T(t)S_i(t)\right\} - \sigma_{3i}tr\left\{W_i^T(t)\tilde{W}_i(t)\right\}.
\end{aligned}$$

Then, combining the adaptive laws (5.14) with (5.21), (5.16) can be rewritten as

$$\begin{aligned}
\dot{V}_i(t) \leq & -\tfrac{K}{2}S_i^2(t) + qh_i\sigma_{2i}\tilde{\gamma}_i(t)\gamma_i(t) + \sigma_{3i}tr\left\{\tilde{W}_i^T(t)W_i(t)\right\} \\
& +0.2785qh_i\gamma_i^*\varepsilon_i + \Sigma_i.
\end{aligned} \tag{5.22}$$

By completion of squares and the Young's inequality in Lemma 2.7 of Chapter 2, we have

$$\begin{aligned}
\tilde{\gamma}_i(t)\gamma_i(t) &= -\tfrac{1}{2}\gamma_i^2(t) - \tfrac{1}{2}\tilde{\gamma}_i^2(t) + \tfrac{1}{2}(\gamma_i^*)^2 \\
&\leq -\tfrac{1}{2}\tilde{\gamma}_i^2(t) + \tfrac{1}{2}(\gamma_i^*)^2
\end{aligned} \tag{5.23a}$$

$$\begin{aligned}
& tr \ \left\{\tilde{W}_i^T(t)W_i(t)\right\} \\
= & \ tr\left\{\tilde{W}_i^T(t)(W_i^* - \tilde{W}_i(t))\right\} \\
= & -tr\left\{\tilde{W}_i^T(t)\tilde{W}_i(t)\right\} + tr\left\{\tilde{W}_i^T(t)W_i^*\right\} \\
\leq & -\tfrac{1}{2}tr\left\{\tilde{W}_i^T(t)\tilde{W}_i(t)\right\} + \tfrac{1}{2}\bar{W}_i
\end{aligned} \tag{5.23b}$$

where the bounded $tr\left\{W_i^{*T}W_i^*\right\} \leq \bar{W}_i$ *has been used.*

Applying (5.23a) and (5.23b) to (5.22), one can obtain that

$$\begin{aligned}
\dot{V}_i(t) \leq & -\tfrac{K}{2}S_i^2(t) - \tfrac{qh_i\sigma_{2i}}{2}\tilde{\gamma}_i^2(t) - \tfrac{\sigma_{3i}}{2}tr\left\{\tilde{W}_i^T(t)\tilde{W}_i(t)\right\} \\
& +\tfrac{qh_i\sigma_{2i}}{2}(\gamma_i^*)^2 + \tfrac{\sigma_{3i}}{2}\bar{W}_i + 0.2785qh_i\gamma_i^*\varepsilon_i + \Sigma_i.
\end{aligned} \tag{5.24}$$

Next, a global Lyapunov function $V(t)$ *is constructed as follows:*

$$V(t) = \sum_{i=1}^{N} V_i(t).$$

In light of (5.24), it then follows that

$$\dot{V}(t) \leq -\zeta_1 V(t) + \zeta_2 \tag{5.25}$$

where the positive parameters ζ_1 and ζ_2 are given as follows

$$\begin{aligned} \zeta_1 &= \min\{K, \min_{1\leq i\leq N} \iota_{1i} q h_i \sigma_{2i}, \min_{1\leq i\leq N} \iota_{2i}\sigma_{3i}\} \\ \zeta_2 &= \sum_{i=1}^{N} [q h_i (\tfrac{\sigma_{2i}}{2}\gamma_i^{*2} + 0.2785\gamma_i^* \varepsilon_i) + \tfrac{\sigma_{3i}}{2}\bar{W}_i + \Sigma_i]. \end{aligned}$$

Thus, by using Lemma 2.6 in Chapter 2, the following inequality holds:

$$V(t) \leq (V(0) - \tfrac{\zeta_2}{\zeta_1})e^{-\zeta_1 t} + \tfrac{\zeta_2}{\zeta_1} \leq V(0) + \tfrac{\zeta_2}{\zeta_1}, \tag{5.26}$$

where the following fact is used for the last inequality

$$V(0) = \sum_{i=1}^{N} [\frac{1}{2} S_i^2(0) + \frac{1}{2\iota_{1i}} \tilde{\gamma}_i(0)^2 + \frac{1}{2\iota_{2i}} tr\{\tilde{W}_i^T(0)\tilde{W}_i(0)\}] \geq 0.$$

Therefore, it is straightforward to show that $S_i(t)$, $\gamma_i(t)$ and $W_i(t)$ are uniformly ultimately bounded if the initial condition is bounded as in (5.15). It is clear that reducing ζ_2, meanwhile increasing ζ_1 will lead to smaller bounds of $S_i(t)$, i.e., $S_i(t)$ can converge to an arbitrary small neighborhood of zero by choosing the design parameters appropriately. Then, according to 4.6 of Chapter 4, $s_i(t)$ and the spacing error $e_i^x(t)$ also converges to a small neighborhood of zero. Simultaneously, the proof of strong string stability is similar to the proof process in Theorem 4.1 of Chapter 4. This completes the proof. □

Remark 5.3 *Since actuator physical constraints can severely degrade the whole vehicle-following platoon, control design for vehicle-following platoon subject to input saturation and nonlinear unmodeled dynamics presents a tremendous challenge. To overcome the effect of input saturation, only one parameter g_m needs be adjusted here, which is much simpler than most existing results such as [20, 76, 175]. It is clear in (5.12) that the vehicle-following platoon system performance is affected by g_m. Therefore, an appropriate value of g_m is important for eliminating the effect of input saturation. It should be pointed out that (5.18) will fail if $g_{u_i}(t) < g_m$.*

Remark 5.4 *In the proof of Theorem 5.1, the strong string stability of the whole vehicle-following platoon is proved following the approach in the Laplace domain under the condition that $S_i(t) = q s_i(t) - s_{i+1}(t)$ converges to zero. However, according to the proof of Theorem 5.1, the sliding mode motion converges to a neighborhood of $S_i(0) = 0$. This is due to actuator saturation nonlinearities and/or unmodeled dynamics, etc. By adjusting the design parameters, this neighborhood of $S_i(0)$ can be made sufficiently small so that the subsequent analysis can be done on the sliding manifold $S_i(t) = 0$. It is further highlighted that the incorporation of zero initial conditions (the MCTH policy below also has zero initial conditions), hence guarantees* *that $s_i(0) = S_i(0) = 0$ and the system starts on the sliding manifold, eliminating completely the reaching phase that normally exists in most sliding mode control schemes. Therefore, even though strong string stability before reaching the sliding manifold is not proven in theory in Theorem 5.1, the proposed control law (5.12) effectively guarantees the strong string stability not only in the transient but also in the steady-state phase of the system response, as is demonstrated by the numerical examples of Section 5.4. Nevertheless, a more efficient and effective notion of and approach to strong string stability when the system is knocked off the sliding surface is a very interesting topic for future research.*

When the input saturation is not considered in the vehicle-following platoon, the following corollary can be obtained.

Corollary 5.1 *Consider the vehicle-following platoon (5.1) with a leader (5.7) under assumption that the initial spacing errors $e_i^x(0)(i \in \mathscr{V}_N)$ are zero. For a sufficiently large positive constant V_{max}, if the initial condition satisfies (5.15), the distributed adaptive NN control laws (5.12)-(5.14) with $g_m = 1$ guarantee that the spacing errors $e_i^x(t)$ in (4.4) of Chapter 4 converge to a small neighborhood around origin by appropriately choosing design parameters, while strong string stability of the vehicle-following platoon also can be guaranteed for $i \in \mathscr{V}_N$ when q satisfies* $0 < q \leq 1$.

5.3.2 Control Strategy II: MCTH Control Law

From a practical point of view, with increasing traffic density (more vehicles on the way), the traffic flow also increases. Therefore, with the aim of overcoming the drawback of low traffic density arising from the use of TCHT policy, in this section, the *MCTH policy* proposed in (4.27) of Chapter 4 is used to design a new adaptive NN control scheme. Furthermore, denote the same sliding mode surfaces $s_i(t)$ and $S_i(t)$ in (4.6) of Chapter 4.

Then, based on the new spacing policy (4.27), the adaptive NN control law is designed as follows:

$$\begin{aligned} u_i(t) &= \frac{1}{h_i g_m} \frac{(v_{i-1}(t) - v_i(t) - \dot{\Xi}(t))^2 S_i(t)}{|(v_{i-1}(t) - v_i(t) - \dot{\Xi}(t)) S_i(t)| + \sigma_{1i}} + \frac{K}{2 q h_i g_m} S_i(t) \\ &\quad + \frac{1}{q h_i g_m} \frac{A_i^2(t) S_i(t)}{|A_i(t) S_i(t)| + \sigma_{1i}} + \frac{1}{g_m} \gamma_i(t) \tanh\left(\frac{S_i(t)}{\varepsilon_i}\right) \\ &\quad + \frac{1}{g_m} \frac{|W_i(t)\xi_i(t)|^2 S_i(t)}{|W_i(t)\xi_i(t) S_i(t)| + \sigma_{1i}} \end{aligned} \tag{5.27}$$

where

$$\begin{aligned} A_i(t) &= \begin{cases} q h_i a_0(t) + q\lambda e_i^x(t) - \dot{e}_{i+1}^x(t) - \lambda e_{i+1}^x(t) \\ \qquad\qquad \text{for } i \in \mathscr{V}_N \backslash N \\ q h_i a_0(t) + q\lambda e_i^x(t), \text{for } i = N. \end{cases} \\ \dot{\Xi}_i(t) &= -\zeta_i[\tilde{e}_i^x(0) + (\zeta_i \tilde{e}_i^x(0) + \dot{\tilde{e}}_i^x(0))t] e^{-\zeta_i t} \\ &\quad + (\zeta_i \tilde{e}_i^x(0) + \dot{\tilde{e}}_i^x(0)) e^{-\zeta_i t}. \end{aligned} \tag{5.28}$$

The adaptive laws for $\gamma_i(t)$ and $W_i(t)$ are the same as (5.14).

Note that if all vehicles' velocities $v_i(t)$ $(i \in \mathscr{V}_N)$ approach the leader's velocity $v_0(t)$ under the MCTH policy (4.27) proposed in Chapter 4, the nonlinear function $f_i(x_i(t), v_i(t), t)$ will converge to a *desired* nonlinear function $F_i(x_0(t), v_0(t), t)$ defined by

$$F_i(x_0(t), v_0(t), t) = f_i\left(\int_0^t v_0(\tau)d\tau - \sum_{j=1}^{i} \delta_j, v_0(t), t\right).$$

The following theorem states that the control law (5.27) guarantees the bounded stability of individual vehicle and *strong string stability* of whole vehicle platoon under the *MCTH policy* (4.27) of Chapter 4.

Theorem 5.2 *Consider the vehicle-following platoon (5.1) with a leader (5.7) subject to input saturation constraint (5.2). For a sufficiently large positive constant* V_{max}, *if the initial condition satisfies (5.15), the distributed adaptive NN control law in (5.27), (5.28), and (5.14) guarantees that the spacing errors* $e_i^x(t)$ *in (4.27) of Chapter 4 converge to a small neighborhood around origin by appropriately choosing design parameters, while strong string stability of the vehicle-following platoon also can be guaranteed for* $i \in \mathscr{V}_N$ *when* q *satisfies* $0 < q \leq 1$.

Proof 5.2 *The proof is immediate and follows along the same lines that are developed for the proof of Theorem 5.1.* □

Remark 5.5 *In comparison with the existing results that also consider adaptive NN control for vehicle-following platoons [116, 138], the main differences in our result are summarized as follows: (a) Actuator saturation nonlinearity is not considered in [116, 138]; (b) as compared with RBFNN used in [116, 138], the key advantage of CP basis function lies in the fact that only one parameter (i.e.,* N_1*, the order of CP) is required to determine the CP basis function [217]; (c) the strong string stability issue as a key issue of platoon control is not considered in [116, 138] which means the vehicles in the platoon may come into collision with each other; (d) zero initial spacing errors are assumed in [116, 138].*

5.4 Simulation Study and Performance Results

In this section, the feasibility of the proposed methods is illustrated via a numerical example and a practical example. All numerical simulations were performed with MATLAB version 2007a and a notebook computer with 1.8GHz Intel (R) Core (TM) i5-3337U CPU and 8.00GB RAM using Windows 10.

5.4.1 Example 1 (Numerical Example)

In this subsection, we will verify the effectiveness of the proposed distributed adaptive neural control based on TCTH policy (4.4) and MCTH policy (4.27) of Chapter 4 for a vehicle platoon of six follower vehicles with input *saturation* as in (5.1) and a leader as in (5.7). In the simulation, the unmodeled dynamic nonlinear function $f_i(x_i(t), v_i(t), t)$, *external disturbances* $w_i(t)$ and the leader's acceleration are given as

$$\begin{aligned} f_i(x_i(t), v_i(t), t) &= \tfrac{0.5x_i^2(t)}{1+x_i^2(t)} + \varphi_i \sin(\tfrac{\pi}{4} + \varphi_i v_i(t)) \\ w_i(t) &= 0.1 \sin(t), i \in \mathscr{V}_N \\ a_0(t) &= \begin{cases} 1.5\, m/s^2, & 2s \leq t < 5s \\ -0.5\, m/s^2, & 6s \leq t < 10s \\ 0\, m/s^2, & otherwise. \end{cases} \end{aligned}$$

Tables 5.1 and 5.2 show the values of the simulation parameters in this section. Furthermore, the input vector of the CNN is $(v_0(t), a_0(t))^T$.

TABLE 5.1
Numerical simulation parameters' values for $i \in \mathscr{V}_N$

Parameter Name	Simulation Values
Standstill Distance	$\delta_i = 0.8m$
Time Headway	$h_i = 1s$
Leader Initial	$x_0(0) = 20m$; $v_0(0) = 1m/s$
The Order of CP	$N_1 = 2$
$\iota_{1i}, \iota_{2i}, \varepsilon_i$	$\iota_{1i} = 20$, $\iota_{2i} = 0.2$, $\varepsilon_i = 0.1$
$\sigma_{ji}, j = 1,2,3$	$\sigma_{1i} = \sigma_{2i} = 0.3$, $\sigma_{3i} = 1$
$\gamma_i(0)$, $W_i(0)$	$\gamma_i(0) = 1$, $W_i(0) = 0_{1\times 5}$

TABLE 5.2
The parameters $x_i(0)$, $v_i(0)$ and φ_i for $i \in \mathscr{V}_N$

i	1	2	3	4	5	6
$x_i(0)$	$18.5m$	$17m$	$15.5m$	$14m$	$12.5m$	$11m$
$v_i(0)$	$1m/s$	$1m/s$	$1m/s$	$1m/s$	$1m/s$	$1m/s$
φ_i	1	0.1	−0.8	0.1	−1.2	0.2

Case I: Zero Initial Spacing Errors with TCTH Policy

According to the definition of TCTH policy (4.4) in Chapter 4, the initial values of $e_i^x(0)$ are zero. For this case, the controller parameters are taken as $K = 5$, $q = 0.85$ and $\lambda = 0.1$. In the following, the comparison between Corollary 5.1 and Theorem 5.1 is studied so as to demonstrate the effectiveness of the proposed single parameter

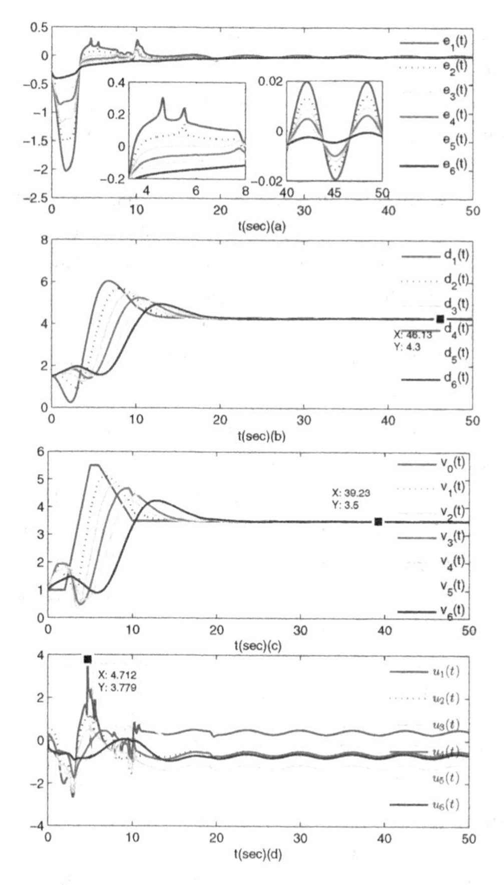

FIGURE 5.3
(a) Spacing error $e_i(t)$; (b) distance $d_i(t)$; (c) velocity $v_i(t)$; (d) control input $u_i(t)$ under Corollary 5.1 without input saturation.

g_m in handling control *saturation*. First of all, when there does not exist input saturation, the simulation results obtained by using Corollary 5.1 are shown in Figure 5.3. It is found from Figure 5.3(a) that the amplitude of the spacing error decreases through the string of vehicles (i.e., $\|e_6(t)\| \leq \cdots, \leq \|e_1(t)\|$), which implies that the *strong string stability* of the whole vehicle platoon is achieved despite the nonlinear unmodeled dynamics and *external disturbances*. Figure 5.3(b) shows that the neighboring-vehicle collisions can be avoided since the distances between consecutive vehicles are always positive. Figures 5.3(b) and 5.3(c) show that neighboring-vehicle spacings of all consecutive vehicles converge to $4.3m$ and the velocities of the vehicles follow the trajectory of the leader and maintain the desired velocity $3.5m/s$ finally. The control input $u_i(t)$ is shown in Figure 5.3(d), where we can find the maximum value of the input is 3.779. Due to the space constraint, the responses of the input $u_i(t)$ will be omitted in the sequel.

Next, given the maximum control effort of 3.7799 in Figure 5.3(d), we consider an input saturation limit of $u_{Mi} = 1.5$. The simulation results under Corollary 5.1 and Theorem 5.1 with $g_m = 0.6$ are shown in Figure 5.4 and Figure 5.5, respectively. With the controller in Corollary 5.1, it can be found from Figures 5.4(a) and 5.4(b) that the input *saturation* causes significant performance degradation and *string instability*, which may cause neighboring-vehicle collisions. However, with the controller in Theorem 5.1, Figures 5.5(a) and (b) show that the collisions can be avoided and the *strong string stability* can be maintained despite the existence of input saturation, demonstrating the effectiveness and advantage of the proposed method. Referring to Figures 5.5(b) and 5.5(c), it is found that the neighboring-vehicle spacings $d_i(t)$ are proportional to vehicle speed $v_i(t)$, and can become very large at high speed, which decreases the capacity of traffic flow.

Case II: Nonzero Initial Spacing Errors with MTCH Policy

According to the definition of MCTH policy (4.27) in Chapter 4, the initial values of $\tilde{e}_i^x(0)$ are non-zero as $\tilde{e}_i^x(0) = 1m, i \in \mathscr{V}_N$ while $e_i^x(0) = 0m, i \in \mathscr{V}_N$ with the MCTH policy (4.27) in Chapter 4. For this case, the controller parameters are taken as $K = 5$, $q = 0.96$, $\lambda = 0.5$ and $\zeta_i = 20$. In addition, the parameter g_m is assumed to be 0.5. Since the input saturation limit $u_{M,i}$ determines the control capacity and saturation degree of the actuator, the *non-zero initial* spacing error condition considered here requires higher input saturation limit, i.e., the input saturation limit u_{Mi} here is chosen as 2 (instead of 1.5 in Figure 5.5 in Case I above). The simulation results are shown in Figure 5.6. Referring to Figure 5.5(a), it is found that Figure 5.6(a) has a faster convergence rate and higher accuracy control performance than Figure 5.5(a) since larger control capacity ($u_{M,i} = 2$ instead of $u_{M,i} = 1.5$) has been used. From Figures 5.6(b) and 5.6(c), by using MCTH policy (4.27) of Chapter 4, the neighboring-vehicle distance ($0.8m$) is much smaller than the one ($4.3m$) in Figures 5.5(b) and 5.5(c), which will increase the traffic density significantly. For the transient phase, it is seen from Figure 5.5(c) and Figure 5.6(c) that the controller (5.27) of Theorem 5.2 has a faster transient response and a higher accuracy control performance than the controller (5.12) of Theorem 5.1.

5.4.2 Example 2 (Practical Example)

Consider a platoon of high speed trains movement on a railway line. The platoon consists of six follower trains and a leader train. The dynamic motion for each follower train is the same as in [101]. After minor transformations, the follower train with *actuator saturation* is given by (5.1) with

$$
\begin{aligned}
f_i(x_i(t), v_i(t), t) &= -c_{i0} - c_{i1} v_i(t) - c_{i2} v_i^2(t) \\
u_i(t) &= \frac{\tau_i(t)}{m_i} \\
w_i(t) &= 0, i \in \mathscr{V}_N
\end{aligned}
$$

where $x_i(t)$, $v_i(t)$, m_i and $\tau_i(t)$ are the position, speed, mass and control force of the ith train, respectively. The leader high speed train's behavior is independent of

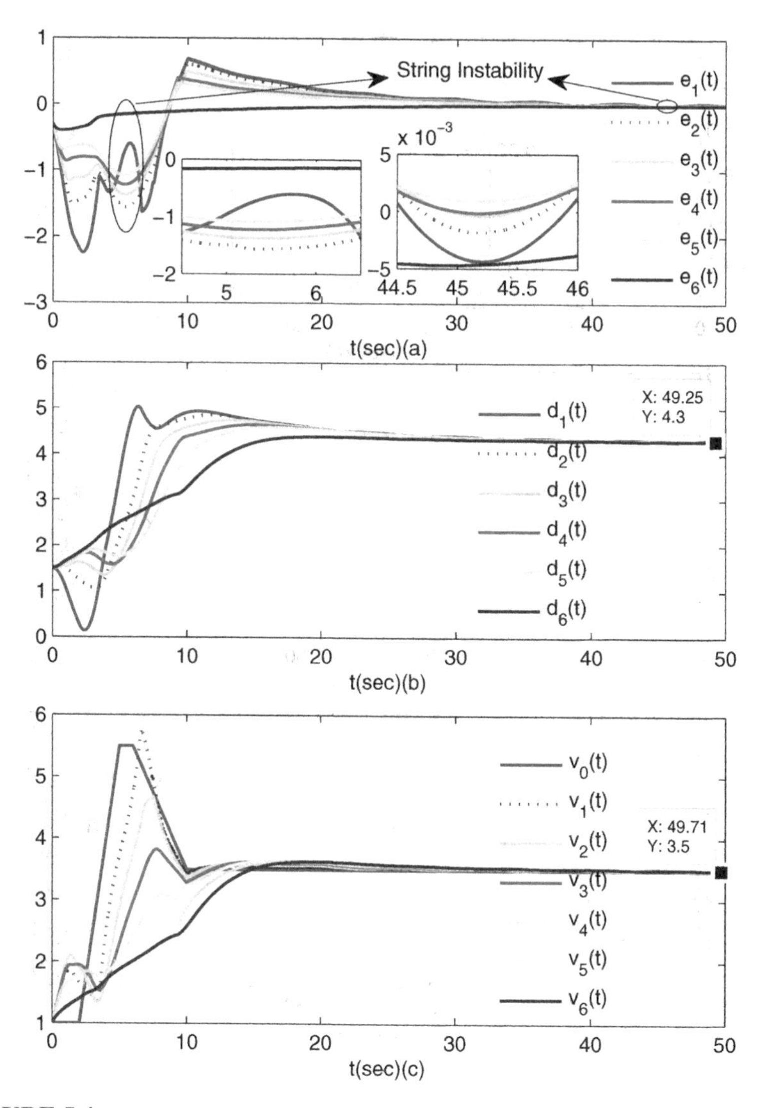

FIGURE 5.4
(a) Spacing error $e_i(t)$; (b) distance $d_i(t)$; (c) velocity $v_i(t)$ under Corollary 5.1 with input saturation.

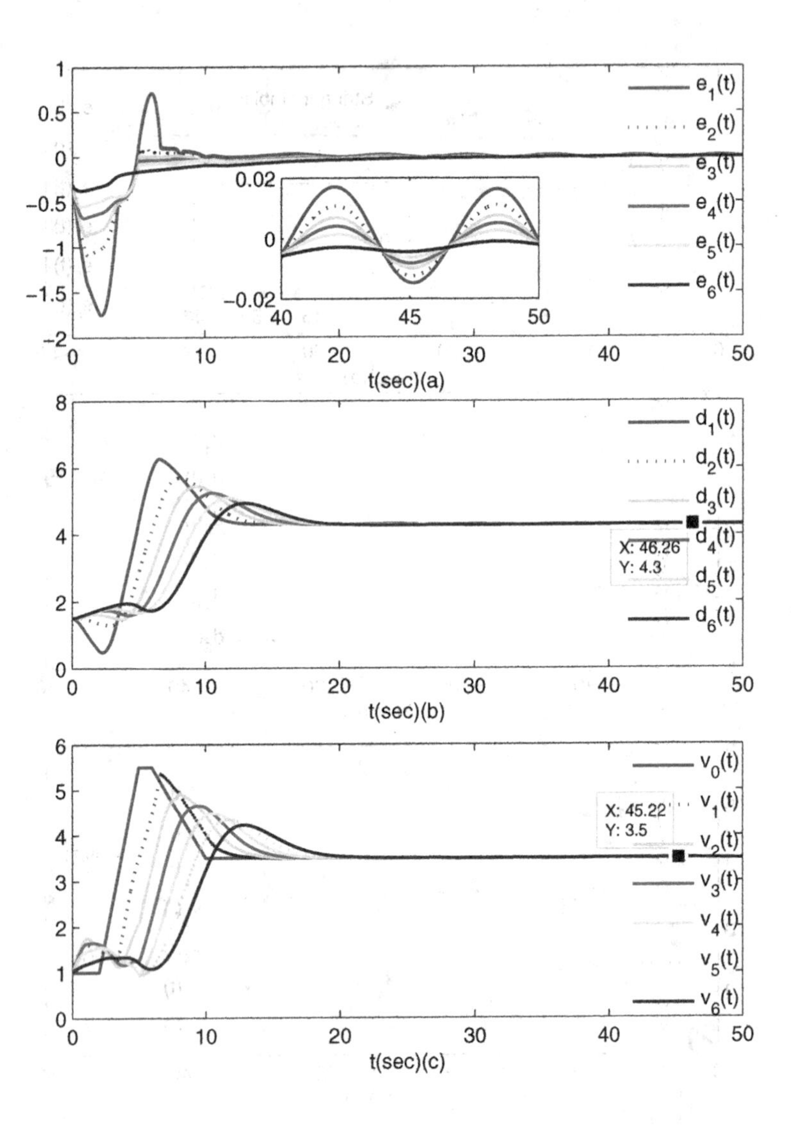

FIGURE 5.5
(a) Spacing error $e_i(t)$; (b) distance $d_i(t)$; (c) velocity $v_i(t)$ under Theorem 5.1 with input saturation.

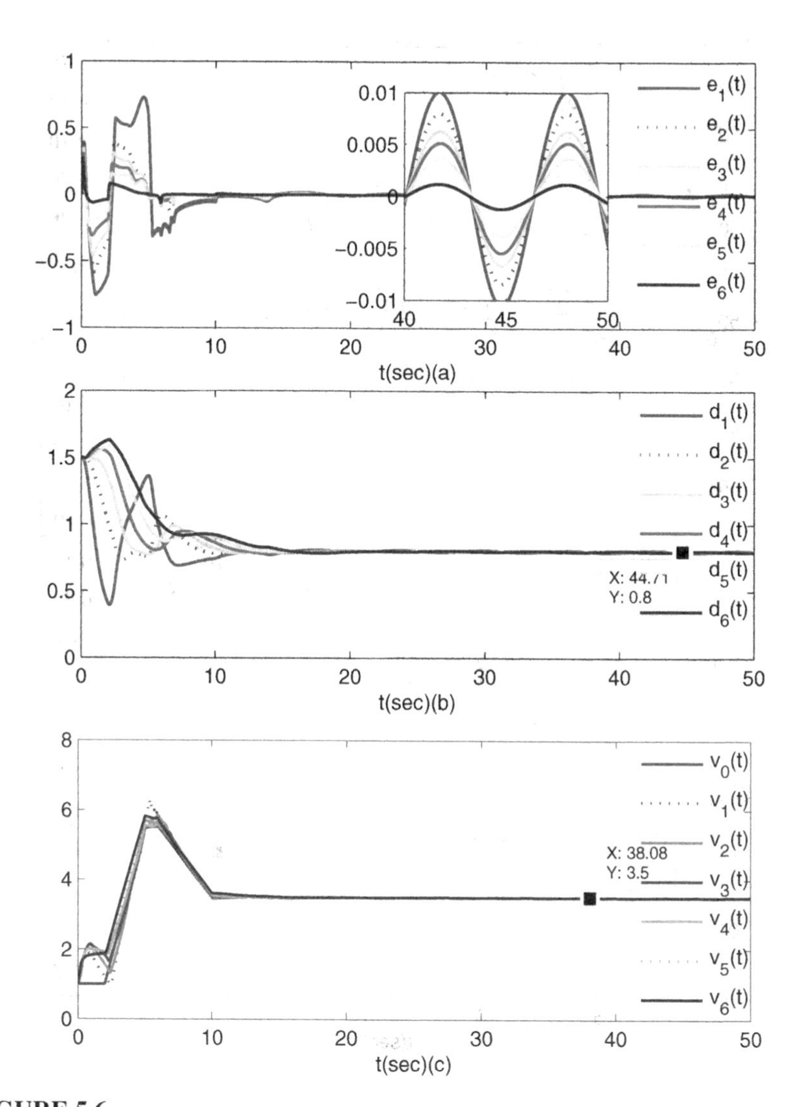

FIGURE 5.6
(a) Spacing error $e_i(t)$; (b) distance $d_i(t)$; (c) velocity $v_i(t)$ under Theorem 5.2 with input saturation.

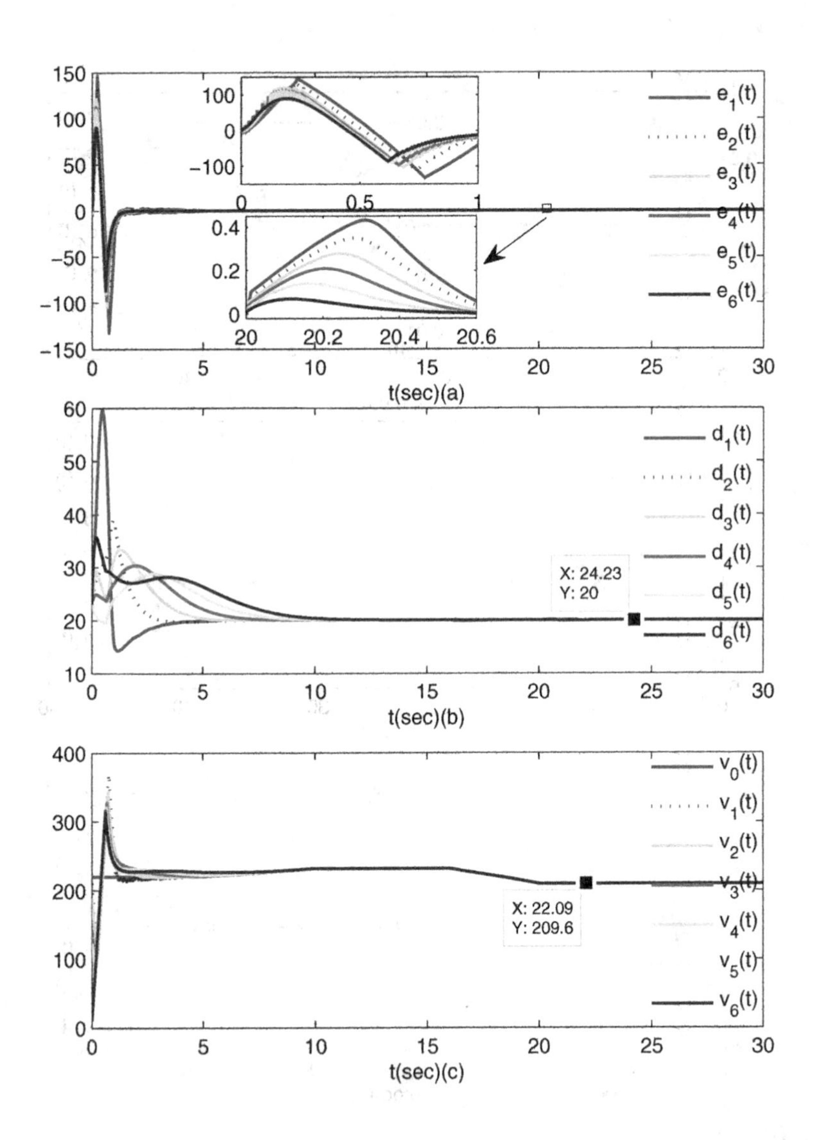

FIGURE 5.7
(a) Spacing error $e_i(t)$; (b) distance $d_i(t)$; (c) velocity $v_i(t)$ under Theorem 5.2 with input saturation.

TABLE 5.3
Parameters of high speed trains for $i \in \mathcal{V}_N(N=6)$

Symbol	Value
m_i	$80 \times 10^4 kg$
c_{i0}	$0.01176N/kg$
c_{i1}	$0.00077616Ns/mkg$
c_{i2}	$1.6 \times 10^{-5} Ns^2/m^2kg$
δ_i, h_i	$20km, 1h$

its followers, and the acceleration $a_0(t)$ in (5.7) keeps changing during the motion process as

$$a_0(t) = \begin{cases} 0.3t\ km/h^2, & 5h \leq t < 10h \\ -0.3t\ km/h^2, & 16h \leq t < 20h \\ 0\ km/h^2, & otherwise. \end{cases}$$

The system parameters of each high speed train and the initial positions and speeds of each high speed train are as in [101] and listed in Table 5.3 and Table 5.4, respectively. The desired spacing between consecutive trains is $\delta_i = 20km$, while the time headway is $h_i = 1h$. In addition, inspired by [217], in order to obtain a good control performance, the input vector of the CNN is normalized as $(v_0(t)/220, a_0(t)/220)^T$. The input saturation limit is taken as $u_{Mi} = 500$, and the control parameters are chosen as $k = 10$, $q = 0.95$, $\lambda = 5$, $\zeta_i = 15$, $g_m = 0.5$. All other design parameters are the same as those in Example 1. According to the definition of $\tilde{e}_i^x(t)$ in (4.27) of Chapter 4, the initial spacing errors $\tilde{e}_i^x(0)$ are non-zero as $\tilde{e}^x(0) = [20, 34, 82.5, 103, 64.5, 213]km$. Under the same *initial conditions* of the adaptive estimations, the proposed controller (5.27) of Theorem 5.2 performs as shown in Figure 5.7. It can be observed that the control method can achieve *strong string stability* in presence of input saturations, nonlinear dynamics and *non-zero initial* spacing errors. It is clear that the distances between two neighboring trains converge to the desired spacing $20km$ and the velocities of the followers converge to that of the leader. It should be highlighted that although the maximum spacing error $150km$

TABLE 5.4
The parameters $x_i(0)$ and $v_i(0)$ for $i \in \mathcal{V}_N \cup \{0\}(N=6)$

i	0	1	2	3	4	5	6
$x_i(0)(km)$	148	118	94	71.5	48.5	24	1
$v_i(0)(km/h)$	220	210	190	140	120	160	10

is much larger than the desired spacing $20km$, there still do not exist collisions in Figure 5.7(b). The main reason for this phenomenon is owed to the use of CTH policy, which also demonstrates the superiority in improving *strong string stability* by using the MCTH policy (4.27) of Chapter 4. All these simulations show the effectiveness and advantage of the proposed method.

5.5 Conclusion

Based on TCTH and MCTH policies, two distributed adaptive NN control schemes are proposed for vehicle-following platoons. The proposed controller can force the followers to track the leader's trajectory while maintaining a *desired safe spacing* simultaneously, even in the presence of input saturation, unknown unmodeled dynamics and *external disturbances*. A simple and straightforward strategy by adjusting only a single parameter is proposed to attenuate the effect of input saturation. CNN is used to approximate the unknown nonlinear function in the followers on-line, and the implementation of CP basis function depends only on the leader's velocity and acceleration. In addition, by adopting the CNN technique, the matching conditions as the existing method are removed. Future research will consider the effect of time-delay, measurement noise and so on.

6

Adaptive Fuzzy Fault-Tolerant Control for Multiple High Speed Trains

A novel adaptive fuzzy fault-tolerant control method is proposed for multiple *high speed trains* (HSTs) in presence of actuator faults, unknown nonlinear dynamics and *external disturbances*. The *fuzzy logic system* (FLS) is used to approximate the unknown functions. Based on proportional and integral (PI)-based sliding mode technique, two adaptive fuzzy control schemes are then proposed to achieve *strong string stability*. In the first scheme, the actuator faults are not considered. In the second scheme, a novel fuzzy *fault-tolerant* control with fewer adjustable parameters is constructed to further reduce the online computational load. By using Lyapunov analysis, the individual stability and *strong string stability* of whole multiple HSTs are proven. In addition, all signals of the closed-loop system are uniformly ultimately bounded and all HSTs track the reference trajectory to a small compact set around zero.

6.1 Introduction

With the rapid explosion of population, by improving rail transit speed and capacity and offering green transportation, HST has recently attracted considerable research and attention [32, 36, 69, 81, 101, 152, 160]. It is noted that all of these results assume that the controller always works efficiently during the operation, and the actuator faults are not taken into account. In practice, however, the controller may lose its efficiency or even completely collapse during operation due to various kinds of reasons, such as overvoltage in traction transformer, overcurrent in traction converter, and overheat in asynchronous motor, and so on [168]. It is known that the actuator faults may cause the system performance deterioration or lead to instability [16, 187, 195, 196, 177], and thus have serious consequence on the safety of the train and the passengers. This is the reason of the necessity to design *fault-tolerant* controllers that are able to tolerate actuator faults and maintain high efficiency and performance. Motivated by this observation, considerable efforts on deriving *fault-tolerant* control algorithm for train systems have been made during the past decades [88, 159, 168, 178]. Besides, automatically controlling the train speed to follow a desired trajectory is one of the demanding control problems of HSTs [36]. If the desired trajectory is regarded as a virtual leader, the multiple HSTs moving on the railway can be viewed as a leader-following vehicle platoon. Since trains in a platoon are coupled, behavior of one train and disturbance on it can affect the other. Hence, *platoon control* must exhibit both individual stability and *string stability* as a group. A broad definition of string stability is from the point of preventing collisions, that is, the real distances between adjacent trains maintain a *desired safe spacing* [2, 27, 142, 143]. A slightly more rigorous definition is that *spacing errors* do not amplify as they propagate along the platoon [2, 55, 57]. Various control approaches have been developed to ensure string stability [2, 55, 57, 93, 120, 142, 143, 212]. Unfortunately, string stability analysis is complicated by the presence of actuator faults and remains an open problem. How to deal with the string stability analysis for HSTs with consideration of unanticipated actuator faults has been a task of major practical interest as well as theoretical significance.

Meanwhile, some previous results such as [88, 189] based on linearized model would limit the validity of the corresponding control strategy in practice because, in practical processes/systems, it is difficult to obtain an accurate model of the train subject to *external disturbances*. Particularly, as train speed increases, the nonlinearities in basic resistance impact on systems dynamics becomes increasingly significant [160]. Furthermore, practical actuator faults often cause unpredictable nonlinear changes in the dynamics of the system [166]. Therefore, it is natural to investigate the *fault-tolerant* control for multiple HSTs in a nonlinear model framework. Nevertheless, the unknown nonlinear dynamic is one of main sources resulting in the instability of systems [109]. To overcome the above restriction, FLSs are known to be particularly powerful tools to control nonlinear systems owning to their universal approximation properties [115]. Another feature of this method is to handle the nonlinear systems without the requirement of matching condition as in [57], where both the follower and leader vehicles have the same type of nonlinear uncertainties. On the other hand, sliding mode control has been applied to reduce the sensitivities to the variations of parameters and *external disturbances* [15, 209]. Since fuzzy sliding mode control indeed alleviates the *chattering phenomenon* in the pure sliding mode control, fuzzy techniques together with the sliding mode control techniques have been applied extensively to systems with nonlinear uncertainties [80, 167]. However, a shortcoming of the above fuzzy sliding mode control method is that it suffers from the explosion of the number of learning parameters because when the number of the fuzzy logic rule base increases, the number of learning parameters increases accordingly. This is particularly serious when the train platoon contains a large number of trains. In this case, the computational cost in the control implementation is not trivial. Therefore, from the viewpoint of practical applications, some efforts will be made in the direction of reducing computational costs.

Motivated by the above discussion, the problem of adaptive fuzzy control for multiple HSTs with unknown *external disturbances* and actuator faults is investigated. The proposed control schemes can guarantee the *strong string stability* of the whole HST platoon and the ultimately uniform bounded of all signals. All HSTs track the reference trajectory to a small neighborhood of origin. The contribution of this chapter is briefly summarized as follows.

(1) The desired trajectory is regarded as a virtual leader, then, the multiple HSTs moving on the railway is viewed as a leader-following vehicle platoon. *Strong string stability* of this platoon is guaranteed to maintain safety spacing between consecutive trains.

(2) By combining adaptive FLS techniques with PI-based sliding mode control method, two novel adaptive fuzzy control schemes are developed to guarantee individual stability and *strong string stability* of whole train platoon even in presence of actuator faults. In addition, the requirement of matching condition as [57], where both the follower and leader vehicles have the same type of nonlinear uncertainties, is removed by introducing FLS approach.

(3) A novel fuzzy structure with minimal computational complexity is proposed based on a novel parameter estimation technique, where the number of learning parameters is independent of the number of the fuzzy rules. Thus, the computational burden of the control algorithm can be reduced significantly, especially when the platoon contains a large number of trains.

6.2 High Speed Train Dynamics and Preliminaries

6.2.1 Model Description of High Speed Train Dynamics

The simplified configuration of an ordered set of N high speed trains running on the railway line is depicted in Figure 6.1, where x_i and v_i denote the position and velocity of the ith high speed train, respectively. The dynamics of train i can be expressed as follows [101, 152]:

$$\begin{aligned} \dot{x}_i &= v_i, i \in \mathscr{V}_N \\ m_i\dot{v}_i &= \tau_i - m_i(c_{i0} + c_{i1}v_i + c_{i2}v_i^2) - m_i(w_{ri} + w_{ci} + w_{ti}) \end{aligned} \tag{6.1}$$

where m_i is the ith train's mass, τ_i is control force of the ith train, w_{ri}, w_{ci}, and w_{ti} are the ramp resistance due to the track slope, the curve resistance due to railway curvature and the tunnel resistance, respectively. In practice, the resistance coefficients c_{i0}, c_{i1}, and c_{i2} are empirical constants obtained from the data fusion using the historical operational data [32]. Correspondingly, the coefficients c_{i0}, c_{i1}, and c_{i2} and other resistances w_{ri}, w_{ci}, and w_{ti} for each train are difficult to obtain accurately.

By letting

$$\begin{aligned} f_i(v_i,t) &= -c_{i0} - c_{i1}v_i - c_{i2}v_i^2 \\ w_i &= -w_{ri} - w_{ci} - w_{ti} \\ u_{ai} &= \frac{\tau_i}{m_i}, \end{aligned}$$

Eq. (6.1) can be rewritten as

$$\begin{aligned} \dot{x}_i &= v_i, i \in \mathscr{V}_N \\ \dot{v}_i &= u_{ai} + f_i(v_i,t) + w_i \end{aligned} \tag{6.2}$$

where $f_i(v_i,t)$ is the basic and aerodynamic resistance [32]. According to practical experience, it is reasonable to assume that v_i is bounded. Thus, since the resistance coefficients c_{i0}, c_{i1}, and c_{i2} cannot be obtained accurately, $f_i(v_i,t)$ is actually an unknown bounded function.

In practical train systems, unanticipated actuator failures may occur, inevitably the case in practice for long-term operation. Then, a discrepancy may exist between the actual control input u_{ai} and the designed control input u_i, and they are related by

$$u_{ai} = \begin{cases} u_i, \text{ fault-free case} \\ \rho_i(t,t_{\rho i})u_i + r_i(t,t_{ri}), \text{ with } i\text{th actuator failure} \end{cases} \tag{6.3}$$

where $0 < \rho_i(t,t_{\rho i}) \leq 1$ indicates the ith actuator's effectiveness and $r_i(t,t_{ri})$ is the uncontrollable actuator bias fault, $t_{\rho i}$ and t_{ri} denote, respectively, the time instant at which the loss of actuation effectiveness fault and the actuation bias fault occur. The various types of actuator faults considered here are listed in Table 6.1.

TABLE 6.1
Various types of actuator faults

Types of Actuator Faults	$\rho_i(t,t_{\rho i})$	$r_i(t,t_{ri})$
Healthy Actuator	1	0
Loss of Effectiveness	<1	0
Actuator Bias Fault	1	Time-Varying
Loss of Effectiveness with Bias Fault	<1	Time-Varying

The time-varying functions $\rho_i(t,t_{\rho i})$, $r_i(t,t_{ri})$ and w_i are physically bounded. Thus, the following assumption is introduced.

Assumption 6.1 *The actuation effectiveness $\rho_i(t,t_{\rho i})$, the actuation bias faults $r_i(t,t_{ri})$ and the disturbances w_i are unknown, but bounded in that there exist a constant ρ_{i0} such that $0 < \rho_{i0} \leq \rho_i(t,t_{\rho i}) \leq 1$, an unknown constant $r_i^* \geq 0$ such that $|r_i(t,t_{ri})| \leq r_i^* < \infty$ and an unknown constant w_i^* such that $|w_i| \leq w_i^* < \infty$. $\rho_i(t,t_{\rho i})$, $r_i(t,t_{ri})$ and w_i are assumed to be unknown, time-varying and undetectable.*

Remark 6.1 *It is noted that similar assumption, as in Assumption 6.1, is commonly imposed in most existing works [17, 18, 108, 161, 177, 178]. Furthermore, $\rho_i(t,t_{\rho i})$ and $r_i(t,t_{ri})$ are unknown and time-varying, and the occurrence instants $t_{\rho i}$ and t_{ri} of the actuator faults are unpredictable. This makes the controller design and strong string stability analysis much more challenging and interesting.*

Remark 6.2 *To achieve the control objectives that each train tracks the desired speed, and meanwhile ensures the safety headway distances of each train with its neighboring trains over the running time, a coordinated cruise control strategy for each train is developed based on potential fields, invariance principle and graph theory in [101]. In this chapter, in order to achieve the same objectives as in [101],*

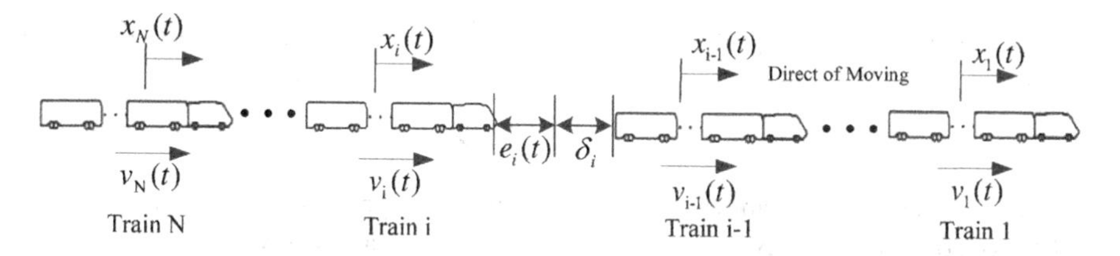

FIGURE 6.1
A simplified configuration of multiple high speed trains in longitudinal motion with constant spacing.

platoon control algorithm is proposed to guarantee strong string stability. This makes the design tools and analysis methods in [101] much different from those in this chapter. In addition, compared to [101], actuator faults make the above two control objectives quite challenging to achieve.

6.2.2 Fuzzy Logic Systems

Fuzzy logic systems (FLSs) are usually used as a tool for modeling nonlinear functions due to their good capabilities in function approximation [107]. The knowledge base for an FLS is comprised of a collection of fuzzy IF-THEN rules of the form:

$$\begin{aligned} R^l: \quad & \text{If } z_1 \text{ is } F_1^l \text{ and } z_2 \text{ is } F_2^l \text{ and } \cdots, \text{ and } z_n \text{ is } F_n^l \\ \text{Then} \quad & y \text{ is } G^l, l = 1, 2, \cdots, M \end{aligned}$$

where $Z = [z_1, \cdots, z_n]^T$ and y are the FLS input and output, respectively. Fuzzy sets F_j^l and G^l are associated with the fuzzy membership functions $\mu_{F_k^l}(z_j)$ and $\mu_{G^l}(y)$, respectively, and M is the number of fuzzy rules. Through singleton fuzzifier, center average defuzzification, and product inference, the FLS can be expressed as

$$y(Z) = \frac{\sum_{l=1}^{M} \bar{y}_l \prod_{k=1}^{n} \mu_{F_k^l}(z_j)}{\sum_{l=1}^{M} [\prod_{k=1}^{n} \mu_{F_k^l}(z_j)]} \tag{6.4}$$

where $\bar{y}_l = \max_{y \in \mathbf{R}} \mu_{G^l}(y)$. Then, define the fuzzy basis functions as

$$\xi_l = \frac{\prod_{k=1}^{n} \mu_{F_k^l}(z_j)}{\sum_{l=1}^{M} [\prod_{k=1}^{n} \mu_{F_k^l}(z_j)]},$$

and by denoting $W^T = [\bar{y}_1, \bar{y}_2, \cdots, \bar{y}_M]$ and $\xi(Z) = [\xi_1(Z), \xi_2(Z), \cdots, \xi_M(Z)]^T$, the FLS (6.4) can be rewritten as

$$y(Z) = W^T \xi(Z).$$

Then, according the above well-known universal approximation property of the FLS, for some sufficiently large integer M_i, there exists an ideal weight vector $W_i^* \in \mathbf{R}^{M_i}$ such that

$$f_i(v_i, t) = W_i^{*T} \xi_i(Z_i) + \varepsilon_i, \, i \in \mathscr{V}_N \tag{6.5}$$

where $Z_i = v_i$, M_i is the number of IF-THEN rules, W_i^* denotes the optimal approximation parameter vector, $\xi_i(Z_i)$ is a basis function and ε_i is the approximation error to be bounded by the constant ε_i^*.

6.2.3 Problem Formulation

The aim of the multiple high speed trains is to track a given desired trajectory (x_0, v_0, a_0) while maintaining a desired safety spacing between two consecutive trains. By viewing the desired trajectory (x_0, v_0, a_0) as a virtual leader (labeled as 0), the original multiple high speed trains can be viewed as leader-following consensus of multi-agent systems. Then, to achieve the above objective, inspired by Chapter 4, the *spacing errors* are defined as

$$\begin{aligned} \tilde{e}_i &= x_{i-1} - x_i - \Delta_{i-1,i} - h_i(v_i - v_0) \\ e_i &= \tilde{e}_i - \Xi_i \\ \Xi_i &= [\tilde{e}_i(0) + (\zeta_i \tilde{e}_i(0) + \dot{\tilde{e}}_i(0))t]e^{-\zeta_i t} \\ \tilde{e}_i(0) &= \tilde{e}_i(t)|_{t=0}, \dot{\tilde{e}}_i(0) = \dot{\tilde{e}}_i(t)|_{t=0} \end{aligned} \tag{6.6}$$

where $\Delta_{i-1,i} > 0$ and $h_i > 0$ are the required ith safety distance and *constant-time headway*, respectively, and ζ_i is a strictly positive constant. In addition, from (6.6), one can imply that

$$e_i(t)|_{t=0} = 0, \, \dot{e}_i(t)|_{t=0} = 0 \tag{6.7}$$

which shows that the *spacing errors* e_i in (6.6) are initially zero for arbitrary initial spacing and velocity errors. By introducing the *constant-time headway* h_i, a systematic improvement of transient performance can be achieved due to the variable spacing during the transient state, while in the steady state, a fixed distance $\Delta_{i-1,i}$ between consecutive trains can be maintained.

The objective of this chapter is to design an adaptive fuzzy-fault tolerant control scheme based on PI sliding mode for multiple *high speed trains* (6.2) with nonlinear uncertainties and disturbances as well as actuator faults described by (6.3) such that not only stable tracking is achieved, but also *strong string stability* is guaranteed. More specifically, the designed control scheme ensures that:

$$(1) \quad \begin{aligned} v_i &\rightarrow v_0 \\ d_i &:= x_{i-1} - x_i \rightarrow \Delta_{i-1,i}; \end{aligned} \tag{6.8}$$

where d_i denotes the real distance of the consecutive trains; (2) all the signals that are involved in the resulting closed-loop system are uniformly ultimately bounded, and the *spacing errors* e_i converge to a small residual set containing the origin; (3) *strong string stability* is guaranteed as [55, 57, 93]

$$|e_N| \leq |e_{N-1}| \leq \cdots \leq |e_1|, \tag{6.9}$$

i.e., the error propagation transfer function $G_i(s) := \frac{E_{i+1}(s)}{E_i(s)}$ satisfies $|G_i(s)| \leq 1$ for all $i = 1, 2, \cdots, N$, where $E_i(s)$ denotes *Laplace transform* of e_i.

6.3 PI-Based Sliding Mode and Coupled Sliding Mode

It is well known that the *sliding mode control* (SMC) method is a robust method that is used to control nonlinear and uncertain systems [209]. As usual in the SMC technique, the control forces the system evolution on a certain surface, which guarantees the accomplishment of the control requirements. In order to achieve the objectives in (6.8), the PI sliding surface in the space of the *spacing errors* e_i is defined as:

$$s_i = K_P e_i + K_I \int_0^t e_i(\tau) d\tau \tag{6.10}$$

where the parameters K_P and K_I are positive scalars representing the proportional and integral constants, respectively. On the other hand, from the definition of the error propagation transfer function $G_i(s)$, the relationship of e_i and e_{i+1} need be constructed. Therefore, similar to [55, 57, 93], the following coupled sliding surface is defined to incorporate the information of e_i and e_{i+1} to realize the *strong string stability* (6.9)

$$S_i = \begin{cases} q s_i - s_{i+1}, i = 1, 2, \cdots, N-1 \\ q s_i, i = N \end{cases} \tag{6.11}$$

where q is a positive parameter which couples both sliding surfaces s_i and s_{i+1}. Since $q > 0$ is a positive scalar, it follows the equivalence of S_i and s_i for all $i = 1, 2, \cdots, N$, that is, when S_i achieves the sliding surface, s_i also achieves the sliding surface at the same time, and vice versa.

6.4 Adaptive Fuzzy Control Design and Stability Analysis

In this section, two adaptive fuzzy controllers will be constructed using proportional and integral *sliding mode control* (PISMC) method and approximating power of FLS. In the first scheme, adaptive fuzzy control in the fault-free case is developed to guarantee the *strong string stability* of all trains, while in the second scheme, an alterative adaptive fuzzy *fault-tolerant* control is investigated, for which learning parameters are employed and the number of such learning parameters is independent of the number of fuzzy rules.

6.4.1 Controller Design for Fault-Free Case

In this subsection, the case of fault-free is investigated. By the definition of (6.3), we can obtain that $u_{ai} = u_i$. To facilitate the control design, define the estimation errors

$$\begin{aligned} \tilde{W}_i &= W_i^* - \hat{W}_i \\ \tilde{\eta}_i &= \eta_i^* - \hat{\eta}_i, i \in \mathscr{V}_N \end{aligned} \tag{6.12}$$

where $\hat{W}_i$ and $\hat{\eta}_i$ are the estimated values of W_i^* and η_i^*, respectively. Here, W_i^* is the ideal fuzzy weight vector in (6.5), while η_i^* will be defined later in (6.21). Then, based on the approximation property and sliding mode techniques, a distributed adaptive fuzzy controller is proposed for each train:

$$u_i = \frac{K_{i0}}{\lambda_i} S_i + \frac{1}{\lambda_i} \frac{A_i^2 S_i}{|A_i S_i| + \delta_i} + \hat{\eta}_i \tanh\left(\frac{S_i}{\varepsilon_i}\right) + \frac{|\hat{W}_i^T \xi_i(Z_i)|^2 S_i}{|\hat{W}_i^T \xi_i(Z_i) S_i| + \delta_i} \tag{6.13}$$

where

$$\begin{cases} A_i = q[K_P(v_{i-1} - v_i - \dot{\Xi}_i) + h_i a_0 + K_I e_i] - K_P \dot{e}_{i+1} - K_I e_{i+1}, \text{for } i \in \mathscr{V}_N \backslash \{N\} \\ A_i = q[K_P(v_{i-1} - v_i - \dot{\Xi}_i) + h_i a_0 + K_I e_i], \text{for } i = N \end{cases}$$
$$\dot{\Xi}_i = -\zeta_i[\tilde{e}_i(0) + (\zeta_i \tilde{e}_i(0) + \dot{\tilde{e}}_i(0))t]e^{-\zeta_i t} + (\zeta_i \tilde{e}_i(0) + \dot{\tilde{e}}_i(0))e^{-\zeta_i t}, i \in \mathscr{V}_N \tag{6.14}$$

and the update laws for $\hat{W}_i$ and $\hat{\eta}_i$ are chosen as

$$\begin{aligned} \dot{\hat{W}}_i &= -\alpha_i \lambda_i \xi_i(Z_i) S_i - \sigma_{1i} \hat{W}_i \\ \dot{\hat{\eta}}_i &= \beta_i \lambda_i S_i \tanh\left(\frac{S_i}{\varepsilon_i}\right) - \sigma_{2i} \hat{\eta}_i, \end{aligned} \tag{6.15}$$

where $\lambda_i = qK_P h_i$, and $K_{i0} > 0$, $K_P > 0$, $K_I > 0$, $\delta_i > 0$, $\varepsilon_i > 0$, $\alpha_i > 0$, and $\beta_i > 0$ are the design parameters. In addition, $\sigma_{1i} \in \mathbf{R}^+$ and $\sigma_{2i} \in \mathbf{R}^+$ are any positive uniform continuous and bounded functions that satisfy

$$\sigma_{ji} > 0, \lim_{t \to \infty} \int_{t_0}^{t} \sigma_{ji}(\tau) d\tau \leq \bar{\sigma}_{ji} < +\infty, j = 1, 2 \tag{6.16}$$

with $\bar{\sigma}_{ji}$ being any finite positive constants.

Now, the first main result is summarized as follows.

Theorem 6.1 *Consider the multiple high speed trains (6.2) without actuator faults. With the adaptive fuzzy control law (6.13), and the adaptive update laws (6.15), the spacing errors e_i in (6.6) converge to a small neighborhood around origin by appropriately choosing design parameters. In addition, strong string stability of the multiple high speed trains can also be guaranteed for all $i \in \mathscr{V}_N$ when q satisfies $0 < q \leq 1$.*

Proof 6.1 *We prove the theorem in two steps.*

(a) Boundedness of S_i: *Construct the following Lyapunov function candidate for the whole platoon:*

$$V_1 = \sum_{i=1}^{N} V_{1i} = \sum_{i=1}^{N} \{\tfrac{1}{2}S_i^2 + \tfrac{1}{2\alpha_i}\tilde{W}_i^T\tilde{W}_i + \tfrac{1}{2\beta_i}\tilde{\eta}_i^2\}$$

where $\tilde{W}_i$ *and* $\tilde{\eta}_i$ *are defined by (6.12). Then, taking the time derivative of* V_{1i} *results in*

$$\dot{V}_{1i} = S_i\dot{S}_i - \tfrac{1}{\alpha_i}\tilde{W}_i^T\dot{\hat{W}}_i - \tfrac{1}{\beta_i}\tilde{\eta}_i^T\dot{\hat{\eta}}_i. \tag{6.17}$$

From the fact $u_{ai} = u_i$ *for the case of actuator fault-free, the derivative of* S_i *can be deduced along (6.2), (6.6), (6.10) and (6.11) as*

$$\begin{aligned}
\dot{S}_i &= q[K_P\dot{e}_i + K_I e_i] - [K_P\dot{e}_{i+1} + K_I e_{i+1}] \\
&= q\{K_P[v_{i-1} - v_i - h_i(u_i + w_i + f_i(v_i,t) - a_0) - \dot{\Xi}_i] + K_I e_i\} \\
&\quad -[K_P\dot{e}_{i+1} + K_I e_{i+1}] \\
&= -\lambda_i[u_i + w_i + f_i(v_i,t)] + A_i, \text{for } i \in \mathscr{V}_N \setminus N \\
\dot{S}_i &= q[K_P\dot{e}_i + K_I e_i] \\
&= -\lambda_i[u_i + w_i + f_i(v_i,t)] + A_i, \text{for } i = N,
\end{aligned} \tag{6.18}$$

where $\lambda_i = qK_P h_i$*, and* A_i *is defined by (6.14).*

Substituting (6.13) into $-\lambda_i u_i S_i$*, one has*

$$\begin{aligned}
-\lambda_i u_i S_i &= -K_{i0}S_i^2 - \tfrac{A_i^2 S_i^2}{|A_i S_i| + \delta_i} - \lambda_i\hat{\eta}_i S_i \tanh(\tfrac{S_i}{\varepsilon_i}) - \lambda_i \tfrac{|\hat{W}_i^T\xi_i(Z_i)|^2 S_i^2}{|\hat{W}_i^T\xi_i(Z_i)S_i| + \delta_i} \\
&\le -K_{i0}S_i^2 - |A_i S_i| - \lambda_i\hat{\eta}_i S_i \tanh(\tfrac{S_i}{\varepsilon_i}) - \lambda_i|\hat{W}_i^T\xi_i(Z_i)S_i| + (\lambda_i + 1)\delta_i.
\end{aligned} \tag{6.19}$$

Using FLS to approximate the unknown function $f_i(v_i,t)$ *as in (6.5), for any* $i \in \mathscr{V}_N$*, we obtain*

$$\begin{aligned}
S_i\dot{S}_i &= S_i\{-\lambda_i[u_i + w_i + W_i^{*T}\xi_i(Z_i) + \varepsilon_i] + A_i\} \\
&= -\lambda_i u_i S_i - \lambda_i W_i^{*T}\xi_i(Z_i)S_i - \lambda_i(w_i + \varepsilon_i)S_i + A_i S_i \\
&\le -K_{i0}S_i^2 - \lambda_i\hat{\eta}_i S_i \tanh(\tfrac{S_i}{\varepsilon_i}) + \lambda_i\eta_i^*|S_i| - \lambda_i\tilde{W}_i^T\xi_i(Z_i)S_i + (\lambda_i + 1)\delta_i
\end{aligned} \tag{6.20}$$

where

$$\eta_i^* \ge w_i^* + \varepsilon_i^*, \tag{6.21}$$

and the following facts are used

$$\begin{aligned}
&- |A_i S_i| + A_i S_i \le 0 \\
&- \lambda_i|\hat{W}_i^T\xi_i(Z_i)S_i| - \lambda_i W_i^{*T}\xi_i S_i \\
&\le \lambda_i\hat{W}_i^T\xi_i(Z_i)S_i - \lambda_i W_i^{*T}\xi_i S_i \\
&= -\lambda_i\tilde{W}_i^T\xi_i(Z_i)S_i.
\end{aligned}$$

By Lemma 2.5 and Lemma 2.7, it follows from (6.15) that

$$\begin{aligned}
&- \lambda_i\tilde{W}_i^T\xi_i(Z_i)S_i - \tfrac{1}{\alpha_i}\tilde{W}_i^T\dot{\hat{W}}_i \\
&= -\lambda_i\tilde{W}_i^T\xi_i(Z_i)S_i - \tilde{W}_i^T(-\lambda_i\xi_i(Z_i)S_i - \tfrac{\sigma_{1i}}{\alpha_i}\hat{W}_i) \\
&= \tfrac{\sigma_{1i}}{\alpha_i}\tilde{W}_i^T\hat{W}_i \\
&\le -\tfrac{\sigma_{1i}}{2\alpha_i}\tilde{W}_i^T\tilde{W}_i + \tfrac{\sigma_{1i}}{2\alpha_i}\|W_i^*\|^2
\end{aligned} \tag{6.22}$$

$$
\begin{aligned}
&- \lambda_i \hat{\eta}_i S_i \tanh(\tfrac{S_i}{\varepsilon_i}) + \lambda_i \eta_i^* |S_i| - \tfrac{1}{\beta_i} \tilde{\eta}_i \dot{\hat{\eta}}_i \\
&= -\lambda_i \hat{\eta}_i S_i \tanh(\tfrac{S_i}{\varepsilon_i}) + \lambda_i \eta_i^* |S_i| - \tilde{\eta}_i(\lambda_i S_i \tanh(\tfrac{S_i}{\varepsilon_i}) - \tfrac{\sigma_{2i}}{\beta_i} \hat{\eta}_i) \\
&= \lambda_i \eta_i^* |S_i| - \eta_i^* \lambda_i S_i \tanh(\tfrac{S_i}{\varepsilon_i}) + \sigma_{2i} \tilde{\eta}_i \hat{\eta}_i \\
&\leq 0.2785 \lambda_i \eta_i^* - \tfrac{\sigma_{2i}}{2\beta_i} \tilde{\eta}_i^2 + \tfrac{\sigma_{2i}}{2\beta_i} \eta_i^{*2}.
\end{aligned}
\tag{6.23}
$$

Substituting (6.20), (6.22) and (6.23) into (6.17) yields

$$
\begin{aligned}
\dot{V}_1 &\leq \sum_{i=1}^{N} \{-K_{i0} S_i^2 - \tfrac{\sigma_{1i}}{2\alpha_i} \tilde{W}_i^T \tilde{W}_i - \tfrac{\sigma_{2i}}{2\beta_i} \tilde{\eta}_i^2 + \tfrac{\sigma_{1i}}{2\alpha_i} \|W_i^*\|^2 \\
&\quad +0.2785 \lambda_i \eta_i^* + \tfrac{\sigma_{2i}}{2\beta_i} \eta_i^{*2} + (\lambda_i + 1)\delta_i\} \\
&\leq -\gamma_1 V_1 + \vartheta_1
\end{aligned}
\tag{6.24}
$$

where the positive parameters γ_1 and ϑ_1 are given as follows

$$
\begin{aligned}
\gamma_1 &= \min\{2K_{i0}, \min_{1 \leq i \leq N} \sigma_{1i}, \min_{1 \leq i \leq N} \sigma_{2i}\} \\
\vartheta_1 &= \sum_{i=1}^{N} [\tfrac{\sigma_{1i}}{2\alpha_i} \|W_i^*\|^2 + 0.2785 \lambda_i \eta_i^* + \tfrac{\sigma_{2i}}{2\beta_i} \eta_i^{*2} + (\lambda_i + 1)\delta_i].
\end{aligned}
$$

According to the comparison principle, it follows from (6.24) that

$$
V_1 \leq (V_1(0) - \tfrac{\vartheta_1}{\gamma_1}) e^{-\gamma_1 t} + \tfrac{\vartheta_1}{\gamma_1} \leq V_1(0) + \tfrac{\vartheta_1}{\gamma_1}, \tag{6.25}
$$

where

$$
V_1(0) = \tfrac{1}{2} \sum_{i=1}^{N} \{S_i^2(0) + \tfrac{1}{\alpha_i} \tilde{W}_i^T(0) \tilde{W}_i(0) + \tfrac{1}{\beta_i} \tilde{\eta}_i^2(0)\}.
$$

Eq. (6.25) implies that all the signals in the closed-loop system are bounded. In particular, we have

$$
\lim_{t \to \infty} \sum_{i=1}^{N} S_i^2 \leq \sqrt{\tfrac{2\vartheta_1}{\gamma_1}}. \tag{6.26}
$$

Using (6.26), we conclude that reducing ϑ_1, meanwhile increasing γ_1 will lead to smaller bounds of S_i, i.e., S_i can converge to a small neighborhood of zero by choosing the design parameters appropriately. It is clear from (6.25) that the bound of S_i can be minimized by selecting appropriately small K_P and large K_{0i}, α_i and β_i (thus large γ_1 and small ϑ_1). However, σ_{1i} and σ_{2i} should be chosen as a tradeoff between the transient and steady-state performance since they are included in both γ_1 and ϑ_1.

(b) Strong String Stability: Since $S_i = qs_i - s_{i+1}$ converges to an arbitrarily small neighborhood of zero, we can get the relationship

$$
q(e_i + \int_0^t \lambda e_i(\tau) d\tau) = e_{i+1} + \int_0^t \lambda e_{i+1}(\tau) d\tau. \tag{6.27}
$$

Recalling (6.7), taking Laplace transform of (6.27) yields

$$
q(E_i(s) + \tfrac{\lambda}{s} E_i(s)) = E_{i+1}(s) + \tfrac{\lambda}{s} E_{i+1}(s).
$$

Then, $G_i(s) = \frac{E_{i+1}(s)}{E_i(s)} = q$. Therefore, when q satisfy $0 < |q| \leq 1$, (6.9) can be satisfied, that is, the strong string stability can be guaranteed. This completes the proof. □

Remark 6.3 *It is worth mentioning that, by introducing the spacing error as (6.6), the distances between consecutive trains will be constant in the steady-state despite the trains running at high speed, which will improve railway capacity and transport efficiency. On the other hand, in the transient state, the constant time headway will improve string stability by using variable spacing.*

Remark 6.4 *The control algorithm in Theorem 6.1 does not involve any linearized analysis for the nonlinear nature of the resistance of the high speed trains, which can satisfy the accuracy requirement of the tracking control for high speed trains in reality. In addition, with the introduction of FLS, the requirement for matching conditions as in [57] is removed. Nevertheless, there are a lot of adaptive parameters that need to be calculated online because the estimations of the ideal weight vector are directly adjusted. This can be particularly serious when there are many trains in the platoon and many rule bases in the fuzzy system are used to approximate the uncertain nonlinear functions $f_i(v_i, t)$. It will result in that the learning time tends to become unacceptably large and time-consuming which is unavoidable when the controller obtained in Theorem 6.1 is implemented.*

6.4.2 Fault-Tolerant Controller Design with Actuator Faults

In control (6.13), actuator faults are not considered, and the online update fuzzy weights $\hat{W}_i$ in (6.15) are a vector, whose dimension depends on the number of fuzzy rules. Since actuator faults can severely degrade the whole train platoon, control design for multiple HSTs subject to actuator faults and *external disturbances* presents a tremendous challenge. This subsection will present an alternative adaptive control to consider the actuator faults and simultaneously reduce the number of adaptive parameters.

The proposed adaptive fuzzy *fault-tolerant* control for train i is of the form:

$$u_i = \frac{K_{i0}}{\lambda_i} S_i + \frac{1}{2\tau_i} S_i \hat{\theta}_i \xi_i^T(Z_i)\xi_i(Z_i) + \frac{\psi_i}{\lambda_i} \frac{A_i^2 S_i}{|A_i S_i| + \delta_i} + \frac{1}{2\tau_i} S_i \hat{\eta}_i \tag{6.28}$$

where $\psi_i \geq \frac{1}{\rho_{i0}} \geq 1$, A_i is the same as (6.14), $K_{i0} > 0$ is the feedback gain, and $\delta_i > 0$ and τ_i are positive constants. The parameters $\hat{\theta}_i$ and $\hat{\eta}_i$ are updated as follows

$$\begin{aligned} \dot{\hat{\theta}}_i &= \frac{\alpha_i \lambda_i}{2\tau_i} \xi_i^T(Z_i)\xi_i(Z_i) S_i^2 - \sigma_{1i}\hat{\theta}_i, \hat{\theta}_i(0) \geq 0 \\ \dot{\hat{\eta}}_i &= \frac{\beta_i \lambda_i}{2\tau_i} S_i^2 - \sigma_{2i}\hat{\eta}_i, \hat{\eta}_i(0) \geq 0 \end{aligned} \tag{6.29}$$

where $\alpha_i > 0$ and $\beta_i > 0$ are adaptation gains, and σ_{1i} and σ_{2i} are the same as (6.16). Denote the parameter errors as:

$$\begin{aligned} \tilde{\theta}_i &= \theta_i^* - \rho_{i0}\hat{\theta}_i \\ \tilde{\eta}_i &= \eta_i^* - \rho_{i0}\hat{\eta}_i \end{aligned} \tag{6.30}$$

where $\theta_i^* = \|W_i^*\|^2$, $\eta_i^* = \bar{\eta}_i^{*2}$ with $\bar{\eta}_i^*$ being defined latter in (6.33).

Then, the following theorem, which guarantees the bounded stability of each train and the *strong string stability* of the whole multiple high speed train system, can be obtained.

Theorem 6.2 *Consider the multiple high speed trains (6.2) with actuator faults (6.3), and suppose Assumption 6.1 is satisfied. With the adaptive fuzzy fault-tolerant control law (6.28), and the adaptive laws (6.29), the spacing errors e_i in (6.6) converge to a small neighborhood around origin by appropriately choosing design parameters. In addition, strong string stability of the multiple high speed trains can also be guaranteed for $i \in \mathscr{V}_N$ when q satisfies $0 < q \leq 1$.*

Proof 6.2 *Construct the following Lyapunov function candidate:*

$$V_2 = \sum_{i=1}^{N} V_{2i} = \sum_{i=1}^{N} \{\tfrac{1}{2}S_i^2 + \tfrac{1}{2\alpha_i\rho_{i0}}\tilde{\theta}_i^2 + \tfrac{1}{2\rho_{i0}\beta_i}\tilde{\eta}_i^2\}$$

where $\tilde{\theta}_i$ and $\tilde{\eta}_i$ are defined by (6.30).

The derivative of V_{2i} for all $i \in \mathscr{V}_N$ can be deduced as

$$\dot{V}_{2i} = S_i\dot{S}_i - \tfrac{1}{\alpha_i}\tilde{\theta}_i^T\dot{\hat{\theta}}_i - \tfrac{1}{\beta_i}\tilde{\eta}_i^T\dot{\hat{\eta}}_i. \tag{6.31}$$

From the fact $u_{ai} = \rho_i(t,t_{\rho i})u_i + r_i(t,t_{ri})$ and after similar manipulation as in (6.18), we obtain

$$\begin{aligned} \dot{S}_iS_i &= -\lambda_i\rho_i(t,t_{\rho i})u_iS_i - \lambda_iW_i^{*T}\xi_i(Z_i)S_i - \lambda_i[r_i(t,t_{ri}) + w_i + \varepsilon_i]S_i + A_iS_i \\ &\leq -\lambda_i\rho_i(t,t_{\rho i})u_iS_i - \lambda_iW_i^{*T}\xi_i(Z_i)S_i + \lambda_i\bar{\eta}_i^*|S_i| + |A_iS_i| \\ &\leq -\lambda_i\rho_i(t,t_{\rho i})u_iS_i + \tfrac{\lambda_i}{2\tau_i}S_i^2\theta_i^*\xi_i^T\xi_i(Z_i) + \tfrac{\lambda_i}{2\tau_i}S_i^2\bar{\eta}_i^{*2} + \lambda_i\tau_i + |A_iS_i| \end{aligned} \tag{6.32}$$

where

$$\begin{aligned} \theta_i^* &= \|W_i^*\|^2, \eta_i^* = \bar{\eta}_i^{*2} \\ \bar{\eta}_i^* &\geq r_i^* + w_i^* + \varepsilon_i^*. \end{aligned} \tag{6.33}$$

It follows from (6.29) that $\dot{\hat{\theta}}_i + \sigma_{1i}\hat{\theta}_i \geq 0$ and $\dot{\hat{\eta}}_i + \sigma_{2i}\hat{\eta}_i \geq 0$, which imply that once we choose $\hat{\theta}_i(0) \geq 0$ and $\hat{\eta}_i(0) \geq 0$, the adaptive parameters $\hat{\theta}_i$ and $\hat{\eta}_i$ will always be non-negative. Then, recalling (6.28), we can obtain

$$\begin{aligned} -\lambda_i\rho_i(t,t_{\rho i})u_iS_i &= -\lambda_iS_i\rho_i(t,t_{\rho i})[\tfrac{K_{i0}}{\lambda_i}S_i + \tfrac{1}{2\tau_i}S_i\hat{\theta}_i\xi_i^T(Z_i)\xi_i(Z_i) \\ &\quad + \tfrac{\psi_i}{\lambda_i}\tfrac{A_i^2S_i}{|A_iS_i|+\delta_i} + \tfrac{1}{2\tau_i}S_i\hat{\eta}_i] \\ &\leq -\rho_{i0}K_{i0}S_i^2 - \tfrac{1}{2\tau_i}\lambda_i\rho_{i0}S_i^2\hat{\theta}_i\xi_i^T(Z_i)\xi_i(Z_i) \\ &\quad -\psi_i\rho_{i0}|A_iS_i| + \psi_i\rho_{i0}\delta_i - \tfrac{1}{2\tau_i}\rho_{i0}\hat{\eta}_i\lambda_iS_i^2. \end{aligned} \tag{6.34}$$

By completion of squares and Lemma 2.7, it follows from (6.29) that

$$\begin{aligned} -\tfrac{1}{\alpha_i}\tilde{\theta}_i\dot{\hat{\theta}}_i &= -\tilde{\theta}_i(\tfrac{\lambda_i}{2\tau_i}\xi_i^T(Z_i)\xi_i(Z_i)S_i^2 - \tfrac{\sigma_{1i}}{\alpha_i}\hat{\theta}_i) \\ &\leq -\tfrac{\lambda_i}{2\tau_i}\tilde{\theta}_i\xi_i^T(Z_i)\xi_i(Z_i)S_i^2 - \tfrac{\sigma_{1i}}{2\alpha_i}\tilde{\theta}_i^2 + \tfrac{\sigma_{1i}}{2\alpha_i}\theta_i^{*2} \\ -\tfrac{1}{\beta_i}\tilde{\eta}_i\dot{\hat{\eta}}_i &= -\tilde{\eta}_i(\tfrac{\lambda_i}{2\tau_i}S_i^2 - \tfrac{\sigma_{2i}}{\beta_i}\hat{\eta}_i) \\ &\leq -\tfrac{\lambda_i}{2\tau_i}\tilde{\eta}_iS_i^2 - \tfrac{\sigma_{2i}}{2\beta_i}\tilde{\eta}_i^2 + \tfrac{\sigma_{2i}}{2\beta_i}\eta_i^{*2}. \end{aligned} \tag{6.35}$$

Recalling the fact $\psi_i \geq \frac{1}{\rho_{i0}} \geq 1$ *and substituting (6.32), (6.34) and (6.35) into (6.31), we can obtain*

$$\begin{aligned} \dot{V}_2 &\leq \sum_{i=1}^{N}\{-K_{i0}\rho_{i0}S_i^2 - \frac{\sigma_{1i}}{2\alpha_i}\tilde{\theta}_i^2 - \frac{\sigma_{2i}}{2\beta_i}\tilde{\eta}_i^2 + \frac{\sigma_{1i}}{2\alpha_i}\theta_i^{*2} \\ &\quad + \frac{\sigma_{2i}}{2\beta_i}\eta_i^{*2} + \psi_i\rho_{i0}\delta_i + \lambda_i\tau_i\} \\ &\leq -\gamma_2 V_1 + \vartheta_1 \end{aligned} \tag{6.36}$$

where the positive parameters γ_2 *and* ϑ_2 *are given as follows*

$$\begin{aligned} \gamma_1 &= \min\{2K_{i0}\rho_{i0}, \min_{1\leq i\leq N}\sigma_{1i}\rho_{i0}, \min_{1\leq i\leq N}\sigma_{2i}\rho_{i0}\} \\ \vartheta_1 &= \sum_{i=1}^{N}[\frac{\sigma_{1i}}{2\alpha_i}\theta_i^{*2} + \frac{\sigma_{2i}}{2\beta_i}\eta_i^{*2} + \psi_i\rho_{i0}\delta_i + \lambda_i\tau_i]. \end{aligned}$$

The result is then established using the same argument as in the proof of Theorem 6.1. □

Remark 6.5 *In comparison with the control algorithm in Theorem 6.1, the actuator faults are taken into consideration and the explosion of learning parameters as in Theorem 6.1 is avoided. In Theorem 6.2, the number of learning parameters is independent of the number of fuzzy rule bases but dependent on the number of the trains in the platoon, which makes the proposed algorithm more efficient in computation.*

6.5 Simulation Study and Performance Results

In this section, we will verify the effectiveness of the proposed adaptive fuzzy controllers obtained in Theorem 6.1 and Theorem 6.2 for a platoon of six high speed trains (i.e., $N = 6$) on the railway line. The parameters of each high speed train are listed in Table 6.2 [101], while the initial positions and speeds of each high speed train containing initial desired trajectories are given out in Table 6.3 [101]. According to the definition of $\tilde{e}_i$ in (6.6), the initial *spacing errors* $\tilde{e}_i(0)$ are non-zero as $\tilde{e}(0) = [20, 34, 82.5, 103, 64.5, 213]km$. The desired acceleration of the given reference trajectory is chosen as

$$a_0(t) = \begin{cases} 0.5t & km/h^2, \quad 2h \leq t < 5h \\ 2.5 & km/h^2, \quad 5h \leq t < 7h \\ -0.5t + 3.5 & km/h^2, \quad 7h \leq t < 10h \\ 0 & km/h^2, \quad otherwise \end{cases}$$

where the train will run in traction, braking and finally cruise phases. In addition, the *external disturbances* are generated by $w_i(t) = 0.1\sin(it + i\pi), i \in \mathcal{V}_N$. On the other hand, define a fuzzy set with 20 rules (i.e., $M_i = 20$) for each train with $Z_i = v_i$, which implies $n = 1$. Then, to construct the fuzzy controller, the fuzzy membership functions are chosen as follows:

$$\mu_{F_1^{l_i}}(v_i) = \exp\left[-\left(\frac{\sin(v_i)+\frac{l_i}{6}}{i\pi}\right)^2\right]$$
$$\text{for } i = 1, \; 2, \; \cdots, 6; \; l_i = 1, 2, \cdots, 20.$$

When there exist actuator faults, the actual input is $u_{ai} = \rho_i(t, t_{\rho i})u_i + r_i(t, t_{ri})$ with $\rho_i(t, t_{\rho i}) = 0.75 + 0.25\sin(0.1it)$ and $r_i(t, t_{ri}) = 2\sin(t)$. Then, we can obtain $\rho_{i0} = 0.5$. Using the fact $\psi_i \geq \frac{1}{\rho_{i0}} \geq 1$, one can obtain $\psi_i \geq 2$.

TABLE 6.2
Parameters of high speed train for $i \in \mathscr{V}_N (N = 6)$

Symbol	Value	Unit
m_i	80×10^4	kg
c_{i0}	0.01176	N/kg
c_{i1}	0.00077616	Ns/mkg
c_{i2}	1.6×10^{-5}	Ns^2/m^2kg
$\Delta_{i-1,i}$	20	km
h_i	1	h

TABLE 6.3
The parameters $x_i(0)$ and $v_i(0)$ for $i \in \mathscr{V}_N \cup \{0\} (N = 6)$

i	0	1	2	3	4	5	6
$x_i(0)(km)$	148	118	94	71.5	48.5	24	1
$v_i(0)(km/h)$	220	210	190	140	120	160	10

6.5.1 Simulation Results of Theorem 6.1

In this subsection, the controller obtained in Theorem 6.1 is applied to the multiple high speed trains (6.2) with or without actuator faults. The controller parameters are taken as $K_{0i} = 50$, $q = 0.5$, $K_P = 0.1$, $K_I = 2$, $\delta_i = 0.3$, $\varepsilon_i = 0.3$, $\zeta_i = 5$, $\alpha_i = 100$, $\beta_i = 10$, $\sigma_{1i} = 0.001e^{-10t}$ and $\sigma_{2i} = 0.1e^{-20t}$. For the case of fault-free actuator, Figure 6.2 illustrates the simulation results of control (6.13) with *initial conditions* $\hat{W}_i(0) = 1_{1\times 20}$ and $\hat{\eta}_i(0) = 1$. It is shown from Figure 6.2(a) that, even in the presence of the non-synchronous disturbances w_i, the *strong string stability* can be guaranteed since the amplitude of the spacing error decreases (i.e., $\|e_6\| \leq \cdots, \leq \|e_1\|$).

Distance and velocity tracking curve using the control law (6.13) given by Theorem 6.1 are presented in Figures 6.2(a) and (b), respectively. Figure 6.2(b) shows that the distances will converge to $\Delta_{i-1,i}$ and the neighboring-vehicle collisions can be avoided not only in the steady-state condition, but also during the initial transient since there do not exist negative distances. It is found that the six trains converge to the desired trajectory in Figure 6.2(c). Good tracking performance can be seen from Figures 6.2(b) and (c) distinctly, despite having non-synchronous disturbances w_i. However, there will be many parameters needed to be estimated in the adaptive fuzzy controller (6.13) given in Theorem 6.1 when $M_i = 20$. The number of $\hat{W}_{ij}$ $(i = 1, \cdots, 6; j = 1, \cdots, 20)$ is 120. For simplicity, only $\hat{W}_{1j}$ is shown in Figure 6.2(e). The sliding surfaces S_i, and the adaptive laws $\hat{W}_{1j}$ $(j = 1, \cdots, 20)$ and $\hat{\eta}_i$ are shown respectively in Figures 6.2(d)-(f), which are all bounded. In addition, the continuous tanh function is applied in the proposed control law, so it is easy to see that the *chattering phenomenon* is almost eliminated fully in Figure 6.2(d). When there exist actuator faults, the simulation results obtained by controller (6.13) are shown in Figure 6.3. Although the *strong string stability* can still be guaranteed in Figure 6.3(a), the actuator faults result in high overshoot and low convergence rate of consensus, which shows the necessity to consider the effect of actuator faults.

6.5.2 Simulation Results of Theorem 6.2

In this subsection, the control (6.28) with adaptive laws (6.29) given in Theorem 6.2 is implemented with parameters $\psi_i = 2$, $\tau_i = 1$ and other parameters are the same as those used in control (6.13) given in Theorem 6.1. Then, the simulation results of control (6.28) with *initial conditions* $\hat{\theta}_i(0) = 1$ and $\hat{\eta}_i(0) = 1$ are shown in Figure 6.4. From Figure 6.4, it is easy to see that not only the *strong string stability* is guaranteed, but also similar tracking performance as compared with Figure 6.2 can be obtained, even in the presence of *external disturbances* and unexpected actuator faults. Furthermore, despite using the same number of fuzzy rules, the number of the adaptive parameters $\hat{\theta}_i$ is only dependent on the number of trains i ($i = 6$ in this case). This can reduce the computation burden significantly (120 in Theorem 6.1 while 6 in Theorem 6.2). All these simulation results show the superiority of the control (6.28).

6.6 Conclusion

By combining *fuzzy logic systems* with PI-based sliding mode techniques, two adaptive fuzzy control schemes are proposed for multiple *high speed trains*. The proposed controller can force all trains to track the reference trajectory while maintaining a *desired safe spacing* simultaneously, even in the presence of actuator faults, unknown nonlinear functions and *external disturbances*. The two control schemes are constructed to achieve individual train stability and *strong string stability* of whole train

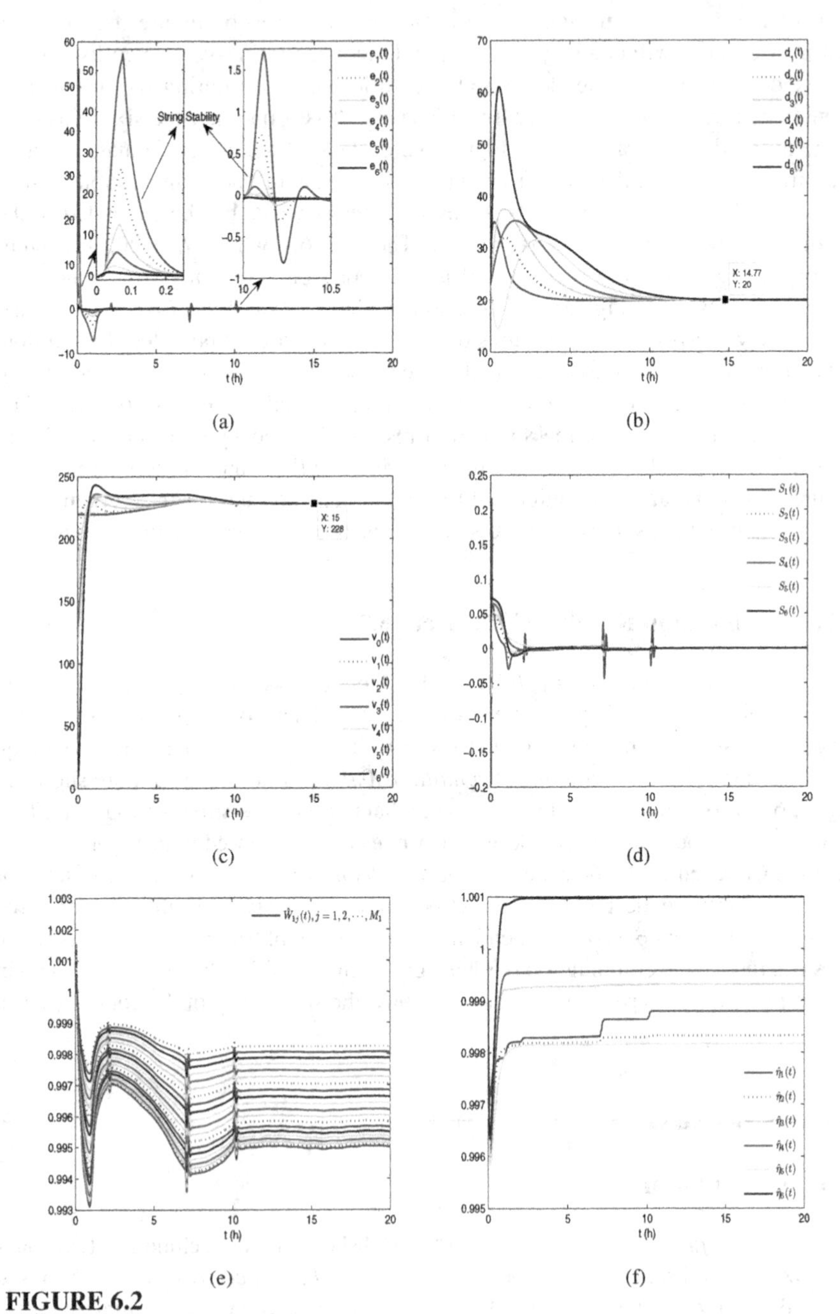

FIGURE 6.2
(a) Spacing error e_i; (b) distance d_i; (c) velocity v_i; (d) sliding surface S_i; (e) response of $\hat{W}_{1j}, j = 1, 2, \cdots, 20$; (f) response of $\hat{\eta}_i$ under Theorem 6.1 without actuator faults.

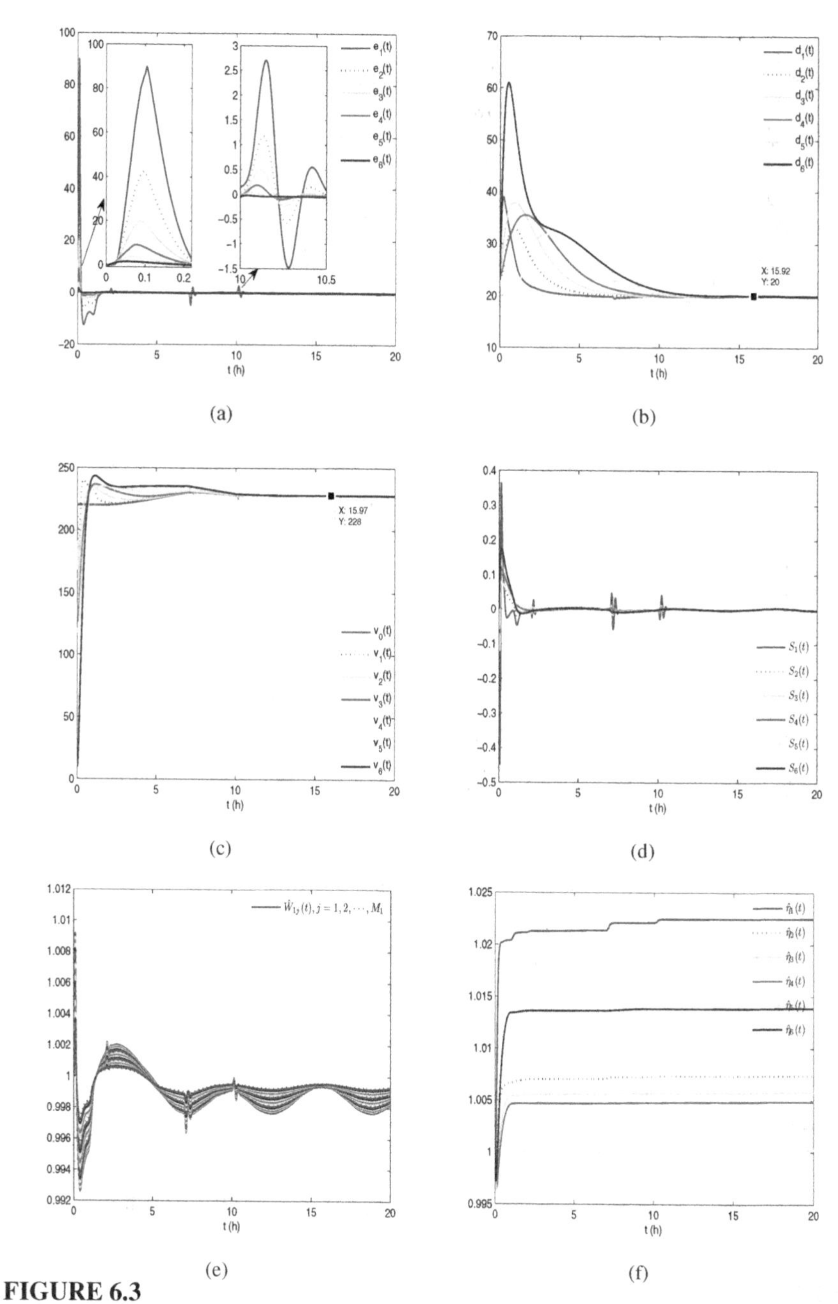

FIGURE 6.3
(a) Spacing error e_i; (b) distance d_i; (c) velocity v_i; (d) sliding surface S_i; (e) response of $\hat{W}_{1j}, j = 1, 2, \cdots, 20$; (f) response of $\hat{\eta}_i$ under Theorem 6.1 with actuator faults.

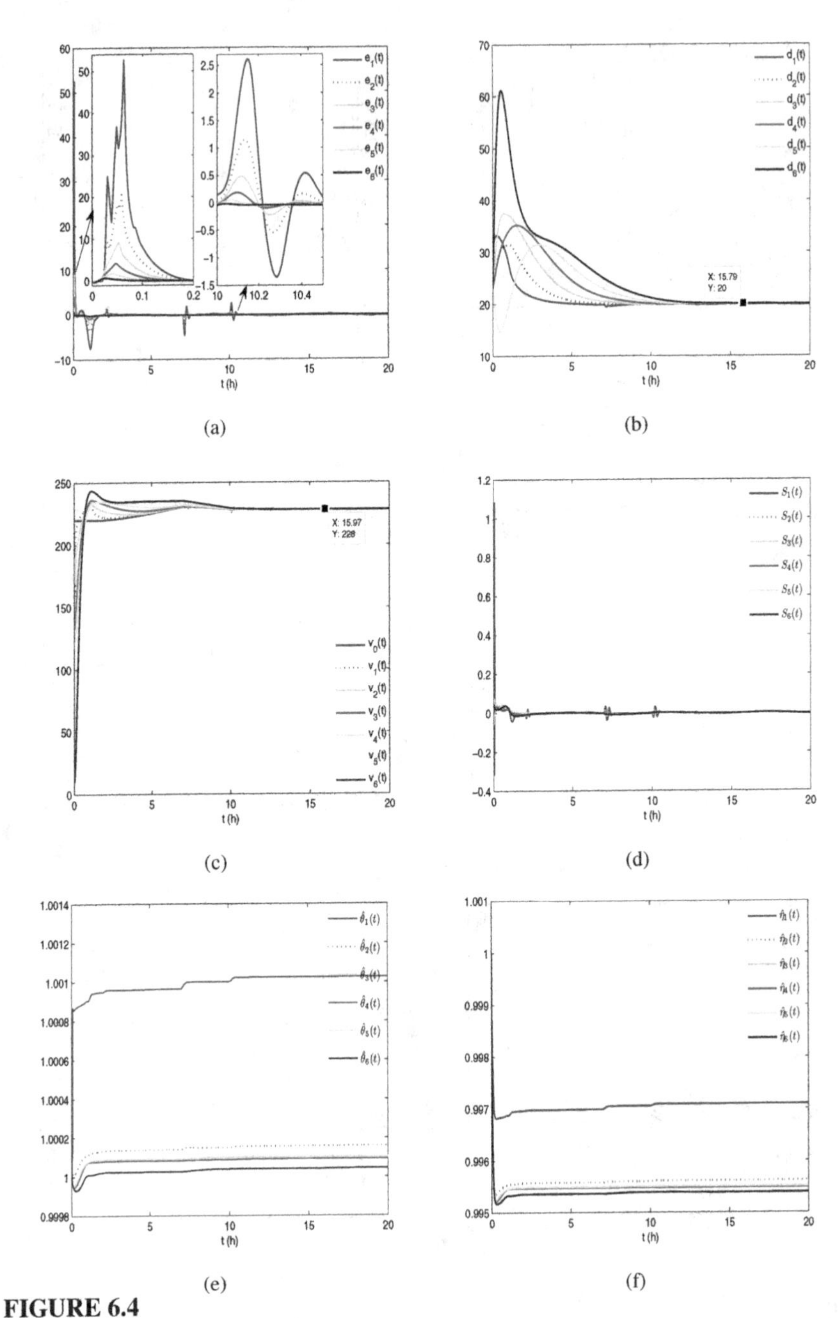

FIGURE 6.4
(a) Spacing error e_i; (b) distance d_i; (c) velocity v_i; (d) sliding surface S_i; (e) response of $\hat{W}_{1j}, j = 1, 2, \cdots, 20$; (f) response of $\hat{\eta}_i$ under Theorem 6.2 with actuator faults.

platoon. Specifically, a novel fuzzy structure with only scalar weight parameters (independent of the number of fuzzy rules) was constructed to reduce the computational costs. The simplicity of the control design renders the developed methods attractive for practical applications. Finally, simulation results are provided to show the effectiveness and advantage of the proposed control schemes.

7

Collision Avoidance for Vehicle Platoon with Input Deadzone

This chapter investigates the adaptive *platoon control* for nonlinear vehicular systems with asymmetric nonlinear input deadzone and neighboring vehicular spacing constraints. Vehicular *platoon control* encounters great challenges from unmodeled dynamic uncertainties, unknown *external disturbances*, unknown asymmetric nonlinear input deadzone and neighboring vehicular spacing constraints. In order to avoid collisions between consecutive vehicles as well as the connectivity breaks owing to limited sensing capabilities, a symmetric barrier Lyapunov function is employed. Then, a neural-network-based terminal *sliding mode control* (TSMC) scheme with minimal learning parameters is developed to maintain neighboring-vehicles keep connectivity and simultaneously avoid collisions. The uniform ultimate boundedness of all signals in the whole vehicular *platoon control* system is proven via Lyapunov analysis.

7.1 Introduction

A vehicular platoon must exhibit both individual stability and stability as a group while avoiding collisions between consecutive vehicles [10]. In order to achieve the stability of the whole vehicular platoon, *string stability* has been a significant topic of the automatic vehicle platoon since mid-1970s [43, 47, 51, 55, 57, 59, 212], where the *spacing errors* are required to attenuate as they propagate upstream from one vehicle to another. This property ensures that any perturbation of the velocity or position of the leader vehicle will not result in amplified fluctuations to the velocity and position of the follower vehicles [214]. However, via simulation examples, [10] has pointed out that the *string stability* introduced in [43, 47, 51, 55, 57, 59, 212] does not provide a formal guarantee of collision avoidance between consecutive vehicles. Another significant issue affecting the *platoon control* for vehicular platoons concerns limited sensing capabilities [6]. Most existing results neglect the communication/sensory limitations, which however are crucial in real-time scenarios [172]. These can be viewed as an issue of neighboring-vehicular spacing constraints. Violation of such constraints may result in collisions or connectivity breaks between consecutive vehicles. Henceforth, how to handle these neighboring-vehicular spacing constraints becomes an important research topic.

The control problem of systems with actuator deadzone nonlinearities is a significant and challenging research topic as such nonlinearities are always present in real plants, particularly in mechanical systems (e.g., the drive train in cars, rolling mills, industrial robots, etc.) [23]. The presence of deadzone nonlinearities may cause undesirable control accuracy, limit cycles, and even instability, and its study has been drawing much interest in the control community for a long time [23, 57, 163, 205]. Recently, the issue of string stability for vehicular platoons with matched velocity nonlinear uncertainties and actuator deadzone nonlinearities under the assumption that the deadzone slopes in the positive and negative regions must be the same has attracted much attention. In order to remove this assumption on the deadzone slopes, asymmetric actuator nonlinear deadzone was investigated in [163, 205] under the assumption that the nonlinear deadzone slopes in the positive and negative regions are strictly monotonous only. This obviously limited application in the real systems. Thus, it is more realistic and reliable to remove the above assumption. Moreover, another well-known major difficulty in the control of vehicular platoons is for unmodeled nonlinear uncertainties and *external disturbances*, which hinders perfect *platoon control* performance. In order to deal with this problem, recently, neural network-based approximation techniques together with the *sliding mode control* techniques have been applied extensively to systems with uncertainties [132, 171].

Inspired by the above observations, it is of both practical and theoretical significance to investigate the problem of adaptive *platoon control* for nonlinear vehicular platoons with asymmetric nonlinear actuator deadzone and neighboring-vehicle spacing constraints, which is however challenging and has not been studied so far. In this work, we present a neural network-based adaptive SMC control scheme for the considered vehicular platoons. In order to address the neighboring-vehicle spacing constraints, we use the *log*-type symmetric barrier Lyapunov function (SBLF) proposed in [76, 148] to avoid collisions and connectivity breaks between consecutive vehicles. To attenuate the negative effects of asymmetric nonlinear actuator deadzone and *unmodeled dynamic nonlinearities*, radial basis function neural network-based (RBFNN) approximation [95] with minimal learning parameters is adopted. It is worth mentioning that the assumption in [163, 205] that the nonlinear *deadzone* slopes in the positive and negative regions are strictly monotonous is removed. We show that the proposed adaptive SMC scheme can guarantee the stability of the whole vehicular platoon and the ultimate uniform boundedness of all closed-loop signals.

7.2 Vehicular Platoon Model and Preliminaries

7.2.1 Vehicular Platoon Description

Consider a vehicular platoon consisting of N follower vehicles indexed by $1, 2, \cdots, N$ and a leader vehicle labeled as 0, respectively. Each follower vehicle i subject to asymmetric unknown nonlinear *deadzone* is characterized by

$$\begin{aligned} \dot{x}_i(t) &= v_i(t), i \in \mathscr{V}_N \\ \dot{v}_i(t) &= \mathrm{Dz}(u_i(t)) + f_i(x_i, v_i, t) + w_i(t) \end{aligned} \tag{7.1}$$

where $(x_i(t), v_i(t)) \in \mathbf{R}^2$ $(i \in \mathscr{V}_N)$ denote the position and velocity of the ith vehicle. For compactness, $(x_i(t), v_i(t)) \in \mathbf{R}^2$, $i \in \mathscr{V}_N$ is abbreviated to (x_i, v_i) in some subsequent formulas. $f_i(x_i, v_i, t)$ represents nonlinear terms of the system, which is assumed to be smooth, continuous and bounded; $u_i(t)$ is the control input and $w_i(t)$ is the *external disturbance* assumed bounded by $|w_i(t)| \leq \bar{w}_i$. The asymmetric unknown nonlinear *deadzone* can be described by [163, 205]

$$\mathrm{Dz}(u_i(t)) = \begin{cases} g_{r,i}(u_i(t)), & \text{if } u_i(t) > b_{r,i} \\ 0, & \text{if } b_{l,i} < u_i(t) < u_{r,i} \\ g_{l,i}(u_i(t)), & \text{if } u_i(t) \leq u_{l,i} \end{cases} \tag{7.2}$$

where $Dz(u_i(t)) \in R$ is the input to the *deadzone*, $b_{l,i}$ and $b_{r,i}$ are the unknown parameters of the *deadzone*, $g_{l,i}(u_i(t))$ and $g_{r,i}(u_i(t))$ are smooth continuous functions, as shown in Figure 7.1. To facilitate control system design later, the following assumption is needed.

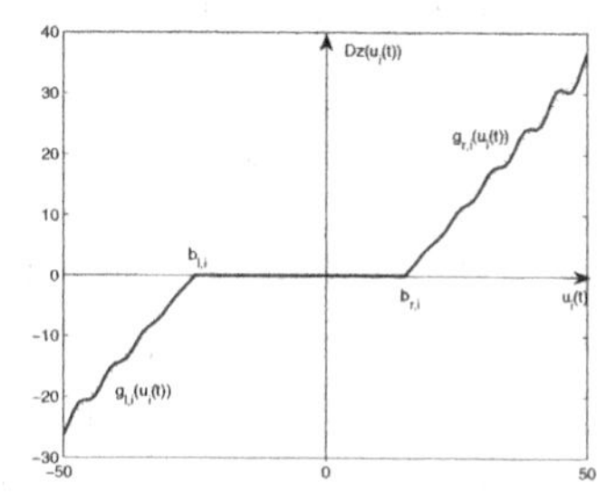

FIGURE 7.1
Asymmetric nonlinear deadzone model.

Assumption 7.1 *[163, 205] The deadzone parameters $b_{r,i}$ and $b_{l,i}$ are unknown bounded constants, but their signs are known, that is, $b_{r,i} > 0$ and $b_{l,i} < 0$.*

Remark 7.1 *Assumption 7.1 on the asymmetric nonlinear deadzone is the same as the assumption used in [163, 205]. However, it is assumed in [163, 205] that the characteristics of the slopes $g_{r,i}(u_i(t))$ and $g_{l,i}(u_i(t))$ are known, while the proposed control method in this chapter will not require the information on the slopes. In addition, the deadzone model considered in [23, 57] is a special case of (7.2).*

The dynamics of the leader vehicle is expressed as follows:

$$\dot{x}_0(t) = v_0(t), \dot{v}_0(t) = a_0(t) \tag{7.3}$$

where $a_0(t)$ is the desired acceleration.

Define

$$\begin{aligned}\Delta u_i(t) &= \mathrm{Dz}(u_i(t)) - u_i(t)\\ &= \begin{cases} u_i(t) - g_{r,i}(u_i(t)), & \text{if } u_i(t) > b_{r,i}\\ u_i(t), & \text{if } b_{l,i} < u_i(t) < u_{r,i}\\ u_i(t) - g_{l,i}(u_i(t)), & \text{if } u_i(t) \le u_{l,i}\end{cases}\end{aligned} \tag{7.4}$$

which will be approximated by RBFNNs in the later design. Thus, the system (7.1) can be rewritten as

$$\begin{aligned}\dot{x}_i(t) &= v_i(t), i \in \mathscr{V}_N\\ \dot{v}_i(t) &= u_i(t) + \Delta u_i(t) + f_i(x_i, v_i, t) + w_i(t).\end{aligned} \tag{7.5}$$

7.2.2 Radial Basis Function Neural Network

Due to the fact that the function $f_i(x_i, v_i, t)$ and $\Delta u_i(t)$ given by (7.1) and (7.4), respectively, are unknown in practical systems, the controller $u_i(t)$ is usually difficult to design. Therefore, the *radial basis function neural network* (RBFNN) is used to approximate these unknown functions. The architecture of the RBFNN comprises an input layer, a hidden layer, and an output layer [95]. In this chapter, the RBFNN is used to approximate any continuous function $f(Z) : \mathbf{R}^p \to \mathbf{R}$ over a compact set to arbitrary accuracy as

$$f(Z) = W^{*T}\xi(Z) + \varepsilon(Z) \tag{7.6}$$

where $Z \in \mathbf{R}^p$ is the input vector of RBFNN, $\varepsilon(Z)$ is the approximation error, $W^* = \left[W_1^*, \cdots, W_{N_1}^*\right]^T \in \mathbf{R}^{N_1}$ is the ideal weight vector of the RBFNN, and N_1 is the number of neurons. It is assumed that there exist ideal constant weights W^* such that $|\varepsilon(Z)| \le \varepsilon^*$ with constant $\varepsilon^* > 0$. $\xi(Z) = [\xi_1(Z), \cdots, \xi_{N_1}(Z)]^T \in \mathbf{R}^{N_1}$ is the Gaussian basis function vector and its kth component $\xi_k(Z)\mathbf{R}$ is described as

$$\xi_k(Z) = \exp\left[-\frac{(Z-\phi_k)^T(Z-\phi_k)}{2\theta_k^2}\right], k = 1, \cdots, N_1 \tag{7.7}$$

where $\phi_k \in \mathbf{R}^p$ is the center of the receptive field and θ_k is the width of the Gaussian function.

7.2.3 Problem Formulation

The control objective is to design a distributed control scheme such that the whole vehicular platoon is maintained arbitrarily accurate while avoiding collisions and connectivity breaks between consecutive vehicles. Defining the spacing error of the ith follower vehicle as

$$\begin{aligned}e_i(t) &= x_{i-1}(t) - x_i(t) - \Delta_{i-1,i},\ i \in \mathscr{V}_N\\ d_i(t) &= x_{i-1}(t) - x_i(t)\end{aligned} \tag{7.8}$$

where $\Delta_{i-1,i}$ represents the desired spacing between two consecutive vehicles, while $d_i(t)$ denotes the spacing between the $(i-1)$th and (i)th vehicles. Collision avoidance is an important issue in the platoon control problem, where $d_i(t)$ must be maintained greater than the collision distance Δ_{col}, i.e., $d_i(t) > \Delta_{col}$. Taking into account the limited communication/sensing distance Δ_{con}, we must have $d_i(t) < \Delta_{con}$. Therefore, the distance $d_i(t)$ should satisfy

$$\Delta_{col} \quad < \quad d_i(t) < \Delta_{con},\ i \in \mathscr{V}_N. \tag{7.9}$$

to avoid collisions and to maintain the communication/sensing connectivity. Furthermore, without loss of generality, assume $0 < \Delta_{col} < d_i(0) < \Delta_{con}, i \in \mathscr{V}_N$. To ensure the feasibility of the desired platoon, we assume that the desired neighboring-vehicle spacing $\Delta_{i-1,i}$ satisfies $0 < \Delta_{col} < \Delta_{i-1,i} < \Delta_{con}, i \in \mathscr{V}_N$. Letting $\Delta_{con} = 2\Delta_{i-1,i} - \Delta_{col}$, then, it follows from (7.9) that we have the following symmetric error bounds

$$\Delta_{col} - \Delta_{i-1,i} \quad < \quad e_i(t) < \Delta_{i-1,i} - \Delta_{col},\ i \in \mathscr{V}_N, \tag{7.10}$$

i.e., $|e_i(t)| < \bar{\Delta}, i \in \mathscr{V}_N$, where $\bar{\Delta} := \Delta_{i-1,i} - \Delta_{col}$. Then, the control goal of this chapter is to maintain a desired neighboring-vehicle spacing and track the leader vehicle while ensuring that all closed-loop signals are bounded and all neighboring-vehicle *spacing errors* $e_i(t)$ evolve within the bounded constraints for all time. More specifically, the following requirements should be satisfied.

(1) Tracking the leader and keeping desired spacing:

$$d_i(t) \quad \rightarrow \quad \Delta_{i-1,i};\ v_i(t) \rightarrow v_0(t), i \in \mathscr{V}_N. \tag{7.11}$$

(2) Avoiding collisions and keeping connectivity:

$$|e_i(t)| \quad < \quad \bar{\Delta}, i \in \mathscr{V}_N. \tag{7.12}$$

To deal with the spacing error constraint requirement in (7.12), the following symmetric barrier Lyapunov function (SBLF) [76, 148] is adopted.

$$V_{b,i}(t) \quad = \quad \frac{k_b}{2} \ln \frac{\bar{\Delta}^2}{\bar{\Delta}^2 - e_i^2(t)} \tag{7.13}$$

where $\ln(\cdot)$ represents the natural logarithm of $(\cdot)$, and Δ is a positive design constant. It is obvious from Figure 7.2 that the SBLF escapes to infinity if $e_i(t) = \bar{\Delta}$. It is also easy to see that $V_{b,i}(t)$ is positive definite and continuous in the region $|e_i(t)| < \bar{\Delta}$.

7.3 Distributed Adaptive NN Control Design

In this section, *terminal sliding mode control* (TSMC), in combination with adaptive control and the RBFNN technique, is used to construct the distributed adaptive

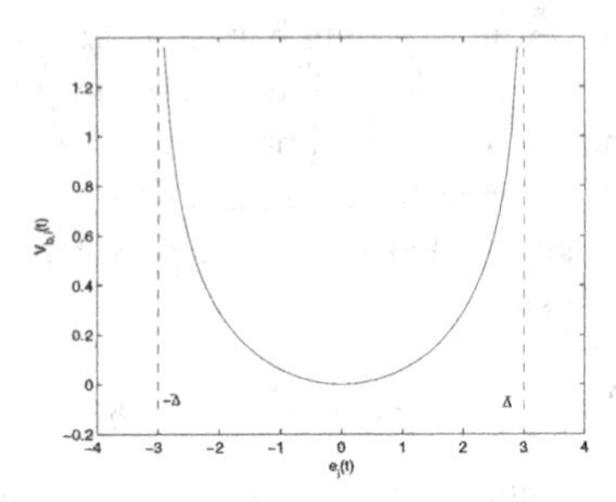

FIGURE 7.2
Schematic illustration of SBLF.

NN control scheme for the vehicle-following platoon in (7.1) and (7.3) subjected to symmetric nonlinear actuator *deadzone* (7.2).

A non-singular TSMC surface is defined as

$$s_i(t) \; = \; \dot{e}_i(t) + \lambda_1 e_i(t) + \lambda_2 |e_i(t)|^{\frac{m}{n}} \operatorname{sgn}(e_i(t)) \tag{7.14}$$

where m, n are positive odd numbers satisfying $m > n$, and λ_1 and λ_2 are positive constants. Then, the derivative of $s_i(t)$ is

$$\dot{s}_i(t) \; = \; \ddot{e}_i(t) + \lambda_1 \dot{e}_i(t) + \lambda_2 \tfrac{m}{n} |e_i(t)|^{\frac{m}{n}-1} \dot{e}_i(t). \tag{7.15}$$

Remark 7.2 *The sliding mode surface (7.14) is different from the one considered in [127], where $m < n$ is required. Then, the fractional power part in (7.15) may raise some doubt on whether its limit exists or not in the singular case. It can be seen in (7.15) that the third term containing $|e_i(t)|^{\frac{m}{n}-1}$ may cause a singularity to occur when $e_i(t) = 0$ if $m < n$. However, when $m > n$ is adopted in (7.14), the singularity problem can be avoided.*

Next, define

$$\begin{aligned} \tilde{\theta}_i(t) &= \theta_i^* - \theta_i(t) \\ \tilde{\eta}_i(t) &= \eta_i^* - \eta_i(t), i \in \mathscr{V}_N \end{aligned} \tag{7.16}$$

where $\theta_i(t)$ and $\eta_i(t)$ are used to estimate θ_i^* and η_i^*, respectively, and θ_i^* and η_i^* are defined later in (7.26). Then, the control law for the ith vehicle in the platoon is established as

$$\begin{aligned} u_i(t) \; = \; & k_i s_i(t) + \tfrac{k_{b,i} e_i(t)}{\Delta_i^{*2} - e_i^2} + \tfrac{1}{2} s_i \theta_i \xi_{1,i}^T(v_i) \xi_{1,i}(v_i) \\ & + \tfrac{1}{2} s_i \eta_i \xi_{2,i}^T(v_i) \xi_{2,i}(v_i) + A_i(t) \end{aligned} \tag{7.17}$$

where $A_i(t) = \ddot{x}_{i-1}(t) + \lambda_1 \dot{e}_i(t) + \lambda_2 \frac{m}{n} |e_i(t)|^{\frac{m}{n}-1} \dot{e}_i(t), i \in \mathscr{V}_N$, and the adaptive laws for $\theta_i(t)$ and $\eta_i(t)$ are designed as

$$\begin{aligned} \dot{\theta}_i &= \tfrac{\alpha_i}{2} \xi_{1,i}^T(v_i) \xi_{1,i}(v_i) s_i^2(t) - \sigma_{1,i}(t) \theta_i(t), \theta_i(0) \geq 0 \\ \dot{\eta}_i &= \tfrac{\beta_i}{2} \xi_{2,i}^T(v_i) \xi_{2,i}(v_i) s_i^2(t) - \sigma_{2,i}(t) \eta_i(t), \eta_i(0) \geq 0 \end{aligned} \tag{7.18}$$

with $k_i > 1$, $k_{b,i}$, α_i and β_i being any positive constants, while ρ_i and φ_i being small positive constants. In addition, $\sigma_{1,i}(t)$ and $\sigma_{2,i}(t)$ are any positive uniformly continuous and bounded functions that satisfy

$$\begin{aligned} \sigma_{j,i}(t) > 0, \lim_{t\to\infty} \int_{t_0}^{t} \sigma_{j,i}(\tau)d\tau &\leq \bar{\sigma}_{j,i} < +\infty \\ & j = 1,2; i \in \mathscr{V}_N \end{aligned} \tag{7.19}$$

with $\bar{\sigma}_{j,i}$ being finite bounds.

The main result is summarized by the following theorem.

Theorem 7.1 *Consider the heterogeneous vehicle platoon (7.1) with a leader (7.3) subject to asymmetric nonlinear actuator deadzone (7.2). For a sufficiently large positive constant V_{max}, suppose that the initial conditions satisfy*

$$\sum_{i=1}^{N} [s_i^2(0) + \tilde{\theta}_i^2(0) + \tilde{\eta}_i^2(0) + \ln \frac{\Delta_i^{\star 2}}{\Delta_i^{\star 2} - e_i^2(0)}] \leq \frac{2V_{max}}{\kappa} \tag{7.20}$$

where $e_i(t)$, $s_i(t)$, $\tilde{\theta}_i(t)$ and $\tilde{\eta}_i(t)$ are defined as in (7.8), (7.14), and (7.16), respectively, and

$$\kappa = max\left\{1, \min_{i\in\mathscr{V}_N}\{\alpha_i\}, \min_{i\in\mathscr{V}_N}\{\beta_i\}, \min_{i\in\mathscr{V}_N}\{k_{b,i}\}\right\}.$$

Then, the distributed adaptive NN control law (7.17)-(7.18) guarantees that the spacing errors $e_i(t)$ in (7.8) satisfy the constraint (7.12).

Proof 7.1 *Substituting (7.5) into (7.15), we can obtain*

$$\begin{aligned} \dot{s}_i(t) &= \ddot{x}_{i-1}(t) - \ddot{x}_i(t) + \lambda_1 \dot{e}_i(t) + \lambda_2 \tfrac{m}{n} |e_i(t)|^{\frac{m}{n}-1} \dot{e}_i(t) \\ &= -u_i(t) - \Delta u_i(t) - f_i(x_i, v_i, t) - w_i(t) + A_i(t) \end{aligned} \tag{7.21}$$

where

$$A_i(t) = \ddot{x}_{i-1}(t) + \lambda_1 \dot{e}_i(t) + \lambda_2 \frac{m}{n} |e_i(t)|^{\frac{m}{n}-1} \dot{e}_i(t).$$

By using RBFNNs, one has

$$\begin{aligned} f_i(x_i, v_i, t) + w_i(t) &= W_{1,i}^{*T} \xi_{1,i}(Z_{1,i}) + \varepsilon_{1,i}(t), \ |\varepsilon_{1,i}(t)| \leq \varepsilon_1 \\ \Delta u_i(t) &= W_{2,i}^{*T} \xi_{2,i}(Z_{2,i}) + \varepsilon_{2,i}(t), \ |\varepsilon_{2,i}(t)| \leq \varepsilon_2 \end{aligned} \tag{7.22}$$

where ε_1 and ε_2 are small unknown positive constants. Taking into account $x_i(t) = x_i(t_0) + \int_{t_0}^{t} v_i(\tau)d\tau$, for the node of the input layer, $Z_{1,i}(t) = Z_{2,i}(t) = v_i(t)$ is chosen as the net input. Then, it follows from (7.22) that (7.21) can be rewritten as

$$\begin{aligned} \dot{s}_i(t) = & -u_i(t) - W_{1,i}^{*T} \xi_{1,i}(v_i) - W_{2,i}^{*T} \xi_{2,i}(v_i) \\ & -\varepsilon_{1,i}(t) - \varepsilon_{2,i}(t) + A_i(t). \end{aligned} \tag{7.23}$$

Now, consider a Lyapunov-Krasovskii functional candidate as follows:

$$\begin{aligned} V(t) &= \sum_{i=1}^{N} [V_i^s(t) + V_i^b(t)] \\ V_i^s(t) &= \tfrac{1}{2} s_i^2(t) + \tfrac{1}{2\alpha_i} \tilde{\theta}_i^2(t) + \tfrac{1}{2\beta_i} \tilde{\eta}_i^2 \\ V_i^b(t) &= \tfrac{k_{b,i}}{2} \ln \frac{\Delta_i^{\star 2}}{\Delta_i^{\star 2} - e_i^2(t)} \end{aligned} \tag{7.24}$$

where α_i, β_i and $k_{b,i}$ are positive design parameters.

By utilizing (7.17) and (7.23), the time derivative of $\frac{1}{2}s_i^2(t)$ is

$$\begin{aligned} s_i(t)\dot{s}_i(t) &\leq -s_i(t)u_i(t) + \tfrac{1}{2}s_i^2(t)\theta_i^*\xi_{1,i}^T(v_i)\xi_{1,i}(v_i) \\ &\quad + \tfrac{1}{2}s_i^2(t)\eta_i^*\xi_{2,i}^T(v_i)\xi_{2,i}(v_i) + s_i^2(t) \\ &\quad + \tfrac{1}{2}[2+\varepsilon_1^2+\varepsilon_2^2] + s_i(t)A_i(t) \\ &= -(k_i-1)s_i^2(t) - \tfrac{k_{b,i}s_i(t)e_i(t)}{\Delta_i^{*2}-e_i^2} + \\ &\quad \tfrac{1}{2}s_i^2[\tilde{\theta}_i\xi_{1,i}^T(v_i)\xi_{1,i}(v_i) + \tilde{\eta}_i\xi_{2,i}^T(v_i)\xi_{2,i}(v_i)] \\ &\quad + \tfrac{1}{2}(2+\varepsilon_1^2+\varepsilon_2^2) \end{aligned} \tag{7.25}$$

where

$$\theta_i^* = \|W_{1,i}^*\|^2,\ \eta_i^* = \|W_{w,i}^*\|^2. \tag{7.26}$$

Then, inspired by [95], a minimal learning parameter mechanism is adopted to estimate θ_i^* and η_i^* instead of estimating the ideal weights $W_{1,i}^*$ and $W_{2,i}^*$ directly. Consequently, the number of adaptive learning parameters used in RBFNN will reduce drastically from $2\sum_{i=1}^N N_i$ to $2N$, which can efficiently solve the explosion of learning parameters.

Recalling the adaptive updated laws in (7.18) and (7.19), we can obtain

$$\begin{aligned} -\tfrac{1}{\alpha_i}\tilde{\theta}_i(t)\dot{\theta}_i(t) &= -\tilde{\theta}_i(t)(\tfrac{1}{2}\xi_{1,i}^T(v_i)\xi_{1,i}(v_i)s_i^2(t) - \tfrac{\sigma_{1,i}(t)}{\alpha_i}\theta_i(t)) \\ &\leq -\tfrac{1}{2}\tilde{\theta}_i(t)\xi_{1,i}^T(v_i)\xi_{1,i}(v_i)s_i^2(t) - \tfrac{\sigma_{1,i}(t)}{2\alpha_i}\tilde{\theta}_i^2(t) \\ &\quad + \tfrac{\bar{\sigma}_{1,i}}{2\alpha_i}\theta_i^{*2} \\ -\tfrac{1}{\beta_i}\tilde{\eta}_i(t)\dot{\eta}_i(t) &= -\tilde{\eta}_i(t)(\tfrac{1}{2}\xi_{2,i}^T(v_i)\xi_{2,i}(v_i)s_i^2(t) - \tfrac{\sigma_{2,i}(t)}{\beta_i}\eta_i(t)) \\ &\leq -\tfrac{1}{2}\tilde{\eta}_i(t)\xi_{2,i}^T(v_i)\xi_{2,i}(v_i)s_i^2(t) - \tfrac{\sigma_{2,i}(t)}{2\beta_i}\tilde{\eta}_i^2(t) \\ &\quad + \tfrac{\bar{\sigma}_{2,i}}{2\beta_i}\eta_i^{*2} \end{aligned} \tag{7.27}$$

The time derivative of $V_i^b(t)$ along (7.14) is

$$\begin{aligned} \dot{V}_i^b(t) &= \tfrac{k_{b,i}}{2}\tfrac{\Delta_i^{*2}-e_i^2}{\Delta_i^{*2}}\tfrac{2\Delta_i^{*2}e_i\dot{e}_i}{(\Delta_i^{*2}-e_i^2)^2} = \tfrac{k_{b,i}e_i\dot{e}_i}{\Delta_i^{*2}-e_i^2} \\ &= \tfrac{k_{b,i}e_i(s_i(t)-k_{b,i}\lambda_1 e_i(t)-k_{b,i}\lambda_2|e_i(t)|^{\frac{m}{n}}sgn(e_i(t))}{\Delta_i^{*2}-e_i^2} \\ &\leq \tfrac{k_{b,i}e_i(t)s_i(t)}{\Delta_i^{*2}-e_i^2} - k_{b,i}\lambda_1\tfrac{e_i^2(t)}{\Delta_i^{*2}-e_i^2} \\ &\leq \tfrac{k_{b,i}e_i(t)s_i(t)}{\Delta_i^{*2}-e_i^2} - k_{b,i}\lambda_1\ln\tfrac{\Delta_i^{*2}}{\Delta_i^{*2}-e_i^2}. \end{aligned} \tag{7.28}$$

It follows from (7.25), (7.27) and (7.28) that

$$\begin{aligned} \dot{V}(t) &= \sum_{i=1}^N[-(k_i-1)s_i^2(t) - k_{b,i}\lambda_1\ln\tfrac{\Delta_i^{*2}}{\Delta_i^{*2}-e_i^2} \\ &\quad + \tfrac{1}{2}(2+\varepsilon_1^2+\varepsilon_2^2) - \tfrac{\sigma_{1,i}(t)}{2\alpha_i}\tilde{\theta}_i^2(t) + \tfrac{\bar{\sigma}_{1,i}}{2\alpha_i}\theta_i^{*2} \\ &\quad - \tfrac{\sigma_{2,i}(t)}{2\beta_i}\tilde{\eta}_i^2(t) + \tfrac{\bar{\sigma}_{2,i}}{2\beta_i}\eta_i^{*2}] \\ &\leq -\chi_1 V(t) + \chi_2 \end{aligned}$$

where the positive parameters χ_1 *and* χ_2 *under the condition* $k_i > 1$ *are given as follows*

$$\begin{aligned} \chi_1 &= \min\{2(k_i-1), \min_{i\in\mathscr{V}_N}\min_{t\to\infty}\sigma_{1,i}(t), \min_{i\in\mathscr{V}_N}\min_{t\to\infty}\sigma_{2,i}(t), 2\lambda_1\} \\ \chi_2 &= \sum_{i=1}^{N}[\tfrac{1}{2}(2+\varepsilon_1^2+\varepsilon_2^2)+\tfrac{\bar{\sigma}_{1,i}}{2\alpha_i}\theta_i^{*2}+\tfrac{\bar{\sigma}_{2,i}}{2\beta_i}\eta_i^{*2}]. \end{aligned} \tag{7.29}$$

Thus, by using Lemma 2.6, the following inequality holds:

$$V(t) \le (V(0)-\tfrac{\chi_2}{\chi_1})e^{-\chi_1 t}+\tfrac{\chi_2}{\chi_1} \le V(0)+\tfrac{\chi_2}{\chi_1}, \tag{7.30}$$

where the following fact is applied in the last inequality.

$$V(0)=\sum_{i=1}^{N}[\frac{1}{2}s_i^2(0)+\frac{1}{2\alpha_i}\tilde{\theta}_i^2(0)+\frac{1}{2\beta_i}\tilde{\eta}_i^2(0)+\frac{k_{b,i}}{2}\ln\frac{\Delta_i^{*2}}{\Delta_i^{*2}-e_i^2(0)}] \ge 0.$$

It is directly obtained that $s_i(t)$, $\tilde{\theta}_i(t)$ *and* $\tilde{\eta}_i(t)$ *must remain bounded. Since* θ_i^* *and* η_i^* *are bounded,* $\theta_i(t)$ *and* $\eta_i(t)$ *are also bounded. Then, one can conclude that all the signals in the closed-loop system remain uniformly ultimately bounded if the initial condition is bounded as in (7.20). As* $t\to\infty$, $|s_i(t)| \le \sqrt{\frac{2\chi_2}{\chi_1}}$, *i.e., the sliding mode surface* $s_i(t)$ *is steered into an adjustable region of zero asymptotically. Then, according to (7.24), the spacing errors* $e_i(t)$ *are also convergent to a neighborhood of zero.* □

Remark 7.3 *The size of spacing errors relies on the selection of design parameters in (7.29). Increasing* k_i, $k_{b,i}$, α_i, *and* β_i *while reducing* $\bar{\sigma}_{1,i}$ *and* $\bar{\sigma}_{2,i}$ *will lead to smaller spacing errors. However, if* χ_1 *is too big and* χ_2 *is too small, the required control energy will be high and may result in saturation. Therefore, in practical applications, these parameters should be chosen carefully according to the required tracking precision and the available control resource [51, 55, 57].*

Remark 7.4 *It is worth mentioning that, by introducing the SBLF, the collision between consecutive vehicles as well as the connectivity breaks owing to limited communication/sensing capabilities can be avoided in Theorem 7.1. Obviously, this method is superior to those methods proposed in [43, 47, 51, 55, 57, 212], where the strong string stability issue is considered but cannot guarantee collision avoidance and also the connectivity keeping.*

7.4 Simulation Study

To validate the effectiveness of the proposed adaptive NN control scheme obtained in Theorem 7.1, computer simulations have been carried out for a high speed train platoon containing twelve followers (i.e., $N = 12$), whose dynamics are described by

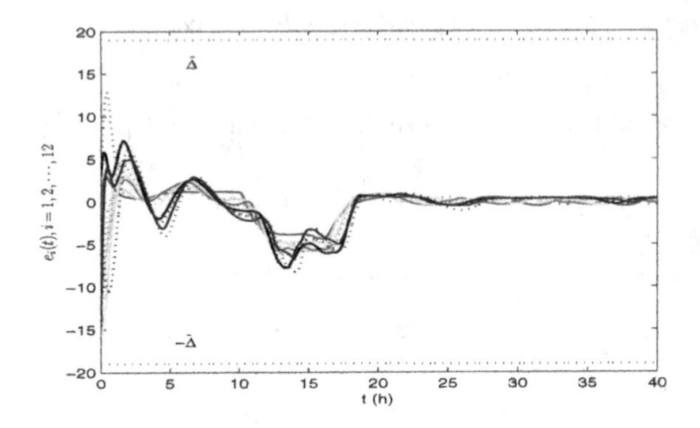

FIGURE 7.3
Responses of spacing errors.

(7.1) with asymmetric *deadzone* (7.2). The nonlinear functions and disturbances are given by

$$\begin{aligned} f_i(x_i, v_i, t) &= -c_{0,i} - c_{1,i}v_i(t) - c_{2,i}v_i^2(t) \\ w_i(t) &= \sin(it + i\pi), i \in \mathscr{V}_N. \end{aligned}$$

The parameters of each high speed train are listed in Table 7.1 [51], while the initial positions and speeds of each high speed train containing *initial conditions* are given out in Table 7.2. Moreover, the desired spacing between consecutive trains is equally set at $\Delta_{i-1,i} = 20km, i \in \mathscr{V}_N$ while the collision and connectivity constraints are given by $\Delta_{col} = 1km$ and $\Delta_{con} = 39km$, respectively, which imply that $\bar{\Delta} = 19km$. According to the definition of $e_i(t)$ in (7.8), the initial spacing $d_i(0)$ between consecutive trains satisfies $\Delta_{col} < d_i(0) < \Delta_{con}$. The desired acceleration of the leader train is chosen as

$$a_0(t) = \begin{cases} 3t & km/h^2, \quad 3h \leq t < 6h \\ -2t & km/h^2, \quad 10h \leq t < 16h \\ 0 & km/h^2, \quad otherwise \end{cases}$$

where the train will run in traction, braking and finally cruise phases. The RBF vectors are given by $\xi_{j,i}(v_i(t)) \in \mathbf{R}^{15}$ (*i.e.*, $N_i = 15; j = 1,2$). The centers for the NN input $Z_i(t) = v_i(t)$ are evenly spaced in $[10, 220]$, and the width is $\theta_k = 1.5$. Furthermore, the controller and adaptive parameters are chosen as $k_{b,i} = 3$, $k_i = 6$, $\lambda_1 = 1$ and $\lambda_2 = 0.2$ with the initial adaptive estimates $\theta_i(0) = \eta_i(0) = 1$. In addition, for simulation purposes, the deadzone parameters are set to

$$\begin{aligned} b_{r,i} &= 2.5, \; b_{l,i} = -1.5 \\ g_{r,i}(u_i(t)) &= 0.5 - 0.3\sin(u_i(t))(u_i(t) - b_{r,i}) \\ g_{l,i}(u_i(t)) &= 0.3 + 0.2\cos(u_i(t))(u_i(t) - b_{l,i}). \end{aligned}$$

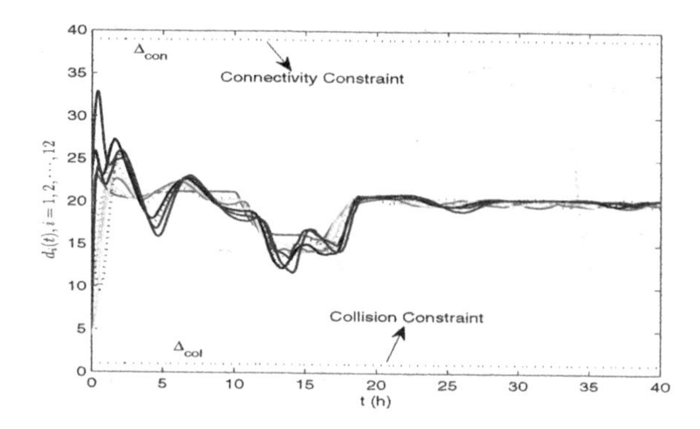

FIGURE 7.4
The spacing between consecutive trains along with the collision and connectivity constraints.

Then, the simulation results obtained by using Theorem 7.1 are shown in Figures 7.3-7.8. It is observed from Figure 7.3 that the spacing error constraints in (7.12) are satisfied despite the presence of asymmetric nonlinear actuator deadzone, unmodeled dynamics and *external disturbances*. The spacing between consecutive trains along with the collision and connectivity constraints are shown in Figure 7.4. From Figure 7.5, it is seen that the followers converge to the trajectory of the time-varying leader in a finite time. Figure 7.6 shows the results of the sliding mode surfaces, while the adaptive parameters $\theta_i(t)$ and $\eta_i(t)$ are shown in Figure 7.7 and Figure 7.8, respectively. All the signals of the whole platoon systems are bounded, which verifies the effectiveness of the proposed control scheme and adaptive laws.

TABLE 7.1
Parameters of high speed train for $i \in \mathcal{V}_N (N = 12)$

Symbol	Value	Unit
m_i	80×10^4	kg
c_{i0}	0.01176	N/kg
c_{i1}	0.00077616	Ns/mkg
c_{i2}	1.6×10^{-5}	Ns^2/m^2kg
$\Delta_{i-1,i}$	20	km
Δ_{con}	39	km
Δ_{col}	1	km

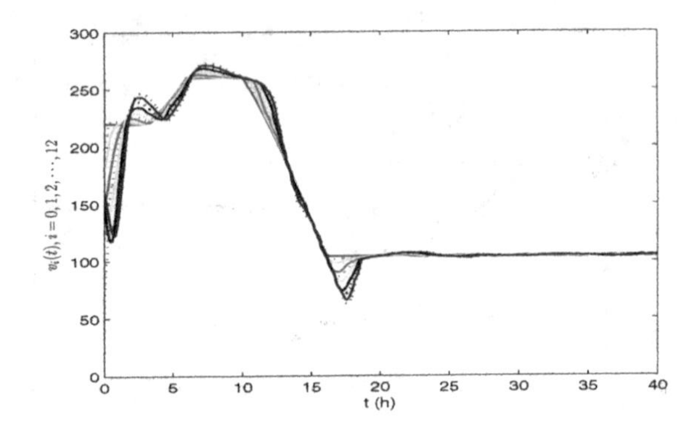

FIGURE 7.5
Velocity tracking trajectories for all trains.

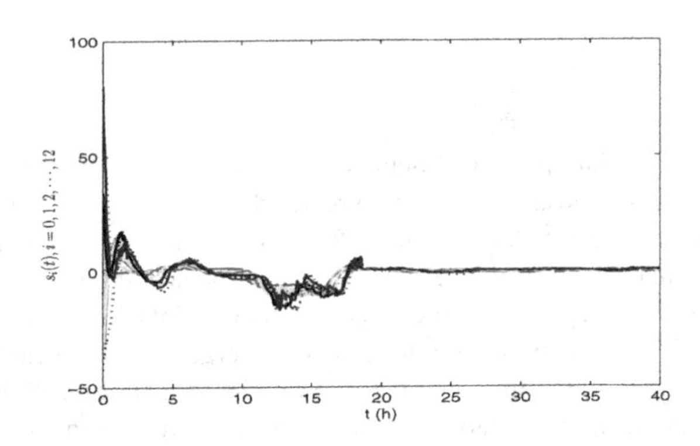

FIGURE 7.6
The responses of all sliding mode $s_i(t)$.

7.5 Conclusion

This chapter addresses the problem of neuro-adaptive terminal *sliding mode control* for a vehicular platoon subject to *external disturbances*, unmodeled dynamics and unknown asymmetric nonlinear actuator deadzone. Based on a symmetric barrier Lyapunov function, the collision between consecutive vehicles as well as the connectivity breaks owing to limited sensing capabilities are avoided. An adaptive NN mechanism with minimal learning parameters makes the online computation burden greatly alleviated. The convergence of the spacing and velocity errors to arbitrarily small compact sets of zero is proven by the Lyapunov theory. Finally, simulation results show the effectiveness of the proposed scheme.

TABLE 7.2
The parameters $x_i(0)(km)$ and $v_i(0)(km/h)$ for $i \in \mathscr{V}_N \cup \{0\}(N = 12)$

i	0	1	2	3	4	5	6	7	8	9	10	11	12
$x_i(0)$	148	128	118	108	100	94	88	83	71.5	48.5	30	24	1
$v_i(0)$	220	200	180	180	150	160	130	145	150	120	160	50	10

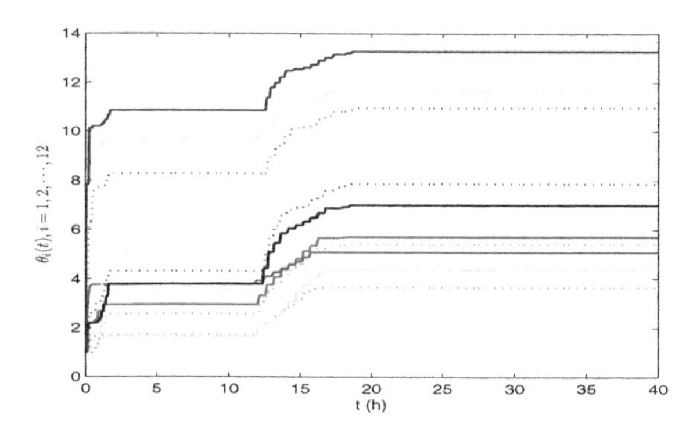

FIGURE 7.7
The responses of adaptive parameters $\theta_i(t)$.

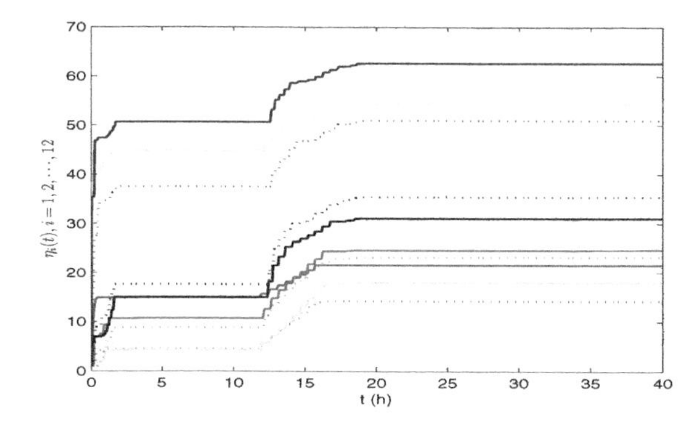

FIGURE 7.8
The responses of adaptive parameters $\eta_i(t)$.

8

Neuro-Adaptive Quantized PID-Based SMC of Vehicular Platoon with Deadzone

This chapter focuses on the problem of neuro-adaptive quantized control for heterogeneous vehicular platoon when the follower vehicles suffer from *external disturbances*, mismatch input quantization and unknown actuator *deadzone*. proportional-integral-derivative-based sliding mode control (PIDSMC) technique is used due to its superior capability to reduce *spacing errors* and chattering and to eliminate the steady-state *spacing errors*. Then, a neuro-adaptive quantized PIDSMC scheme with minimal learning parameters is designed not only to guarantee the *strong string stability* of the whole vehicular platoon and ultimate uniform boundedness of all adaptive law signals, but also to attenuate the negative effects caused by *external disturbances*, mismatch input quantization and unknown actuator *deadzone*. Furthermore, different from Chapters 3-7, a new modified constant time headway (MCTH) policy for third-order model is proposed to not only increase traffic density, but also address the negative effect of *non-zero initial* spacing, velocity and acceleration errors in this chapter. Compared with most existing methods, the proposed method does not linearize the system model and neither does it require precise knowledge of the system model.

8.1 Introduction

Awfully crowded urban traffic has become a serious problem in the world due to the increase in the number of vehicles. Increasing attention has been paid to the field of vehicular platoons as they help relieve traffic congestion [27]. Since vehicles in a platoon are dynamically coupled, disturbances acting on one vehicle may affect other vehicles. Therefore, an important concept regarding vehicular platoons is *string stability*, i.e., *spacing errors* do not amplify as they propagate upstream from one vehicle to another vehicle [2, 55, 57, 93, 142, 143, 164]. If the *spacing errors* amplify as they propagate upstream (i.e., *string instability*), it not only will likely provide poor ride quality but also could result in collision [55, 57, 164, 186]. Therefore, the issue of how to guarantee string stability has received tremendous attention. Interested readers are referred to the seminal works [43, 44, 46, 47, 120, 162, 201, 212, 213]. In most existing results on string stability, by reducing a nonlinear system to a linear one via feedback linearization is extensively used to simplify the control problem [43, 44, 46, 47, 120, 162, 212, 213]. However, feedback linearization is usually rather difficult for implementation in practice due to the numerical complexity of the control scheme, the difficulty in measuring various parameters (such as aerodynamic drag, the rolling resistance moment and air resistance), and environmental noises, etc. Therefore, some nonlinear platoon control approaches have been developed to control the nonlinear vehicular platoon directly in recent research works [116, 138, 201]. Distributed control protocols based on backstepping technique were proposed to guarantee *string stability* in [138, 201]. Nevertheless, the determination of virtual control terms and their time derivatives require tedious and complex analysis by using backstepping technique [133]. In [116], an neuro-adaptive sliding mode control method was developed for a car-following system, but the issue of the string stability was ignored, which might cause neighboring-vehicle collisions. Henceforth, it is more appealing to develop a more feasible and efficient nonlinear control method to overcome the above drawbacks. Technical difficulty arises from the significant nonlinear and time-varying behavior of the system model and simultaneously guaranteeing string stability.

Note that one common feature in the above-mentioned literature is that they all assume that the signal transmission required by the system can be performed with infinite precision. However, signal quantisation problem is often encountered in engineering practice and quantization plays an important role in information exchange within the neighboring-vehicle communication network, while quantization is a common source of errors in feedback control systems, which may deteriorate the performance of the closed-loop systems [86]. Therefore, quantized control has been an active research topic in the literature [74, 86, 209]. Most of the existing results assume that the quantization adjustable parameters are time-invariant and identical, such as [74]. However, the mismatch of the quantization adjustable parameters may result in instability of the closed-loop systems [209]. Therefore, recently, much effort has been focused on investigating the mismatch of the quantization adjustable parameters [86, 209]. Note that the aforementioned quantized control approaches are commonly based on an assumption that the actuator is ideal (i.e., the actuator input is completely equal to the actuator output). However, it is well known that the *deadzone* input nonlinearity is commonly encountered in actuators in real control systems [23] and is a very important actuator nonlinearity [121]. The presence of this nonlinear input characteristic could severely deteriorate the achievable control performance, leading to undesirable control accuracy, limit cycles, and even instability [79]. So far, a large number of publications have been devoted to developing actuator *deadzone* nonlinearity rejection techniques, such as the adaptive method [79], neural network/fuzzy-logic approximation method [121], the inverse *deadzone* nonlinearity technique [79], and so on. However, in presence of input quantization, the actual control signal is quantized before transmitting to actuator. For this reason, it is not clear whether the previously mentioned techniques are still effective. How to construct an adaptive control scheme for vehicular platoon that effectively attenuates the negative effects of input quantization and actuator *deadzone* has been a task of major practical interest as well as theoretical significance.

The PIDSMC technique has its superior capability to eliminate steady-state spacing error [30], while neural networks (NNs) have the capability to approximate any continuous functions over a compact set to arbitrary accuracy [95, 106, 122]. Since unmodeled nonlinear dynamics and *external disturbances* often exist in vehicular platoon, we are interested in investigating the *strong string stability* issue by combining PIDSMC technique with NN approximation method. Therefore, this chapter devotes to developing a neuro-adaptive quantized PIDSMC for heterogeneous vehicular platoon with both input quantization and unknown actuator *deadzone*. There are two major challenging difficulties. One difficulty is how to achieve string stability without backstepping technique when there exist simultaneous *non-zero initial* spacing, velocity and acceleration errors, nonlinearities, and *external disturbances*. The other difficulty is that the actual input is subject to strong coupling of the quantization effects and the unknown asymmetric *deadzone*. The main contributions of the chapter are summarized as follows:

(1) Compared with the existing results [43, 213], our proposed method does not *feedback linearize* the system mode, and the negative effects of mismatch input quantization and unknown asymmetric *deadzone* are considered simultaneously.

(2) The strong coupling between mismatch input quantization and unknown asymmetric *deadzone* is first decoupled by some nonlinear decomposition methods. Then, the common nonlinear control techniques such as PIDSMC technique and adaptive NN approximation technique can be used to reject the negative effects of mismatch input quantization and unknown asymmetric *deadzone*. The proposed control schemes can guarantee the *strong string stability* of the whole heterogeneous vehicular platoon and the ultimate uniform boundedness of all closed-loop signals.

(3) In order to improve the road capacity and avoid traffic congestion, and simultaneously remove the common restriction of zero initial spacing, velocity and acceleration errors in [43, 201, 213], a new *modified constant time headway* (MCTH) policy is proposed. This policy is an obvious extension compared with those in Chapters 3-6.

(4) It is known that good approximated performance requires more NN nodes. However, increasing the number of NN nodes will result in unacceptably long learning time. This is particularly time-consuming when the vehicular platoon contains a large number of vehicles. To overcome this drawback, an optimized adaptation method proposed in [95, 106, 122] is incorporated into the adaptation mechanism. Then, the number of adaptive parameters will not increase as the number of NN nodes increases and thus the online computation burden can be greatly alleviated. Another important point worth noting is that neuro-adaptive technique allows us to remove a restrictive condition used in [57] that the nonlinear uncertainties of both the leader and follower vehicles should match each other.

8.2 Vehicle-Following Platoon Model and Preliminaries

8.2.1 Vehicle-Following Platoon Description

Suppose a vehicular platoon composed of N follower vehicles and a leader vehicle (labeled as 0) running in a horizontal environment shown in Figure 8.1. Let $(x_i, v_i, a_i) \in \mathbf{R}^3$ $(i \in \mathscr{V}_N \cup \{0\})$ denote respectively the position, velocity and acceleration of the ith vehicle. Each follower vehicle i is subject to both unknown *deadzone* and input quantization as characterized by the following third-order nonlinear time-variant dynamics [43, 44, 46, 47, 120, 162, 201, 212, 213]:

$$\begin{aligned} \dot{x}_i &= v_i, \dot{v}_i = a_i, i \in \mathscr{V}_N \\ \dot{a}_i &= f_i(v_i, a_i, t) + \frac{1}{\tau_i} u_i^d + w_i \\ f_i(v_i, a_i, t) &= -\frac{1}{\tau_i}\left(a_i + \frac{\rho_{mi} H_i c_i}{2m_i} v_i^2 + \frac{d_{mi}}{m_i}\right) - \frac{\rho_{mi} H_i c_i v_i a_i}{m_i} \\ u_i^d &= \mathscr{D}_i(u_i^q),\ u_i^q = \mathscr{Q}_i(u_i),\ u_i = \frac{\eta_i}{m_i} \end{aligned} \tag{8.1}$$

where m_i is the vehicle mass, η_i is the engine/brake input, ρ_{mi} is the specific mass of the air, c_i is the coefficient of aerodynamic drag, H_i is the cross-sectional area, $\frac{\rho_{mi} H_i c_i}{2m_i}$ is the air resistance, d_{mi} is the mechanical drag, τ_i is the engine time constant, and w_i is the *external disturbance*, which satisfies $|w_i| \leq w_i^M$ with w_i^M being an unknown parameter. Note that u_i instead of η_i used here makes the control independent of the vehicle masses.

The unknown input *deadzone* nonlinearity $\mathscr{D}_i(u_i^q)$ can be described by [79]

$$u_i^d = \begin{cases} n_i^r(u_i^q - b_i^r), & \text{if } u_i^q \geq b_i^r \\ 0, & \text{if } -b_i^l < u_i^q < b_i^r \\ n_i^l(u_i^q + b_i^l), & \text{if } u_i^q \leq -b_i^l \end{cases} \tag{8.2}$$

where n_i^r and n_i^l stand for the right and the left slope of the *deadzone* characteristic. The parameters b_i^r and b_i^l represent the breakpoints of the input nonlinearity. In (8.1) and (8.2), u_i^q represents the quantized input, which is taken as follows

$$\mathscr{Q}_i(u_i) \triangleq \mu_i^d q(u_i/\mu_i^c) \tag{8.3}$$

where the function $q(\cdot) \triangleq round(\cdot)$ rounds toward the nearest integer, and $q(u_i/\mu_i^c)$ denotes the finite precision implementation of the control input. The time-varying parameters $\mu_i^c > 0$ and $\mu_i^d > 0$ are called *quantization sensitivity parameters.* In addition, the quantizer operator $\mathscr{Q}_i(\cdot)$ in (8.3) can further be rewritten as

$$\begin{aligned} \mathscr{Q}_i(u_i) &= \mu_i q_{\mu_i^c}(u_i) \\ q_{\mu_i^c}(u_i) &= \mu_i^c q(u_i/\mu_i^c) \end{aligned} \tag{8.4}$$

where $\mu_i = \frac{\mu_i^d}{\mu_i^c}$ is defined as the *mismatch ratio* and is time-varying.

Remark 8.1 *Since the vehicle dynamics (8.1) have been investigated widely in the existing literature, some necessary explanations are given as follows.*

- ⋆ *It is noted that the existing literature on platoon control does not consider the effects of input quantization and actuator deadzone, which may result in oscillations, delays, inaccuracy and severe deterioration in performance [23]. In addition, in the presence of input quantization and unknown actuator deadzone, the problem of platoon control becomes more challenging and much more difficult than those in [43, 44, 46, 47, 120, 162, 201, 212, 213].*

- ⋆ *In most existing results such as [43, 44, 46, 47, 120, 162, 212, 213], in order to simplify the platoon control problem, the original nonlinear vehicle dynamic (8.1) is feedback linearized as* $\dot{a}_i = -\frac{1}{\tau_i}a_i + \frac{1}{\tau_i}u_i + w_i$ *by introducing a feedback linearization controller* $\eta_i = u_i m_i + \rho_{mi} H_i c_i v_i^2/2 + d_{mi} + \tau_i \rho_{mi} H_i c_i v_i a_i$. *However, in practical engineering, the mechanical drag* d_{mi} *and the air resistance* $\frac{\rho H_i c_i}{2m_i}$ *are always uncertain. So, in the presence of input quantization and unknown actuator deadzone, the above feedback linearization approach will not apply.*

- ⋆ *Distributed control protocols based on backstepping technique [138, 201] have a drawback is that the determination of virtual control terms and their time derivatives requires tedious and complex analysis [133].*

- ⋆ *A common assumption of the control schemes proposed in [43, 44, 46, 47, 120, 138, 162, 201, 212, 213] is that the initial spacing, velocity and acceleration errors are assumed to be zero. However,* non-zero *initial errors may cause string instability and thus may result in collision with each other [57, 164].*

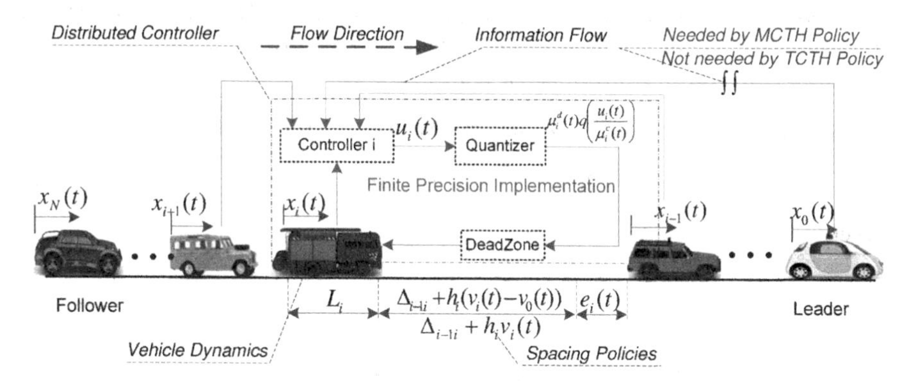

FIGURE 8.1
Controller configuration for a car-following platoon on a straight lane.

The dynamics of the leader is expressed as follows:

$$\dot{x}_0 = v_0, \dot{v}_0 = a_0 \tag{8.5}$$

where a_0 is the desired time-varying acceleration.

Remark 8.2 *It is important to mention that the nonlinear function $f_i(v_i, a_i, t)$ in (8.1) is more general than the one considered in [57], where the nonlinear function $f_i(x_i, v_i, a_i, t)$ should satisfy some Lipschitz condition. Another drawback of the scheme developed in [57] is that the considered second-order vehicle system is assumed to be subject to matched nonlinear uncertainties only, that is, both the follower and leader vehicles have the same type of nonlinear uncertainties $f_i(x_i, v_i, a_i, t)$.*

8.2.2 Nonlinear Actuator Decomposition

Note that *deadzone* $\mathscr{D}_i(u_i^q)$ in (8.2) and quantization $q_{\mu_i^c}(u_i)$ in (8.4) make the controller design difficult. To overcome this difficult, nonlinear decomposition strategies are used to decompose $\mathscr{D}_i(u_i^q)$ and $q_{\mu_i^c}(u_i)$ into proper forms. Before giving the nonlinear decomposition strategies, and to facilitate control design, the following assumptions are made.

Assumption 8.1 *[79] The deadzone parameters n_i^r, n_i^l, b_i^r and b_i^l are positive but unknown. Although the lower bounds of n_i^r and n_i^l are not known a priori, it is reasonable to assume that there exist some unknown constants $\zeta_i > 0$ such that $0 < \zeta_i \leq n_i^r$ and $0 < \zeta_i \leq n_i^l$ for all $i \in \mathscr{N}$.*

Assumption 8.2 *The mismatch ratio μ_i between the quantization adjustable parameters μ_i^d and μ_i^c is bounded as $\mu_i^m \leq \mu_i \leq \mu_i^M$ for any $t > 0$, where μ_i^m and μ_i^M are unknown positive constants.*

Remark 8.3 *The mismatch ratio was first introduced in [86]. The time-varying property of the mismatch ratio μ_i can be used to model the mismatch problem caused by the discrepancy in quantization sensitivity initialization as discussed in [86] and by the non-synchronous adjustment of the quantization adjustable parameters μ_i^d and μ_i^c [209]. Assumption 8.2 is an obvious extension compared to the previous result in [86], where μ_i is a time-invariant constant.*

Similar to [79], the *deadzone* input (8.2) can be rewritten as follows

$$\mathscr{D}(u_i^q) \quad = \quad n_i u_i^q + \Delta n_i \tag{8.6}$$

where

$$\begin{aligned} n_i &= \begin{cases} n_i^l, & \text{if } u_i^q \leq 0 \\ n_i^r, & \text{if } u_i^q > 0 \end{cases} \\ \Delta n_i &= \begin{cases} -n_i^r b_i^r, & \text{if } u_i^q \geq b_i^r \\ -n_i u_i^q, & \text{if } -b_i^l < u_i^q < b_i^r \\ n_i^l b_i^l, & \text{if } u_i^q \leq -b_i^l \end{cases} \end{aligned}$$

with Δn_i being bounded as $|\Delta n_i| \leq n_i^M$ and $n_i^M > 0$ an unknown constant. In addition, Assumption 8.1 implies $0 < \zeta_i \leq n_i$.

To overcome the difficulty arising from the quantizer $\mathscr{Q}_i(u_i)$ in (8.3), we decompose the quantizer $q_{\mu_i^c}(u_i)$ in (8.4) into a linear part and a nonlinear part as follows:

$$q_{\mu_i^c}(u_i) = u_i + \Delta u_i \tag{8.7}$$

where the nonlinear part is given by

$$\Delta u_i \triangleq q_{\mu_i^c}(u_i) - u_i. \tag{8.8}$$

Then, based on the quantizer characteristics (8.4), each component of Δu_i at time t is bounded by half of the quantization level μ_c, i.e.,

$$\|\Delta u_i\|_\infty \leq 0.5 \max_{t\geq 0} \mu_i^c$$

which implies that larger μ_i^c leads to larger quantization error.

By substituting (8.4) and (8.7) into (8.6), under Assumption 8.2, the unknown input *deadzone* nonlinearity $\mathscr{D}_i(u_i^q)$ is finally decomposed into the following form:

$$\begin{aligned} \mathscr{D}_i(u_i^q) &= n_i \mu_i u_i + \Delta N_i \\ \Delta N_i &= n_i \mu_i \Delta u_i + \Delta n_i \end{aligned} \tag{8.9}$$

where

$$\begin{aligned} 0 &< \zeta_i \mu_i^m \leq n_i \mu_i \leq \mu_i^M \max\{n_i^l, n_i^r\} \\ |\Delta N_i| &\leq \frac{1}{2} \mu_i^M \max\{n_i^l, n_i^r\} \max_{t\geq 0} \mu_i^c + n_i^M \triangleq N_i^M. \end{aligned}$$

It is worth pointing out that the value of the control gain $n_i \mu_i$ is unknown since ζ_i, μ_i^m, μ_i^M and $\max\{n_i^l, n_i^r\}$ are unknown. Taking this into consideration, an online adaptive updating law Ψ_i is used to estimate the reciprocal $\Psi_i^* = \frac{1}{\zeta_i \mu_i^m}$ of the lower bound of the control gain in (8.9).

8.2.3 Radial Basis Function Neural Network

Combining *radial basis function neural network* (RBFNN) technique with adaptive control technique is widely adopted to deal with unknown nonlinear functions due to the inherent approximation capability of RBFNN [95, 106, 122]. RBFNN is composed of three layers, namely, the input layer, the neuron layer (hidden layer) and the output layer. Given any unknown continuous function $f(Z) : \mathbf{R}^p \to \mathbf{R}$ defined on the compact set $\Sigma_Z \subseteq \mathbf{R}^p$ and any precision $\varepsilon^* > 0$, there exists an RBFNN $f_{NN}(Z) = W^{*T}\xi(Z)$ such that

$$f(Z) = W^{*T}\xi(Z) + \varepsilon(Z) \tag{8.10}$$

where $Z \in \mathbf{R}^p$ is the input vector of RBFNN, $\varepsilon(Z)$ is the approximation error satisfying $|\varepsilon(Z)| \leq \varepsilon^*$, $W^* = \left[W_1^*, \cdots, W_{N_1}^*\right]^T \in \mathbf{R}^{N_1}$ is the ideal weight vector of the RBFNN, and N_1 is the number of neurons. $\xi(Z) = [\xi_1(Z), \cdots, \xi_{N_1}(Z)]^T \in \mathbf{R}^{N_1}$ is the Gaussian basis function vector and its kth componentwise $\xi_k(Z)$ is described as

$$\xi_k(Z) = \exp\left[-\frac{(Z-\phi_k)^T(Z-\phi_k)}{2\theta_k^2}\right], k = 1, \cdots, N_1 \tag{8.11}$$

where $\phi_k \in \mathbf{R}^p$ is the center of the receptive field and θ_k is the width of the Gaussian function.

8.3 Neuro-Adaptive Quantized PIDSMC Design and Strong String Stability Analysis

In this section, a PID-based sliding mode control (PIDSMC), in combination with adaptive control and RBFNN technique, is used to construct the distributed adaptive NN control schemes for the vehicle-following platoon in (8.1) and (8.5). A new *modified constant time headway* (MCTH) policy is also proposed to improve the traffic density and simultaneously remove the common assumption that the initial spacing, velocity and acceleration errors are all zero.

8.3.1 MCTH Policy and Control Problem

In a platoon as in Figure 8.1, each vehicle is required to follow its predecessor and maintain a desired spacing. The *traditional constant time headway* (TCTH) policy is known to have the advantage of improving string stability and has been used extensively to achieve this objective. However, a drawback of the TCTH policy is its low traffic density since the desired neighboring-vehicle spacing varies with vehicle velocity. On the other hand, the previous work [57] has shown that *non-zero initial* spacing and velocity errors may result in *string instability*. Therefore, in order to overcome the above two difficulties, a new *MCTH policy* is proposed as follows

$$\begin{aligned} e_i &= \tilde{e}_i - \Omega_i \\ \tilde{e}_i &= x_{i-1} - x_i - L_i - \Delta_{i-1,i} - h_i(v_i - v_0) \\ \Omega_i &= \{\tilde{e}_i(0) + [\pi_i \tilde{e}_i(0) + \dot{\tilde{e}}_i(0)]t + \tfrac{1}{2}[\pi_i^2 \tilde{e}_i(0) \\ &\quad +2\pi_i \dot{\tilde{e}}_i(0) + \ddot{\tilde{e}}_i(0)]t^2\} e^{-\pi_i t} \end{aligned} \tag{8.12}$$

where L_i, $\Delta_{i-1,i}$ and h_i are the length of the ith vehicle, the required ith standstill distance and *constant-time headway*, respectively; and π_i is a strictly positive constant. Then, we can obtain

$$e_i(0) = 0, \; \dot{e}_i(0) = 0, \; \ddot{e}_i(0) = 0. \tag{8.13}$$

This is an obvious extension compared with that in [57] where $e_i(0) = 0$ and $\dot{e}_i(0) = 0$ are ensured.

Defining the real spacing between two consecutive vehicles as

$$d_{i-1,i} = x_{i-1} - x_i - L_i.$$

Then, the control objective of this chapter is to utilize RBFNN, adaptive control technique and PIDSMC method to determine a distributed adaptive NN sliding mode controller u_i and adaptive updating laws so that

$$d_{i-1,i} \to \Delta_{i-1,i}; \; v_i \to v_0; \; a_i \to a_0. \tag{8.14}$$

Simultaneously, for the vehicle-following platoon control problem, *strong string stability* requires that [57, 93, 201]

$$|e_N| \leq |e_{N-1}| \leq \cdots \leq |e_1| \tag{8.15}$$

i.e., the error propagation transfer function $G_i(s) := \frac{E_{i+1}(s)}{E_i(s)}$ satisfies $|G_i(s)| \leq 1$ for all $i \in \mathscr{V}_N$, where $E_i(s)$ denotes *Laplace transform* of e_i. It is highlighted that, according to (8.12), $E_i(s)$ (and hence $G_i(s)$) includes the effects of initial errors $\tilde{e}(0)$, $\dot{\tilde{e}}(0)$ and $\ddot{\tilde{e}}(0)$.

8.3.2 PID-Based Sliding Mode Control Design

Since PIDSMC technique covers the treatment of both transient and steady-state responses, PIDSMC has superior capability to reduce *spacing errors*, chattering, and to eliminate the steady-state *spacing errors* [30]. Define a PIDSMC surface as follows

$$s_i = K_P e_i + K_I \int_0^t e_i(\tau)d\tau + K_D \dot{e}_i \tag{8.16}$$

where K_P, K_I and K_D are the proportional, integral and derivative gains, respectively, while K_P provides an overall control action proportional to the error signal through the all-pass gain factor, K_I can be used to reduce the steady-state *spacing errors*, and K_D will improve the transient response.

In order to guarantee *strong string stability*, and similar to [57, 93], a coupled sliding surface, coupling the sliding surfaces s_i with s_{i+1}, is defined as follows

$$S_i = \begin{cases} \lambda s_i - s_{i+1}, & i \in \mathscr{V}_N \setminus \{N\} \\ \lambda s_i, & i = N \end{cases} \tag{8.17}$$

where λ is a positive parameter. When s_i's reach the sliding surface, S_i's also reach sliding surface at the same time because λ is a strictly positive parameter.

Note that the nonlinear functions $f_i(v_i, a_i, t)$ in (8.1) may be unknown due to the unavailability of the air resistance, mechanical drag, etc. According to (8.10), the RBFNN technique is adopted to compensate for these unknown functions as follows

$$f_i(v_i, a_i, t) = W_i^{*T} \xi_i(Z_i(t)) + \varepsilon_i(Z_i(t)) \tag{8.18}$$

where $W_i^* = [W_{i,1}^*, \cdots, W_{i,N_i}^*]^T \in \mathbf{R}^{N_i}$ are the ideal weight vectors of RBFNNs, N_i is the number of neurons, $Z_i(t)$ is selected as $Z_i(t) = [v_i(t), a_i(t)]^T \in \mathbf{R}^2$, and $\xi_i(Z_i(t)) \in \mathbf{R}^{N_i}$ are the corresponding radial basis function vectors whose form is selected to be the same as that in (8.11), while $\varepsilon_i(Z_i(t))$ denote the approximation errors that are bounded as $|\varepsilon_i(Z_i(t))| \leq \varepsilon_i^M$ with ε_i^M an unknown positive scalar.

Next, define

$$\begin{aligned} \tilde{\Psi}_i &= \Psi_i^* - \Psi_i \\ \tilde{W}_i &= \hat{W}_i^* - W_i \\ \tilde{\Xi}_i &= \Xi_i^* - \Xi_i, i \in \mathscr{V}_N \end{aligned} \tag{8.19}$$

where Ψ_i, W_i and Ξ_i are used to estimation Ψ_i^*, $\hat{W}_i^*$ and Ξ_i^*, respectively, and Ψ_i^*, $\hat{W}_i^*$ and Ξ_i^* are unknown constants as follows

$$\begin{aligned} \Psi_i^* &= \frac{1}{\zeta_i \mu_i^m} \\ \hat{W}_i^* &= \|W_i^*\|^2 = W_i^{*T} W_i^* \\ \Xi_i^* &\geq \varepsilon_i^M + \frac{N_i^M}{\tau_i} + w_i^M, i \in \mathscr{V}_N. \end{aligned} \tag{8.20}$$

with τ_i and w_i^M introduced below (8.1), and Ψ_i^* and N_i^M introduced below (8.9). Then, the control law for the ith vehicle in the platoon is established as

$$\begin{aligned} u_i &= \Psi_i \bar{u}_i \\ \bar{u}_i &= \tfrac{\tau_i K_i}{\lambda h_i K_D} S_i + \tfrac{\tau_i}{\lambda h_i K_D} \tfrac{F_i^2 S_i}{|F_i S_i| + \rho_i} + \tau_i \Xi_i \tanh(\tfrac{S_i}{\varphi_i}) \\ &\quad + \tfrac{\tau_i}{2\delta_i^2} S_i W_i \xi_i^T(Z_i) \xi_i(Z_i) \end{aligned} \tag{8.21}$$

where

$$\begin{cases} F_i = \lambda [K_P \dot{e}_i + K_I e_i + K_D(a_{i-1} - a_i - \ddot{\Omega}_i) + h_i \dot{a}_0] \\ \qquad -K_P \dot{e}_{i+1} - K_I e_{i+1} - K_D \ddot{e}_{i+1}, \text{for } i \in \mathscr{V}_N \backslash \{N\} \\ F_i = \lambda [K_P \dot{e}_i + K_I e_i + K_D(a_{i-1} - a_i - \ddot{\Omega}_i) \\ \qquad + h_i \dot{a}_0], \text{for } i = N \end{cases} \tag{8.22}$$

with Ω_i as in (8.12), and the following adaptive laws designed as

$$\begin{aligned} \dot{W}_i &= \frac{\alpha_i \lambda h_i K_D}{2\delta_i^2} S_i^2 \xi_i^T \xi_i - \sigma_{1i} W_i, \ W_i(0) \geq 0 \\ \dot{\Xi}_i &= \beta_i \lambda h_i K_D S_i \tanh(\frac{S_i}{\varphi_i}) - \sigma_{2i} \Xi_i, \ \Xi_i(0) \geq 0 \\ \dot{\Psi}_i &= \frac{\gamma_i \lambda h_i K_D}{\tau_i} S_i \bar{u}_i - \sigma_{3i} \Psi_i, \ \Psi_i(0) \geq 0. \end{aligned} \tag{8.23}$$

Here, K_i, λ, K_P, K_I, K_D, δ_i, α_i, β_i, and γ_i are any positive constants, while ρ_i and φ_i are small positive constants. For simplicity, some design parameters such as λ, K_P, K_I, and K_D are assumed the same for all vehicles. In addition, σ_{1i}, σ_{2i}, and σ_{3i} are any positive uniformly continuous and bounded functions that satisfy

$$\sigma_{ji}(t) > 0, \ \lim_{t \to \infty} \int_{t_0}^{t} \sigma_{ji}(\tau) d\tau \ \leq \ \bar{\sigma}_{ji} < +\infty, \ j = 1,2,3$$

with $\bar{\sigma}_{ji}$ being finite bounds.

8.3.3 Stability Analysis

The main result is summarized by the following theorem.

Theorem 8.1 *Consider the heterogeneous vehicle platoon (8.1) with a leader (8.5) subject to input quantization (8.4) and nonlinear actuator deadzone (8.6). For a sufficiently large positive constant V_{max}, suppose that the initial conditions satisfy*

$$\sum_{i=1}^{N} [S_i^2(0) + \tilde{W}_i^2(0) + \tilde{\Xi}_i^2(0) + \tilde{\Psi}_i^2(0)] \ \leq \ \tfrac{2V_{max}}{\kappa} \tag{8.24}$$

where S_i, $\tilde{\Psi}_i$, $\tilde{W}_i$ and $\tilde{\Xi}_i$ are as in (8.17) and (8.19), and

$$\kappa \ = \ max\left\{1, \min_{i \in \mathscr{V}_N}\{\alpha_i\}, \min_{i \in \mathscr{V}_N}\{\beta_i\}, \min_{i \in \mathscr{V}_N}\{\tfrac{\gamma_i}{\zeta_i \mu_i^m}\}\right\}.$$

Then, the distributed adaptive NN control law (8.21)-(8.23) guarantees that the spacing errors e_i in (8.12) converge to a small neighborhood around the origin by appropriately choosing the design parameters. Furthermore, strong string stability of the vehicle-following platoon can also be guaranteed when $0 < \lambda \leq 1$.

Proof 8.1 *Choose the Lyapunov function candidate as follows:*

$$\begin{aligned} V &= \sum_{i=1}^{N} [V_i^s + V_i^e] \\ V_i^s &= \frac{1}{2} S_i^2 \\ V_i^e &= \frac{1}{2\alpha_i} \tilde{W}_i^2 + \frac{1}{2\beta_i} \tilde{\Xi}_i^2 + \frac{\zeta_i \mu_i^m}{2\gamma_i} \tilde{\Psi}_i^2 \end{aligned} \tag{8.25}$$

where $\tilde{\Psi}_i$, $\tilde{W}_i$ and $\tilde{\Xi}_i$ are defined as in (8.19).

Differentiating V_i^s and V_i^e yields

$$\begin{aligned} \dot{V}_i^s &= S_i \dot{S}_i \\ \dot{V}_i^e &= -\frac{1}{\alpha_i} \tilde{W}_i \dot{W}_i - \frac{1}{\beta_i} \tilde{\Xi}_i \dot{\Xi}_i - \frac{\zeta_i \mu_i^m}{\gamma_i} \tilde{\Psi}_i \dot{\Psi}_i. \end{aligned} \tag{8.26}$$

Then, along the trajectories of (8.1), (8.5) and (8.9), as well (8.12), (8.16) and (8.17), one can obtain that

$$\begin{aligned} \dot{S}_i &= \lambda [K_P \dot{e}_i + K_I e_i + K_D \ddot{e}_i] - K_P \dot{e}_{i+1} - K_I e_{i+1} \\ &\quad - K_D \ddot{e}_{i+1} \\ &= \lambda \{K_P \dot{e}_i + K_I e_i + K_D [a_{i-1} - a_i - \ddot{\Omega}_i \\ &\quad - h_i (f_i(v_i, a_i, t) + \frac{n_i \mu_i}{\tau_i} u_i + \frac{1}{\tau_i} \Delta N_i + w_i - \dot{a}_0)]\} \\ &\quad - K_P \dot{e}_{i+1} - K_I e_{i+1} - K_D \ddot{e}_{i+1} \\ &= -\lambda h_i K_D [\frac{n_i \mu_i}{\tau_i} u_i + f_i(v_i, a_i, t) + \frac{1}{\tau_i} \Delta N_i + w_i] \\ &\quad + F_i, \ i \in \mathscr{V}_N \backslash \{N\} \\ \dot{S}_i &= \lambda \{K_P \dot{e}_i + K_I e_i + K_D [a_{i-1} - a_i - \ddot{\Omega}_i \\ &\quad - h(f_i(v_i, a_i, t) + \frac{n_i \mu_i}{\tau_i} u_i + w_i - \dot{a}_0)]\} \\ &= -\lambda h_i K_D [\frac{n_i \mu_i}{\tau_i} u_i + \frac{1}{\tau_i} \Delta N_i \\ &\quad + f_i(v_i, a_i, t) + w_i] + F_i, \ i = N \end{aligned} \tag{8.27}$$

where F_i and $\dot{\Omega}_i$ are defined as (8.22).

Recalling (8.18), (8.27) can further be rewritten as

$$\begin{aligned} \dot{S}_i &= -\lambda h_i K_D [\frac{n_i \mu_i}{\tau_i} u_i + \frac{1}{\tau_i} \Delta N_i + w_i \\ &\quad + W_i^{*T} \xi_i(Z_i) + \varepsilon(Z_i)] + F_i, i \in \mathscr{V}_N. \end{aligned} \tag{8.28}$$

In (8.18), instead of estimating the ideal weights W_i^* directly, a minimal learning parameter mechanism similar to the ones in [95, 106] is adopted to only estimate the constants $\hat{W}_i^*$ defined in (8.20). Then, the number of adaptive parameters updated online in RBFNN will reduce drastically from $\sum_{i=1}^{N} N_i$ to N, which can efficiently solve the explosion of learning parameters and the computational burden of the algorithm can be drastically reduced and thus is convenient to implement in platoon control.

Using Lemma 2.8 in Chapter 2, it follows

$$\begin{aligned} -\lambda \ \ & h_iK_DS_iW_i^{*T}\xi_i(Z_i) \\ \leq \ \ & \frac{\lambda h_iK_DS_i^2}{2\delta_i^2}\|W_i^*\|^2\xi_i^T(Z_i)\xi_i(Z_i)+\frac{\lambda h_iK_D}{2}\delta_i^2 \end{aligned} \tag{8.29}$$

where δ_i is a positive design parameter.

Substituting (8.28) and (8.29) into $\dot{V}_i^s$, we have

$$\begin{aligned} \dot{V}_i^s &= S_i\dot{S}_i \\ &= -\lambda h_iK_DS_i[\frac{n_i\mu_i}{\tau_i}u_i+W_i^{*T}\xi_i(Z_i)+\varepsilon(Z_i)+\frac{\Delta N_i}{\tau_i} \\ &\quad +w_i]+S_iF_i \\ &\leq -\lambda h_iK_D[S_i\frac{n_i\mu_i}{\tau_i}u_i-\frac{1}{2\delta_i^2}S_i^2\|W_i^*\|^2\xi_i^T(Z_i)\xi_i(Z_i) \\ &\quad -\frac{1}{2}\delta_i^2]-\lambda h_iS_i[\varepsilon(Z_i)+\frac{\Delta N_i}{\tau_i}+w_i]+S_iF_i. \end{aligned} \tag{8.30}$$

It follows from (8.23) that $\dot{W}_i+\sigma_{1i}W_i\geq 0$, which means that $W_i\geq e^{-\int_0^t\sigma_{1i}(\tau)d\tau}W_i(0)$. This means that once we select $W_i(0)\geq 0$, the adaptive law W_i will always be non-negative. Similarly, Ψ_i and Ξ_i are non-negative as well. Then, recalling (8.21), we can obtain

$$\begin{aligned} & -\frac{\lambda h_iK_Dn_i\mu_i}{\tau_i}S_iu_i \\ &= -\lambda h_iK_DS_i\frac{n_i\mu_i}{\tau_i}\Psi_i\Big[\frac{\tau_iK_i}{\lambda h_iK_D}S_i+\frac{\tau_i}{\lambda h_iK_D}\frac{F_i^2S_i}{|F_iS_i|+\rho_i} \\ &\quad +\tau_i\Xi_i\tanh(\frac{S_i}{\varphi_i})+\frac{\tau_i}{2\delta_i^2}S_iW_i\xi_i^T(Z)\xi_i(Z)\Big] \\ &\leq -\frac{\lambda h_iK_D\zeta_i\mu_i^m}{\tau_i}S_i(\Psi_i-\Psi_i^*+\Psi_i^*)\bar{u}_i \\ &= -\frac{\lambda h_iK_D\zeta_i\mu_i^m}{\tau_i}S_i(-\tilde{\Psi}_i+\Psi_i^*)\bar{u}_i \\ &= \frac{\lambda h_iK_D\zeta_i\mu_i^m}{\tau_i}S_i\tilde{\Psi}_i\bar{u}_i-\frac{\lambda h_iK_D}{\tau_i}S_i\bar{u}_i. \end{aligned} \tag{8.31}$$

It is worth noticing that

$$\begin{aligned} -\frac{\tau_iF_i^2S_i^2}{|F_iS_i|+\rho_i} &\leq -\tau_i|F_iS_i|+\tau_i\rho_i \\ -S_i[\varepsilon(Z_i)+\frac{\Delta N_i}{\tau_i}+w_i] &\leq \Xi_i^*|S_i| \end{aligned}$$

where Ξ_i^* is defined as in (8.20). Then, by substituting (8.21) and (8.31) into (8.30), we have

$$\begin{aligned} \dot{V}_i^s &\leq \frac{\lambda h_iK_D\zeta_i\mu_i^m}{\tau_i}S_i\tilde{\Psi}_i\bar{u}_i+\frac{\lambda h_iK_D}{2\delta_i^2}S_i^2\|W_i^*\|^2\xi_i^T(Z_i)\xi_i(Z_i) \\ &\quad -\frac{\lambda h_iK_D\mu}{\tau_i}S_i\bar{u}_i+\lambda h_iK_D\Xi_i^*|S_i|+\frac{\lambda h_iK_D}{2}\delta_i^2+S_iF_i \\ &\leq \frac{\lambda h_iK_D\zeta_i\mu_i^m}{\tau_i}S_i\tilde{\Psi}_i\bar{u}_i-[\lambda h_iK_D\Xi_iS_i\tanh(\frac{S_i}{\varphi_i}) \\ &\quad +K_iS_i^2+\frac{F_i^2S_i^2}{|F_iS_i|+\rho_i}+\frac{\lambda h_iK_D}{2\delta_i^2}S_i^2W_i\xi_i^T(Z_i)\xi_i(Z_i)] \\ &\quad +\frac{\lambda h_iK_D}{2\delta_i^2}S_i^2\|W_i^*\|^2\xi_i^T(Z_i)\xi_i(Z_i)+\lambda h_iK_D\Xi_i^*|S_i| \\ &\quad +\frac{\lambda h_iK_D}{2}\delta_i^2+S_iF_i \\ &\leq \frac{\lambda h_iK_D\zeta_i\mu_i^m}{\tau_i}S_i\tilde{\Psi}_i\bar{u}_i+\lambda h_iK_D\Xi_i^*|S_i|+\rho_i-K_iS_i^2 \\ &\quad -\lambda h_iK_D\Xi_iS_i\tanh(\frac{S_i}{\varphi_i})+\frac{\lambda h_iK_D}{2\delta_i^2}S_i^2\tilde{W}_i\xi_i^T(Z_i)\xi_i(Z_i) \\ &\quad +\frac{\lambda h_iK_D}{2}\delta_i^2 \end{aligned} \tag{8.32}$$

where the fact $S_iF_i-|S_iF_i|\leq 0$ has been used.

Using the adaptive laws (8.23) in $\dot{V}_i^e$, it yields that

$$\begin{aligned}\dot{V}_i^e = & -\frac{\lambda h_i K_D}{2\delta_i^2} S_i^2 \tilde{W}_i \xi_i^T \xi_i + \frac{\sigma_{1i}}{\alpha_i}\tilde{W}_i W_i \\ & -\lambda h_i K_D S_i \tilde{\Xi}_i \tanh(\frac{S_i}{\varphi_i}) + \frac{\sigma_{2i}}{\beta_i}\tilde{\Xi}_i \Xi_i \\ & -\frac{\lambda h_i K_D \zeta_i \mu_i^m}{\tau_i}\tilde{\Psi}_i S_i \bar{u}_i + \frac{\sigma_{3i}\zeta_i\mu_i^m}{\gamma_i}\tilde{\Psi}_i\Psi_i.\end{aligned} \tag{8.33}$$

Combining (8.32) with (8.33), one can obtain that

$$\begin{aligned}& \dot{V}_i^s + \dot{V}_i^e \\ = & -K_i S_i^2 + \lambda h_i K_D \Xi_i^* |S_i| + \frac{\sigma_{2i}}{\beta_i}\tilde{\Xi}_i \Xi_i \\ & -\lambda h_i K_D S_i \Xi_i^* \tanh(\frac{S_i}{\varphi_i}) + \frac{\lambda h_i K_D}{2}\delta_i^2 \\ & +\frac{\sigma_{3i}\zeta_i\mu_i^m}{\gamma_i}\tilde{\Psi}_i\Psi_i + \frac{\sigma_{1i}}{\alpha_i}\tilde{W}_i W_i + \rho_i \\ \leq & -K_i S_i^2 + \frac{\sigma_{1i}}{\alpha_i}\tilde{W}_i W_i + \frac{\sigma_{2i}}{\beta_i}\tilde{\Xi}_i \Xi_i \\ & +\frac{\sigma_{3i}\zeta_i\mu_i^m}{\gamma_i}\tilde{\Psi}_i\Psi_i + \rho_i + \frac{\lambda h_i K_D}{2}\delta_i^2 \\ & +0.2785\lambda h_i K_D \varphi_i\end{aligned} \tag{8.34}$$

where Lemma 2.5 in Chapter 2 has been used.

Based on Lemma 2.7 in Chapter 2, one has

$$\begin{aligned}\frac{\sigma_{1i}}{\alpha_i}\tilde{W}_i W_i = & \frac{\sigma_{1i}}{\alpha_i}\tilde{W}_i(\hat{W}_i^* - \tilde{W}_i) \\ \leq & -\frac{\sigma_{1i}}{2\alpha_i}\tilde{W}_i^2 + \frac{\sigma_{1i}}{2\alpha_i}\hat{W}_i^{*2}.\end{aligned} \tag{8.35}$$

Similarly, one can also obtain that

$$\begin{aligned}\frac{\sigma_{2i}}{\beta_i}\tilde{\Xi}_i\Xi_i \leq & -\frac{\sigma_{2i}}{2\beta_i}\tilde{\Xi}_i^2 + \frac{\sigma_{2i}}{2\beta_i}\Xi_i^{*2} \\ \frac{\sigma_{3i}}{\gamma_i}\tilde{\Psi}_i\Psi_i \leq & -\frac{\sigma_{3i}}{2\gamma_i}\tilde{\Psi}_i^2 + \frac{\sigma_{3i}}{2\gamma_i}\Psi_i^{*2}.\end{aligned} \tag{8.36}$$

Then, from (8.25) and making use of (8.34)-(8.36), we have

$$\begin{aligned}\dot{V} = & \sum_{i=1}^{N}[-K_i S_i^2 - \frac{\sigma_{1i}}{2\alpha_i}\tilde{W}_i^2 - \frac{\sigma_{2i}}{2\beta_i}\tilde{\Xi}_i^2 - \\ & \frac{\sigma_{3i}\zeta_i\mu_i^m}{2\gamma_i}\tilde{\Psi}_i^2 + 0.2785\lambda h_i K_D\varphi_i + \frac{\sigma_{1i}}{2\alpha_i}\hat{W}_i^{*2} \\ & +\rho_i + \frac{\sigma_{2i}}{2\beta_i}\Xi_i^{*2} + \frac{\sigma_{3i}\zeta_i\mu_i^m}{2\gamma_i}\Psi_i^{*2} + \frac{\lambda h_i K_D}{2}\delta_i^2] \\ \leq & -\chi_1 V + \chi_2\end{aligned}$$

where the positive parameters χ_1 and χ_2 are given as follows

$$\begin{aligned}\chi_1 = & \min\{2K_i, \min_{i\in\mathscr{V}_N}\sigma_{1i}, \min_{i\in\mathscr{V}_N}\sigma_{2i}, \min_{i\in\mathscr{V}_N}\sigma_{3i}\} \\ \chi_2 = & \sum_{i=1}^{N}[\rho_i + 0.2785\lambda h_i K_D\varphi_i + \frac{\sigma_{1i}}{2\alpha_i}\hat{W}_i^{*2} \\ & +\frac{\sigma_{2i}}{2\beta_i}\Xi_i^{*2} + \frac{\sigma_{3i}\zeta_i\mu_i^m}{2\gamma_i}\Psi_i^{*2} + \frac{\lambda h_i K_D}{2}\delta_i^2].\end{aligned} \tag{8.37}$$

Thus, by using Lemma 2.6 in Chapter 2, the following inequality holds:

$$V \leq (V(0) - \frac{\chi_2}{\chi_1})e^{-\chi_1 t} + \frac{\chi_2}{\chi_1} \leq V(0) + \frac{\chi_2}{\chi_1}, \quad (8.38)$$

where the following fact has been used for the last inequality.

$$V(0) = \sum_{i=1}^{N}[\frac{1}{2}S_i^2(0) + \frac{1}{2\alpha_i}\tilde{W}_i^2(0) + \frac{1}{2\beta_i}\tilde{\Xi}_i^2(0) + \frac{\zeta_i\mu_i^m}{2\gamma_i}\tilde{\Psi}_i^2(0)] \geq 0.$$

Then, it is directly obtained that S_i, $\tilde{W}_i$, $\tilde{\Xi}_i$ and $\tilde{\Psi}_i$ must remain bounded. Because of the boundedness of $\hat{W}_i^*$, Ξ_i^* and Ψ_i^*, we conclude that W_i, Ξ_i and Ψ_i are also bounded. Based on the above analysis, one can conclude that all the signals in the closed-loop system remain uniformly ultimately bounded if the initial condition is bounded as in (8.24). As $t \to \infty$, $|S_i| \leq \sqrt{\frac{2\chi_2}{\chi_1}}$, i.e., the sliding surface S_i is steered into an adjustable region of zero asymptotically. Then, according to (8.16) and (8.17), s_i and the spacing error e_i are also convergent to a neighborhood of zero.

In the sequel, the strong string stability of the whole vehicle-following platoon is proved following the approach in [57, 93]. Since $S_i = \lambda s_i - s_{i+1}$, when S_i converges to an arbitrarily small neighborhood of zero by choosing the design parameters appropriately, we have

$$\begin{aligned} \lambda & \ [K_P e_i + K_I \textstyle\int_0^t e_i(\tau)d\tau + K_D\dot{e}_i] \\ = & \ K_P e_{i+1} + K_I \textstyle\int_0^t e_{i+1}(\tau)d\tau + K_D\dot{e}_{i+1}. \end{aligned} \quad (8.39)$$

Taking Laplace transform of (8.39) and using (8.13), we have

$$\begin{aligned} \lambda & \ [K_P E_i(s) + K_I s E_i(s) + \tfrac{K_D}{s}E_i(s)] \\ = & \ K_P E_{i+1}(s) + K_I s E_{i+1}(s) + \tfrac{K_D}{s}E_{i+1}(s), \end{aligned}$$

i.e., $G_i(s) = \frac{E_{i+1}(s)}{E_i(s)} = \lambda$. When $0 < \lambda \leq 1$, $|G_i(s)| \leq 1$ is guaranteed and strong string stability of the whole vehicular platoon as in (8.15) is also guaranteed. Theorem 8.1 is thus proved. □

Remark 8.4 *The size of spacing errors relies on the selection of design parameters in (8.37). Increasing K_i, α_i, β_i, and γ_i while reducing K_D, φ_i, ρ_i and δ_i will lead to smaller spacing errors. In addition, σ_{1i}, σ_{2i}, and σ_{3i} should also be chosen small to reduce the spacing errors. However, if χ_1 is too big and χ_2 is too small, the required control energy will be high and may result in saturation. Therefore, in practical applications, these parameters should be chosen carefully according to the required tracking precision and the available control resource.*

Remark 8.5 *As a conclusion of Theorem 8.1, the following two points need be highlighted. First, by requiring additional communication load in leader's velocity, acceleration and jerk, the traffic density ($\Delta_{i-1,i}$) of the MCTH policy (8.12) is much higher than that of the TCTH policy ($\Delta_{i-1,i} + h_i v_i$) as shown in Figure 8.1. This is very important to improve road capacity and avoid traffic congestion. Second, the* *feedback linearization approach developed in [43, 213] cannot be applied directly to the nonlinear vehicular system (8.1) due to the uncertainty of mechanical drag and air resistance (not to mention the input quantization and unknown actuator deadzone). Furthermore, the control schemes developed in [138, 201] cannot guarantee strong string stability when there exist non-zero initial condition uncertainties, input quantization, and unknown actuator deadzone, which may cause collisions between consecutive vehicles.*

Following a similar process as in the proof of Theorem 8.1, when the effects of input *quantization* and actuator deadzone are not considered as in [43, 138, 201, 213], we have the following corollary.

Corollary 8.1 *Consider the heterogeneous vehicle platoon (8.1) with a leader (8.5). For a sufficiently large positive constant V_{max}, if the initial conditions satisfy (8.24), the adaptive NN control law (8.40) below guarantees that the spacing errors e_i in (8.12) converge to a small neighborhood around the origin by appropriately choosing design parameters, while strong string stability of the vehicle-following platoon also can be guaranteed when $0 < \lambda \leq 1$.*

$$
\begin{aligned}
u_i(t) &= \frac{\tau_i}{\lambda h_i K_D}\frac{F_i^2(t)S_i(t)}{|F_i(t)S_i(t)|+\rho_i} + \tau_i \Xi_i(t)\tanh\left(\frac{S_i(t)}{\varphi_i}\right) \\
&\quad + \frac{\tau_i}{2\delta_i^2} S_i(t) W_i(t) \xi_i^T(Z)\xi_i(Z) + \frac{\tau_i K_i}{\lambda h_i K_D} S_i \\
\dot{W}_i(t) &= \frac{\alpha_i \lambda h_i K_D}{2\delta_i^2} S_i^2(t)\xi_i^T(t)\xi_i(t) - \sigma_{1i}(t)W_i(t) \\
\dot{\Xi}_i(t) &= \beta_i \lambda h_i K_D S_i(t)\tanh\left(\frac{S_i(t)}{\varphi_i}\right) - \sigma_{2i}(t)\Xi_i(t)
\end{aligned} \tag{8.40}
$$

where $F_i(t)$ is defined in (8.22).

For convenience, this controller is called as *Standard Controller.*

8.4 Simulation Study

To validate the effectiveness of the proposed adaptive quantized NN control, computer simulations are carried out for a heterogenous vehicular platoon containing ten followers, i.e., $N = 10$. In the simulations, the parameters of the heterogeneous vehicular platoon are the same as those in [43]. For clarity, the detailed parameters of all follower vehicles are listed in Table 8.1 and Table 8.3. The quantized parameters and non-symmetric deadzone parameters are given in Table 8.2. The mismatch ratio μ_i in (8.4) is assumed as $\mu_i \in [5, 30]$ with a small $\mu_i^c \in [0.05, 0.3]$. The RBF vectors are given by $\xi_i(Z_i) \in \mathbf{R}^{15}$ $(i.e., N_i = 15)$. The centers ϕ_k for the NN input $Z_i = [v_i, a_i]$ are randomly spaced in $[-2, 5] \times [3.5, 4]$, and the width is $\theta_k = 1.5$, $k = 1, 2, \cdots, 15$. In addition, the *external disturbances* and the desired acceleration of the leader are

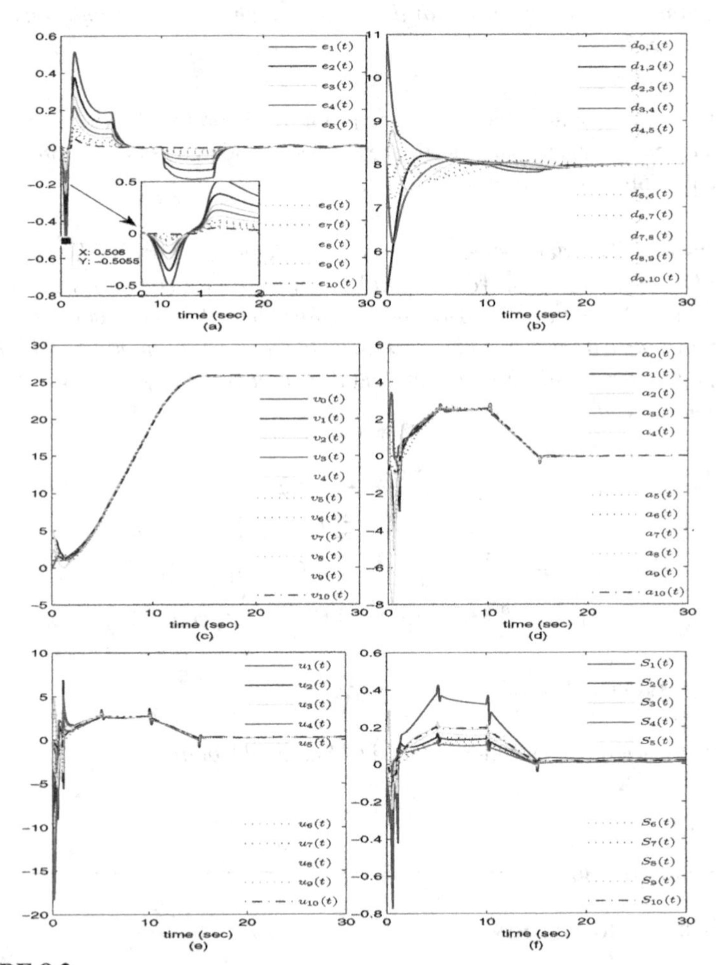

FIGURE 8.2
(a) Spacing error $e_i(t)$; (b) position $x_i(t)$; (c) velocity $v_i(t)$; (d) acceleration $a_i(t)$; (e) control input $u_i(t)$; (f) sliding surface $S_i(t)$ under Corollary 8.1 without input quantization and actuator deadzone.

generated by

$$w_i = 0.1\sin(t), i \in \mathcal{V}_N$$
$$a_0 = \begin{cases} 0.5t \; m/s^2, & 1s \le t < 5s \\ 2.5 \; m/s^2, & 5s \le t < 10s \\ -0.5t + 7.5 \; m/s^2, & 10s \le t < 15s \\ 0 \; m/s^2, & otherwise. \end{cases}$$

The controller design parameters and adaptive gains are listed in Table 8.2. In addition, the initial spacings, velocities and accelerations of the leader and the followers are listed in Table 8.4. From *MCTH policy* in (8.12), the initial errors $\tilde{e}_i(0)$, $\dot{\tilde{e}}_i(0)$ and $\ddot{\tilde{e}}_i(0)$ are nonzero.

TABLE 8.1
Simulation parameters' values of vehicle dynamics for $i \in \mathcal{V}_N$

Parameter Name	Simulation Values
Desired vehicle spacing	$\Delta_{i-1,i} = 8m$
Specific mass of the air	$\rho_{mi} = 1N/m^3$
Cross-sectional area	$H_i = 2.2m^2$
Drag coefficient	$c_i = 0.35$
Mechanical drag	$d_{mi} = 150N$
Headway time	$h_i = 1s$

TABLE 8.2
Controller and adaptive designed parameters for $i \in \mathcal{V}_N$

Parameter Name	Simulation Values
Quantizer parameters	$\mu_i^c = 0.05(1+5e^{-0.25t})$, $\mu_i^d = 0.01$
Deadzone parameters	$n_i^r = 0.7$, $n_i^l = b_i^r = 1$, $b_i^l = 2$
Controller parameters	$K_i = 45$, $\lambda = 0.75$, $\pi_i = 5$
PID parameters	$K_P = 10$, $K_I = 12$, $K_D = 1$
Initial adaptive values	$W_i(0) = 1, \Xi_i(0) = 1, \Psi_i(0) = 1$
Adaptive parameters 1	$\alpha_i = \beta_i = 10$, $\gamma_i = 0.6$
Adaptive parameters 2	$\sigma_{1i} = \sigma_{3i} = 0.1e^{-10t}$, $\sigma_{2i} = 0.001e^{-10t}$
Tuning parameters	$\delta_i = 1$, $\varphi_i = 0.5$, $\rho_i = 0.5$

In the following, the comparison between Corollary 8.1 and Theorem 8.1 is studied. First of all, when there do not exist input quantization and actuator deadzone, the simulation results obtained by using the standard controller (8.40) of Corollary 8.1 are shown in Figure 8.2, where the adaptive updating laws are omitted due to

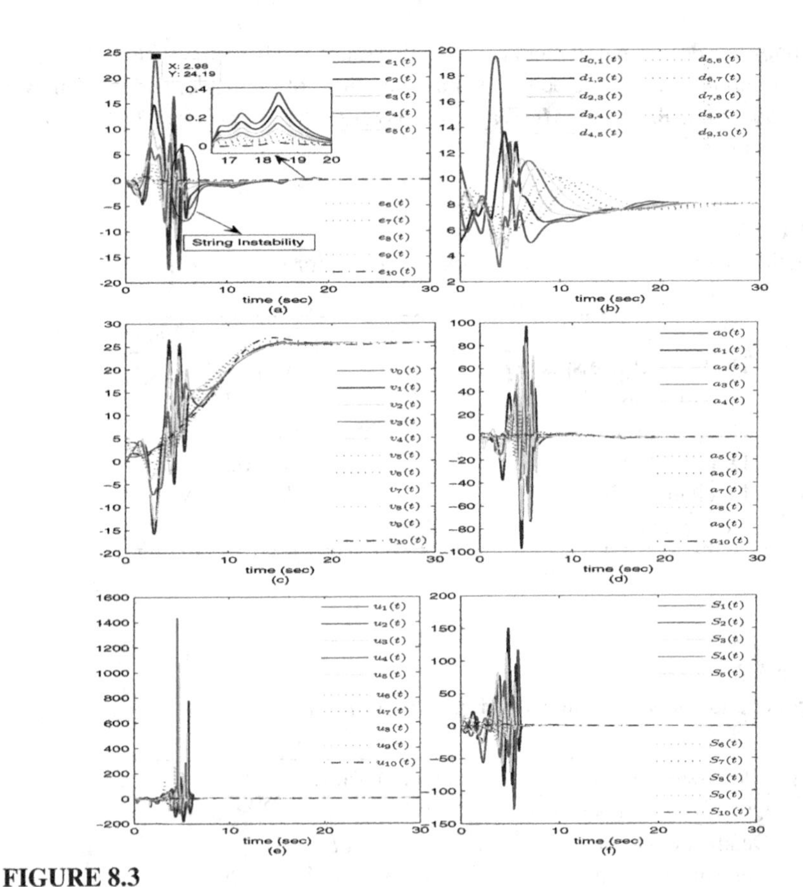

FIGURE 8.3
(a) Spacing error $e_i(t)$; (b) position $x_i(t)$; (c) velocity $v_i(t)$; (d) acceleration $a_i(t)$; (e) control input $u_i(t)$; (f) sliding surface $S_i(t)$ under Corollary 8.1 with input quantization and actuator deadzone.

TABLE 8.3
The parameters m_i, τ_i and L_i for $i \in \mathcal{V}_N$ [43]

i	1	2	3	4	5	6	7	8	9	10
$m_i(kg)$	1500	1600	1550	1650	1500	1400	1550	1550	1700	1600
$\tau_i(s)$	0.1	0.3	0.2	0.4	0.25	0.4	0.3	0.15	0.35	0.2
$L_i(m)$	4	4.5	5	5	4.5	3.5	4	4	5	4.5

TABLE 8.4
The initial values $x_i(0)(m)$, $v_i(0)(m/s)$ and $a_i(0)(m/s^2)$ for $i \in \mathcal{V}_N \cup \{0\}$

i	0	1	2	3	4	5	6	7	8	9	10
$x_i(0)$	150	135	125.5	112.5	99.5	87	75.5	63.5	51.5	38.5	26
$v_i(0)$	1	4	2	0	5	3	1	2	1	3	2
$a_i(0)$	0	1	5	2	1	3	1	2	1	2	−1

limited space. Figure 8.2(a) shows the performance of *strong string stability* of the platoon as the amplitude of the spacing error decreases through the platoon (i.e., $\|e_{10}(t)\| \leq \cdots \leq \|e_1(t)\|$). It follows from Figure 8.2(b) that neighboring-vehicle collisions are also avoided as there do not exist any crossed and/or overlapped positions. Figures 8.2(c) and (d) show that the velocities and accelerations of all follower vehicles eventually track those of the leader, respectively. The control input $u_i(t)$ and the sliding surface $S_i(t)$ are shown in Figures 8.2(e) and (f), respectively. It can be seen that the case where control inputs are smooth and chattering is indeed eliminated. However, when there exist input quantization and actuator deadzone, the *strong string stability* cannot be guaranteed as is shown in Figure 8.3(a), which could result in collisions in Figure 8.3(b). It can also be observed from Figure 8.3 that the performance of the whole platoon by using (8.40) deteriorates significantly in the presence of input quantization and actuator deadzone as the *spacing errors* in Figure 8.3(a) is much larger than those in Figure 8.2(a). These simulation results demonstrate the need for effective control schemes to handle such input quantization and actuator deadzone nonlinearities.

Next, the simulation results using (8.21)-(8.23) of Theorem 8.1 are shown in Figure 8.4. It can be found from Figure 8.4 that, even in the presence of input quantization and actuator deadzone, *strong string stability* can still be guaranteed and the performance is very similar to those in Figure 8.2. This shows that the control scheme proposed in Theorem 8.1 is superior in its ability to tolerate input quantization and actuator deadzone as compared with the standard controller in Corollary 8.1. More specifically, the maximum values of spacing errors in Figure 8.2(a), Figure 8.3(a) and Figure 8.4(a) are $0.51m$, $24.19m$ and $1.89m$, respectively, while the desired spacing between two consecutive vehicles is $\Delta_{i-1,i} = 8m$ only. This shows the likelihood of collisions when the standard controller is used in the presence of input quantization

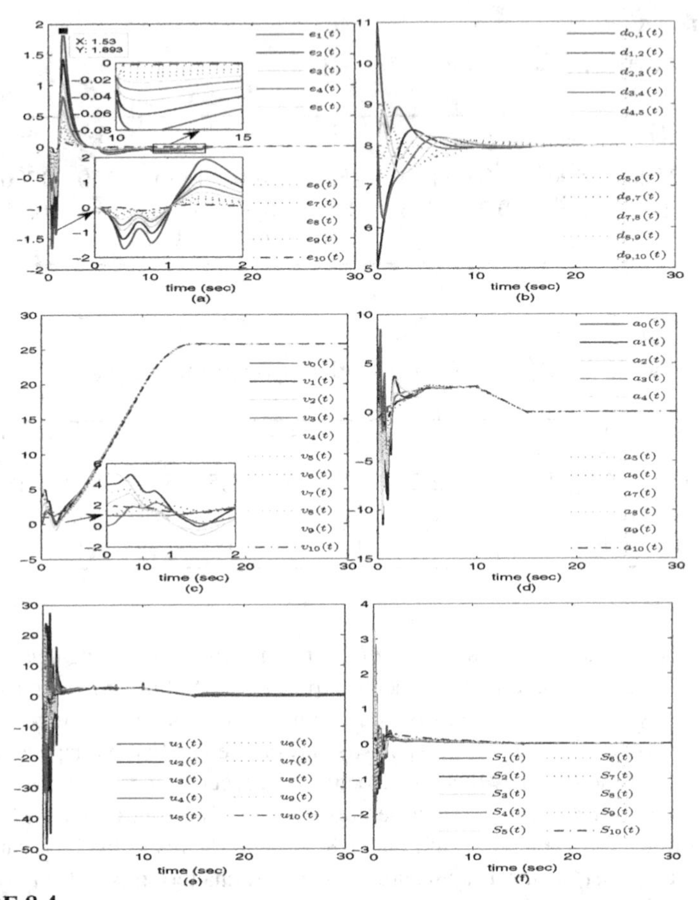

FIGURE 8.4
(a) Spacing error $e_i(t)$; (b) position $x_i(t)$; (c) velocity $v_i(t)$; (d) acceleration $a_i(t)$; (e) control input $u_i(t)$; (f) sliding surface $S_i(t)$ under Theorem 8.1 with input quantization and actuator deadzone.

and actuator deadzone. It is worth mentioning that although the maximum spacing error $24.19m$ is much larger than the desired spacing $8m$, there is still no collision in Figure 8.3(b). The main reason for this phenomenon is due to the use of the CTH policy, which also demonstrates the superiority of improving *strong string stability* by using the *MCTH policy* (8.12).

8.5 Conclusion

This chapter addresses the problem of neuro-adaptive quantized PIDSMC for heterogeneous vehicular platoon subject to *external disturbances*, mismatch input quantization and unknown actuator deadzone. To improve the traffic density and remove the common assumption that the initial spacing, velocity and acceleration errors must be zero, a new MCTH policy is proposed. In addition, an optimized adaptation algorithm included in the adaptive NN mechanism makes the online computation burden greatly reduced. The *strong string stability* of the whole vehicular platoon and the convergence of the spacing, velocity and acceleration errors to arbitrarily small compact sets of zero is proven by the Lyapunov theory. Finally, the simulation results show the effectiveness and advantages of the proposed scheme.

9

Low-Complexity Control of Vehicular Platoon with Asymmetric Saturation

An approximation-free adaptive PID-based *sliding mode control* (PIDSMC) scheme is designed for nonlinear vehicle platoon subject to *asymmetric actuator saturation*, capable of guaranteeing, for any initial system condition, *strong string stability* of the whole vehicular platoon. It is shown that under the proposed scheme, the whole vehicle platoon can tolerate the *asymmetric actuator saturation* and *unmodeled dynamic nonlinearities*, and the scheme also retains the main advantages of the PIDSMC technique. These advantages include robustness and capability to reduce the spacing errors and chattering and to eliminate the steady-state *spacing errors*. Moreover, *adaptive compensation* instead of approximation approach such as neuro-network and fuzzy logic approaches is adopted to attenuate the negative effects caused by *asymmetric actuator saturation* and unmodeled dynamic nonlinearities. Compared with most existing methods, the proposed method does not linearize the system model and neither requires precise knowledge of the system model.

9.1 Introduction

In recent years, intelligent vehicle control is increasingly receiving deserved attention from the control communities due to increased traffic congestion. The objective of intelligent vehicle control is to maintain a *desired safe spacing* between consecutive vehicles as well as effectively increase traffic throughput, which thus enhance the safety and improve the traffic flow [120]. To achieve the above objective, much effort has been spent on various control laws for vehicular platoons [43, 57, 51, 93, 120, 138, 212]. The literature in [43, 120, 212] simplified the *platoon control* problem by reducing a nonlinear system to a linear one via feedback linearization technique. Obviously, feedback linearization approach would have limitation to practical application because accurate models are extremely difficult to obtain in practical processes/systems. *String stability* of a type of second order vehicle systems was guaranteed based on an adaptive coupled *sliding mode control* technique by requiring zero initial spacing and velocity errors in [93]. In order to remove the above requirement on zero initial errors, [57] proposed a *modified constant time headway* policy but under the unrealistic assumption that the nonlinearities of the consecutive vehicles (including the leader) should satisfy certain *Lipschitz* constraints and should exactly match each other. By considering the fact that different kinds of vehicles may require different model structures and parameters, powerful approximation capabilities of neural networks (NNs) and fuzzy logic systems (FLSs) for control of vehicular platoons have been adopted in [51, 138]. However, for NNs or FLSs, the learning is slow since all the weights are updated during each learning cycle [138]. Therefore, the effectiveness of NNs and FLSs is limited in problems requiring a large amount of on-line learning parameters.

Actuator saturation is one of the most important non-smooth nonlinearities and is often encountered in engineering practice because actuators usually have limitations on their magnitude due to the physical characteristics of the actuator or safety considerations [35]. It often severely limits the system performance, give rise to undesirable inaccuracy and even instability [19, 205]. Therefore, the phenomenon of input saturation has to be considered for practical systems. Consequently, the analysis and the design methods for nonlinear saturation systems have attracted extensive attention during the past decades [19, 35]. Asymmetric actuator nonlinear saturation was first studied in [19], where the nonlinear functions of the *asymmetric actuator saturation* are required to be strictly monotonous. This obviously limits its application in the real systems. Furthermore, there are still few results for the control of nonlinear vehicular platoon by taking *actuator saturation* into account in the controller design and analysis. How to construct a control scheme to guarantee string stability of the vehicular platoon and simultaneously attenuate the negative effect of the *actuator saturation* is a tremendous challenge. On the other hand, *sliding mode control* (SMC) technique is known as an efficient way to be robust against model uncertainties and insensitive to the bounded disturbances [209], while in practice, 90% of the controllers used for industrial control are proportional-integral-derivative (PID) controllers [5], known for their excellent ability in achieving tracking, robustness and disturbance rejection. Taking these into account, the PID-based SMC (*PIDSMC*) method is employed to analyze and control vehicle platoons under *actuator saturation* in this chapter.

This chapter pursues the *platoon control* for nonlinear vehicular platoons subjected to *asymmetric actuator saturation*. The engineering motivation is that *platoon control* is able to achieve *strong string stability* without involving feedback linearization. Another objective is to make full use of *adaptive compensation* instead of approximation approaches such as NNs and FLSs. Thus, our approach is quite low in complexity since approximation approaches are not adopted despite there being mismatched nonlinearities, unknown bounded *external disturbances* and *asymmetric actuator saturation*. The assumption in [19] that the nonlinear functions of the *asymmetric actuator saturation* are required to be strictly monotonous is removed in this chapter. Therefore, the type of *asymmetric actuator saturation* considered in this chapter is more general than the one considered in [19]. Based on the *MCTH policy* proposed in Chapter 8, an adaptive *PIDSMC* scheme is proposed to guarantee the *strong string stability* of the whole nonlinear vehicular platoon and the ultimate uniform boundedness of all closed-loop signals through Lyapunov stability analysis.

9.2 Vehicular Platoon Description

Consider a leader-follower vehicular platoon of N follower vehicles and a leader vehicle (indexed by 0), moving in a one-dimensional line. The position, velocity and acceleration of the leader vehicle are denoted by $(x_0, v_0, a_0) \in \mathbf{R}^3$, respectively. Then, the motion of the leader is expressed as follows:

$$\dot{x}_0 = v_0, \dot{v}_0 = a_0 \tag{9.1}$$

where a_0 is the desired time-varying acceleration. Furthermore, $(x_i, v_i, a_i) \in \mathbf{R}^3$ $(i \in \mathscr{V}_N)$ denote the position, velocity and acceleration of the $(i)^{\text{th}}$ vehicle whose dynamics are described by

$$\begin{aligned} \dot{x}_i &= v_i, \dot{v}_i = a_i, i \in \mathscr{V}_N \\ \dot{a}_i &= \text{sat}(u_i) + f_i(x_i, v_i, a_i, t) + w_i \end{aligned} \tag{9.2}$$

where $f_i(x_i, v_i, a_i, t)$ is its intrinsic nonlinear dynamics; $u_i \in \mathbf{R}$ is the control input action on vehicle i; w_i is an unknown time-varying external disturbance. Considering actuator asymmetric saturation constraints as shown in Figure 9.1, the control input $sat(u_i)$ is defined by

$$\text{sat}(u_i(t)) = \begin{cases} u_{r,imax}, & \text{if } u_i(t) > b_{r,i} \\ g_{r,i}(u_i(t)), & \text{if } 0 \le u_i(t) \le b_{r,i} \\ g_{l,i}(u_i(t)), & \text{if } b_{l,i} \le u_i(t) < 0 \\ u_{l,imin}, & \text{if } u_i(t) < b_{l,i} \end{cases} \tag{9.3}$$

where $b_{r,i} > 0$ and $b_{l,i} < 0$ are saturation amplitudes, $u_{r,imax} > 0$ and $u_{l,imin} < 0$ are unknown constants, while $g_{r,i}(u_i(t))$ and $g_{l,i}(u_i(t))$ are unknown bounded nonlinear functions. The difference between the desired and actual control signals can be described as

$$\begin{aligned} \Delta u_i &\triangleq \text{sat}(u_i(t)) - u_i(t) \\ &= \begin{cases} u_{r,imax} - u_i(t), & \text{if } u_i(t) > b_{r,i} \\ g_{r,i}(u_i(t)) - u_i(t), & \text{if } 0 \le u_i(t) \le b_{r,i} \\ g_{l,i}(u_i(t)) - u_i(t), & \text{if } b_{l,i} \le u_i(t) < 0 \\ u_{l,imin} - u_i(t), & \text{if } u_i(t) < b_{l,i} \end{cases} \end{aligned} \tag{9.4}$$

which cannot be implemented because of the saturation and will be compensated by an adaptive law in the later design. Thus, the follower dynamics (9.2) can be rewritten as

$$\begin{aligned} \dot{x}_i &= v_i, \dot{v}_i = a_i, i \in \mathscr{V}_N \\ \dot{a}_i &= u_i + \Delta u_i + f_i(x_i, v_i, a_i, t) + w_i. \end{aligned} \tag{9.5}$$

Remark 9.1 *Third-order vehicle systems have been investigated widely in the existing literature, such as in [43, 138, 212]. However, in order to simplify the considered* *problems, feedback linearized approach has been used widely in [43, 212]. In addition, the considered nonlinear vehicle dynamic (9.2) will be more general than those considered in [43, 138, 212] since the nonlinear dynamics $f_i(x_i, v_i, a_i, t)$ and the non-symmetric saturation constraints are considered simultaneously. In addition, the nonlinear function $f_i(x_i, v_i, a_i, t)$ in (9.2) is also more general than those considered in [57], where the nonlinear function $f_i(x_i, v_i, a_i, t)$ should satisfy some Lipschitz condition and the follower and leader vehicles have the same type of nonlinear uncertainties $f_i(x_i, v_i, a_i, t)$, which means the considered vehicle platoon is subjected to matched nonlinear uncertainties.*

Remark 9.2 *The type of actuator saturation is inspired by [205], where a type of asymmetric nonlinear actuator deadzone is considered. Then, [19] investigated the case of actuator saturation. It should be pointed out that in [19, 205], the nonlinear functions $g_{r,i}(u_i(t))$ and $g_{l,i}(u_i(t))$ are assumed strictly monotonous. In this chapter, this assumption is removed by using an adaptive compensation technique.*

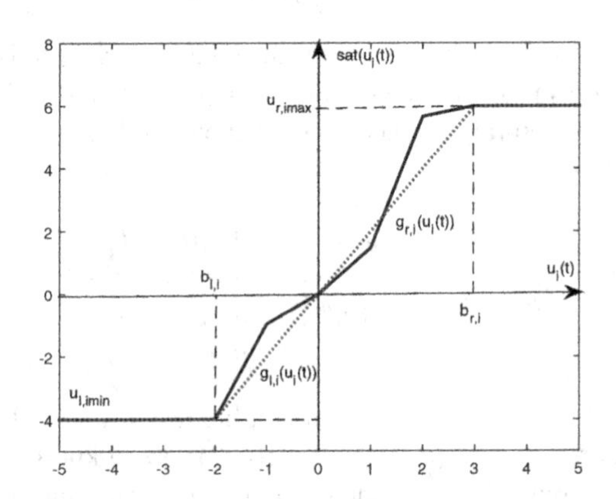

FIGURE 9.1
Asymmetric actuator saturation constraint.

For the development of control laws, the following assumption is made.

Assumption 9.1 *There exist constants $\bar{f}_i$, $\bar{w}_i$ and $\bar{\Delta}_i$ such that $f_i(x_i, v_i, a_i, t)$, w_i and Δu_i satisfied $|f_i(x_i, v_i, a_i, t)| \leq \bar{f}_i$, $|w_i| \leq \bar{w}_i$ and $|\Delta u_i| \leq \bar{\Delta}_i$, respectively.*

Remark 9.3 *In Assumption 9.1, the nonlinear uncertainties $f_i(x_i, v_i, a_i, t)$ and the external disturbances w_i are required to be bounded, which is practical in reality. In addition, in practice, the energy of control input is always finite and the objective is to propose a method for finding an effective bounded controller to bring the system into the sliding mode, thus it is reasonable to assume that u_i and hence Δu_i is bounded. It should be noted that all these bounds $\bar{f}_i$, $\bar{w}_i$ and $\bar{\Delta}_i$ are not required in the implementation of the proposed control design. They are used only for analytical purposes.*

9.3 Adaptive PIDSMC Design and Strong String Stability Analysis

In this section, adaptive PIDSMC is proposed for the vehicle-following platoon in (9.1) and (9.2). Inspired by [57], a similar *modified constant time headway* (MCTH) policy is introduced to improve the traffic density and simultaneously attenuate the negative effect of *non-zero initial* spacing, velocity and acceleration errors.

9.3.1 Control Problem

The control objective of this chapter is to utilize an *adaptive compensation* technique to determine a distributed adaptive PIDSMC law u_i and adaptive updating laws such that all the signals that are involved in the resulting closed-loop system are uniformly ultimately bounded. In addition, based on *MCTH policy* defined in (8.12) of Chapter 8, the *strong string stability* as defined in (8.15) of Chapter 8 is guaranteed.

9.3.2 PID-Based Sliding Mode Control Design

In order to avoid using backstepping technique as in [138], a PID-based sliding manifold and a coupling the sliding surfaces s_i the same as (8.16) and (8.17) of Chapter 8 are defined. Next, define

$$\begin{aligned} \tilde{\Xi}_i &= \Xi_i^* - \Xi_i \\ \tilde{\Psi}_i &= \Psi_i^* - \Psi_i, i \in \mathscr{V}_N \end{aligned} \tag{9.6}$$

where Ψ_i and Ξ_i are used to estimation Ψ_i^* and Ξ_i^*, respectively, and Ψ_i^* and Ξ_i^* are unknown constants as follows

$$\begin{aligned} \Psi_i^* &= \bar{f}_i + \bar{w}_i \\ \Xi_i^* &= \bar{\Delta}_i, i \in \mathscr{V}_N. \end{aligned} \tag{9.7}$$

with $\bar{f}_i$, $\bar{w}_i$ and $\bar{\Delta}_i$ introduced in Assumption 9.1. Then, the control law for the $(i)^{\text{th}}$ vehicle in the platoon is established as

$$\begin{aligned} u_i = & \frac{K_i}{\lambda h_i K_D} S_i + \frac{1}{\lambda h_i K_D} \frac{F_i^2 S_i}{|F_i S_i| + \rho_i} \\ & + \Xi_i \tanh\left(\frac{S_i}{\varphi_{1i}}\right) + \Psi_i \tanh\left(\frac{S_i}{\varphi_{2i}}\right) \end{aligned} \tag{9.8}$$

where

$$\begin{cases} F_i = \lambda[K_P \dot{e}_i + K_I e_i + K_D(a_{i-1} - a_i - \ddot{\Omega}_i) + h_i \dot{a}_0] \\ \quad -K_P \dot{e}_{i+1} - K_I e_{i+1} - K_D \ddot{e}_{i+1}, \text{for } i \in \mathscr{V}_N \backslash \{N\} \\ F_i = \lambda[K_P \dot{e}_i + K_I e_i + K_D(a_{i-1} - a_i - \ddot{\Omega}_i) + h_i \dot{a}_0] \\ \qquad \qquad \text{for } i = N \end{cases} \tag{9.9}$$

with Ω_i as in (8.12) of Chapter 8, and the following adaptive laws designed as

$$\dot{\Xi}_i = \alpha_i \lambda h_i K_D S_i \tanh\left(\frac{S_i}{\varphi_{1i}}\right) - \sigma_{1i} \Xi_i, \ \Xi_i(0) \geq 0$$

$$\dot{\Psi}_i = \beta_i \lambda h_i K_D S_i \tanh(\frac{S_i}{\varphi_{2i}}) - \sigma_{2i}\Psi_i,\ \Psi_i(0) \geq 0. \tag{9.10}$$

Here, K_i, λ, K_P, K_I, K_D, α_i and β_i are any positive constants, while ρ_i and $\varphi_{ji}(j = 1,2)$ are small positive constants. In addition, σ_{1i} and σ_{2i} are any positive uniformly continuous and bounded functions that satisfy

$$\sigma_{ji}(t) > 0,\ \lim_{t\to\infty} \int_{t_0}^{t} \sigma_{ji}(\tau)d\tau \ \leq\ \bar{\sigma}_{ji} < +\infty,\ j = 1,2$$

with $\bar{\sigma}_{ji}$ being finite bounds.

Based on the above control strategy, an immediate result is the following sufficient condition for *strong string stability* for vehicle platoon.

Theorem 9.1 *Consider the vehicle platoon containing a leader (9.1) and N followers (9.2) subjected to asymmetric nonlinear actuator saturation (9.3) with Assumption 9.1. For any non-zero initial spacing, velocity and acceleration errors, with the distributed adaptive control law (9.8)-(9.10), the signals involved in the vehicle platoon will be uniformly ultimately bounded and the strong string stability can also be guaranteed when* $0 < \lambda \leq 1$.

Proof 9.1 *Let us consider the following Lyapunov function candidate*

$$V = \sum_{i=1}^{N} [\tfrac{1}{2}S_i^2 + \tfrac{1}{2\alpha_i}\tilde{\Xi}_i^2 + \tfrac{1}{2\beta_i}\tilde{\Psi}_i^2] \tag{9.11}$$

where $\tilde{\Psi}_i$ *and* $\tilde{\Xi}_i$ *are defined as in (9.6).*

The derivative of the function (9.11) is

$$\dot{V} = \sum_{i=1}^{N} [S_i\dot{S}_i - \tfrac{1}{\alpha_i}\tilde{\Xi}_i\dot{\Xi}_i - \tfrac{1}{\beta_i}\tilde{\Psi}_i\dot{\Psi}_i]. \tag{9.12}$$

Differentiating S_i *along the trajectories of (9.1), (9.5), as well (8.12), (8.16) and (8.17) of Chapter 8 yields*

$$\begin{aligned}
\dot{S}_i &= \lambda\{K_P\dot{e}_i + K_I e_i + K_D[a_{i-1} - a_i - \ddot{\Omega}_i \\
&\quad - h_i(f_i(v_i,a_i,t) + u_i + \Delta u_i + w_i - \dot{a}_0)]\} \\
&\quad - K_P\dot{e}_{i+1} - K_I e_{i+1} - K_D\ddot{e}_{i+1} \\
&= -\lambda h_i K_D[u_i + f_i(x_i,v_i,a_i,t) + \Delta u_i + w_i] + F_i \\
&\qquad\qquad i \in \mathscr{V}_N \setminus \{N\} \\
\dot{S}_i &= \lambda\{K_P\dot{e}_i + K_I e_i + K_D[a_{i-1} - a_i - \ddot{\Omega}_i \\
&\quad - h(f_i(v_i,a_i,t) + u_i + \Delta u_i + w_i - \dot{a}_0)]\} \\
&= -\lambda h_i K_D[u_i + \Delta u_i + f_i(x_i,v_i,a_i,t) + w_i] + F_i \\
&\qquad\qquad i = N
\end{aligned} \tag{9.13}$$

where F_i *is defined in (9.9).*

Recalling Assumption 9.1 yields

$$\begin{aligned}
&-S_i[f_i(x_i,v_i,a_i,t) + w_i] \leq \Xi_i^*|S_i| \\
&-S_i\Delta u_i \leq \Psi_i^*|S_i|
\end{aligned} \tag{9.14}$$

where Ξ_i^* *and* Ψ_i^* *are defined as in (9.7).*

Then, substituting (9.8), (9.13) and (9.14) into $S_i\dot{S}_i$ yields

$$
\begin{aligned}
S_i\dot{S}_i &= -\lambda h_i K_D S_i[u_i+\Delta u_i+w_i+f_i(x_i,v_i,a_i,t)]+S_iF_i \\
&\leq -\lambda h_i K_D S_i u_i+\lambda h_i K_D(\Psi_i^*+\Xi_i^*)|S_i|+|S_iF_i| \\
&\leq -K_iS_i^2-\lambda h_i K_D[\Xi_i\tanh(\tfrac{S_i}{\varphi_{1i}})+\Psi_i\tanh(\tfrac{S_i}{\varphi_{2i}})]S_i \\
&\quad +\rho_i+\lambda h_i K_D(\Psi_i^*+\Xi_i^*)|S_i|
\end{aligned}
\tag{9.15}
$$

where the fact $-\frac{F_i^2S_i^2}{|F_iS_i|+\rho_i}\leq -|F_iS_i|+\rho_i$ *has been used.*

By using (9.10), we can obtain that $\dot{\Xi}_i+\sigma_{1i}\Xi_i\geq 0$*, which implies that* $\Xi_i\geq e^{-\int_0^t\sigma_{1i}(\tau)d\tau}\Xi_i(0)$*. This means that once we select* $\Xi_i(0)\geq 0$*, the adaptive law* Ξ_i *will always be non-negative. Similarly,* Ψ_i *is also non-negative. In addition, by using the adaptive laws (9.10), we can obtain the following results*

$$
\begin{aligned}
&-\tfrac{1}{\alpha_i}\tilde{\Xi}_i\dot{\Xi}_i-\tfrac{1}{\beta_i}\tilde{\Psi}_i\dot{\Psi}_i \\
&= -\lambda h_i K_D S_i\tilde{\Xi}_i\tanh(\tfrac{S_i}{\varphi_{1i}})+\tfrac{\sigma_{1i}}{\alpha_i}\tilde{\Xi}_i\Xi_i \\
&\quad -\lambda h_i K_D S_i\tilde{\Psi}_i\tanh(\tfrac{S_i}{\varphi_{2i}})+\tfrac{\sigma_{2i}}{\beta_i}\tilde{\Psi}_i\Psi_i \\
&= -\lambda h_i K_D S_i\Xi_i^*\tanh(\tfrac{S_i}{\varphi_{1i}})+\lambda h_i K_D S_i\Xi_i\tanh(\tfrac{S_i}{\varphi_{1i}}) \\
&\quad -\lambda h_i K_D S_i\Psi_i^*\tanh(\tfrac{S_i}{\varphi_{2i}})+\lambda h_i K_D S_i\Psi_i\tanh(\tfrac{S_i}{\varphi_{2i}}) \\
&\quad +\tfrac{\sigma_{1i}}{\alpha_i}\tilde{\Xi}_i\Xi_i+\tfrac{\sigma_{2i}}{\beta_i}\tilde{\Psi}_i\Psi_i.
\end{aligned}
\tag{9.16}
$$

Combining (9.15) with (9.16), one can obtain that

$$
\begin{aligned}
\dot{V} &= \sum_{i=1}^{N}[-K_iS_i^2+\lambda h_i K_D(\Psi_i^*+\Xi_i^*)|S_i|+\tfrac{\sigma_{1i}}{\alpha_i}\tilde{\Xi}_i\Xi_i \\
&\quad -\lambda h_i K_D S_i[\Xi_i^*\tanh(\tfrac{S_i}{\varphi_{1i}})+\Psi_i^*\tanh(\tfrac{S_i}{\varphi_{2i}})] \\
&\quad +\tfrac{\sigma_{2i}}{\beta_i}\tilde{\Psi}_i\Psi_i+\rho_i] \\
&\leq -K_iS_i^2+\tfrac{\sigma_{1i}}{\alpha_i}\tilde{\Xi}_i\Xi_i+\tfrac{\sigma_{2i}}{\beta_i}\tilde{\Psi}_i\Psi_i+\rho_i \\
&\quad +0.2785\lambda h_i K_D(\varphi_{1i}+\varphi_{2i})
\end{aligned}
\tag{9.17}
$$

where Lemma 2.5 has been used.

Based on Young's inequality in Lemma 2.7 of Chapter 2, one has

$$
\begin{aligned}
\frac{\sigma_{1i}}{\alpha_i}\tilde{\Xi}_i\Xi_i &\leq -\frac{\sigma_{1i}}{2\alpha_i}\tilde{\Xi}_i^2+\frac{\bar{\sigma}_{1i}}{2\alpha_i}\Xi_i^{*2} \\
\frac{\sigma_{2i}}{\beta_i}\tilde{\Psi}_i\Psi_i &\leq -\frac{\sigma_{2i}}{2\beta_i}\tilde{\Psi}_i^2+\frac{\bar{\sigma}_{2i}}{2\beta_i}\Psi_i^{*2}.
\end{aligned}
\tag{9.18}
$$

Then, from (9.11) and making use of (9.17)-(9.18), we have

$$
\begin{aligned}
\dot{V} &= \sum_{i=1}^{N}[-K_iS_i^2-\tfrac{\sigma_{1i}}{2\alpha_i}\tilde{\Xi}_i^2-\tfrac{\sigma_{2i}}{2\beta_i}\tilde{\Psi}_i^2 \\
&\quad +\tfrac{\bar{\sigma}_{1i}}{2\alpha_i}\Xi_i^{*2}+\tfrac{\bar{\sigma}_{2i}}{2\beta_i}\Psi_i^{*2}+0.2785\lambda h_i K_D(\varphi_{1i}+\varphi_{2i})+\rho_i] \\
&\leq -\chi_1V+\chi_2
\end{aligned}
$$

where the positive parameters χ_1 and χ_2 are given as follows

$$\begin{aligned}\chi_1 &= \min\{2K_i, \min_{t\to\infty}\min_{i\in\mathscr{V}_N}\sigma_{1i}, \min_{t\to\infty}\min_{i\in\mathscr{V}_N}\sigma_{2i}\} \\ \chi_2 &= \sum_{i=1}^{N}[\rho_i + 0.2785\lambda h_i K_D(\varphi_{1i}+\varphi_{2i}) + \tfrac{\bar{\sigma}_{1i}}{2\alpha_i}\Xi_i^{*2} + \tfrac{\bar{\sigma}_{2i}}{2\beta_i}\Psi_i^{*2}].\end{aligned} \tag{9.19}$$

Thus, by using Lemma 2.6, the following inequality holds:

$$V \le (V(0) - \tfrac{\chi_2}{\chi_1})e^{-\chi_1 t} + \tfrac{\chi_2}{\chi_1} \le V(0) + \tfrac{\chi_2}{\chi_1}, \tag{9.20}$$

where the fact $V(0) \geq 0$ has been used for the last inequality. Then, it is directly obtained that S_i, $\tilde{\Xi}_i$ and $\tilde{\Psi}_i$ must remain bounded. Therefore, one can conclude that all the signals in the closed-loop system remain uniformly ultimately bounded. Then, according to (8.16) and (8.17) of Chapter 8, the spacing error e_i can also be bounded. On the other hand, the proof of strong string stability is by following similar steps as in the proof of Theorem 8.1 in Chapter 8. Theorem 9.1 is thus proved. □

Remark 9.4 *In the derivation of Theorem 9.1, the adaptive compensation technique instead of neural networks (NNs) and fuzzy logic systems (FLSs) as in [51, 138] is introduced to attenuate the negative effects of asymmetric nonlinear actuator saturation, nonlinear uncertainties and external disturbances. Therefore, the disadvantage of requiring large number of on-line learning parameters for NNs or FLSs is removed in Theorem 9.1. At the same time, the assumption made in [57] that the nonlinearities of the consecutive vehicles (including the leader) should satisfy the Lipschitz constraints and should exactly match each other is also removed.*

9.4 Simulation Results

Numerical simulations for a vehicle platoon containing ten followers (i.e., $N = 10$) are executed to validate the main results obtained in the previous section. Consider that the dynamics of vehicles on road is described by (9.2) with $f_i(x_i, v_i, a_i, t)$ and $\text{sat}(u_i(t))$ given by

$$\begin{aligned} f_i(v_i, a_i, t) &= -\frac{1}{\tau_i}\left(a_i + \frac{\rho_{mi}H_i c_i}{2m_i}v_i^2 + \frac{d_{mi}}{m_i}\right) - \frac{\rho_{mi}H_i c_i v_i a_i}{m_i} \\ \text{sat}(u_i(t)) &= \begin{cases} 10, & \text{if } u_i(t) > 5 \\ (1-0.3\cos(u_i))u_i(t), & \text{if } 0 \le u_i(t) \le 5 \\ (1+0.3\sin(u_i))u_i(t), & \text{if } -10 \le u_i(t) < 0 \\ -15, & \text{if } u_i(t) < -10. \end{cases} \end{aligned}$$

Obviously, the assumption made in [19, 205] will not be satisfied. Therefore, the considered saturation will be more general than those considered in [19, 205]. In the simulations, the parameters of the vehicle platoon are the same as those in [43]. For clarity, the detailed parameters of all follower vehicles are listed in Table 9.1 and

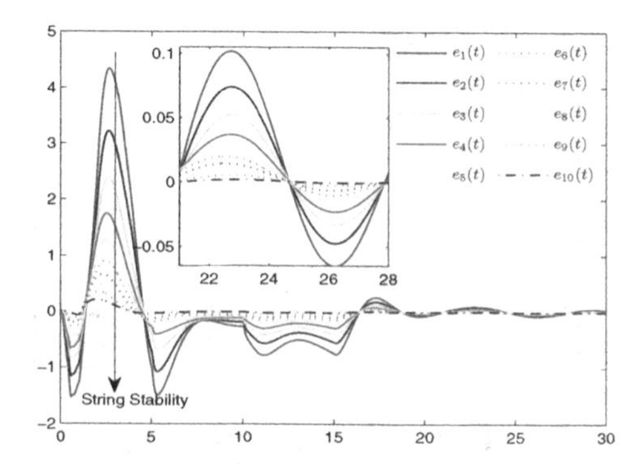

FIGURE 9.2
Responses of spacing error $e_i(t)$.

Table 9.3. In addition, the *external disturbances* and the desired acceleration of the leader are generated by

$$\begin{aligned} w_i &= 0.1\sin(t), i \in \mathscr{V}_N \\ a_0 &= \begin{cases} 0.5t\ m/s^2, & 1s \le t < 5s \\ 2.5\ m/s^2, & 5s \le t < 10s \\ -0.5t + 7.5\ m/s^2, & 10s \le t < 15s \\ 0\ m/s^2, & otherwise. \end{cases} \end{aligned}$$

The controller design parameters and adaptive gains are listed in Table 9.2. In addition, the initial spacings, velocities and accelerations of the leader and the followers are listed in Table 9.4. From *MCTH policy* in (8.12), the initial errors $\tilde{e}_i(0)$, $\dot{\tilde{e}}_i(0)$ and $\ddot{\tilde{e}}_i(0)$ are non-zero. Our simulation results are presented in Figures 9.2-9.9. The spacing error responses of the vehicles over time are shown in Figure 9.2, where the *strong string stability* can be guaranteed despite the asymmetric nonlinear actuator saturations. It follows from Figure 9.3 that neighboring-vehicle collisions are avoided as there do not exist negative distances. It is observed from Figure 9.4 and Figure 9.5 that the velocities and accelerations of all follower vehicles eventually track those of the leader. The saturation control input $\text{sat}(u_i(t))$ and the sliding surface $S_i(t)$ are shown in Figure 9.6 and Figure 9.7, respectively. Figure 9.8 and Figure 9.9 show that signals $\Xi_i(t)$ and $\Psi_i(t)$ are bounded. These simulation results validate the effectiveness of the proposed control scheme.

9.5 Conclusion

This chapter addresses the problem of adaptive PIDSMC for third-order vehicle platoon subject to *external disturbances* and unknown asymmetric actuator saturation.

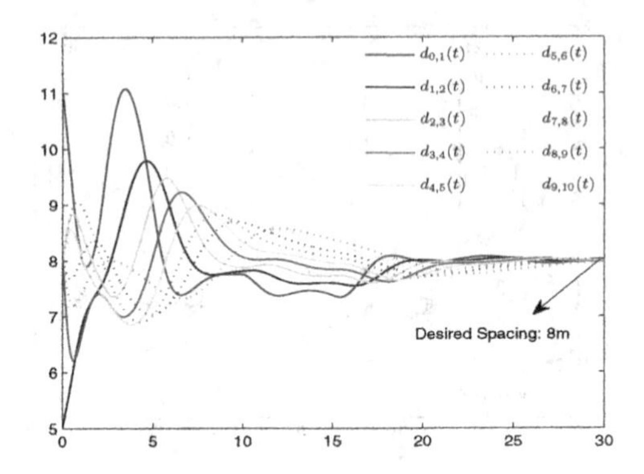

FIGURE 9.3
The real distance $d_{i-1,i}$ between consecutive vehicles.

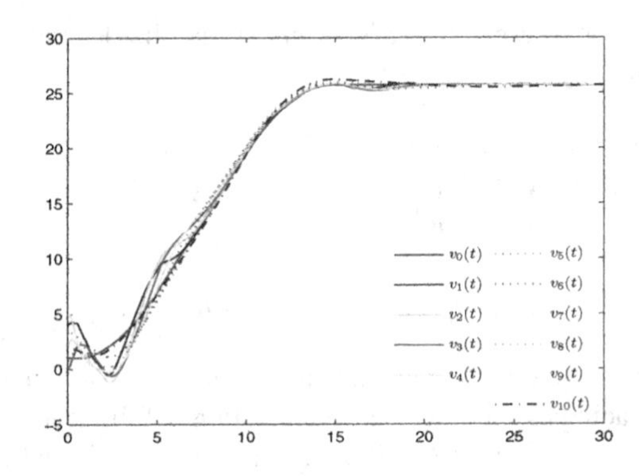

FIGURE 9.4
Responses of velocity $v_i(t)$.

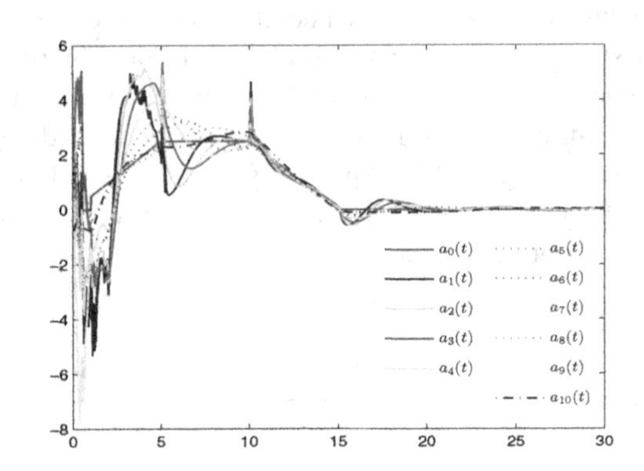

FIGURE 9.5
Responses of acceleration $a_i(t)$.

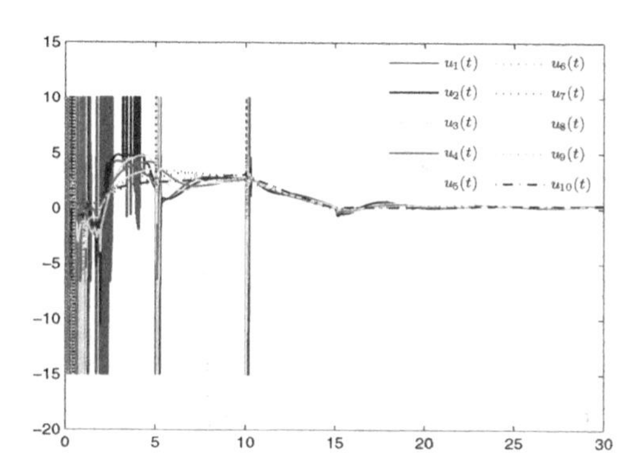

FIGURE 9.6
Responses of control input $u_i(t)$.

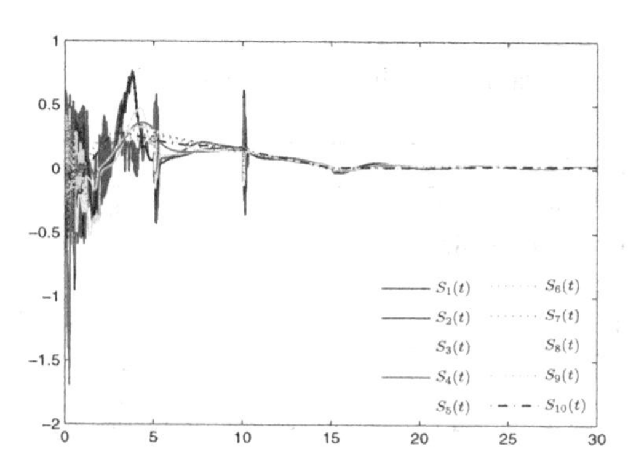

FIGURE 9.7
Responses of sliding surface $S_i(t)$.

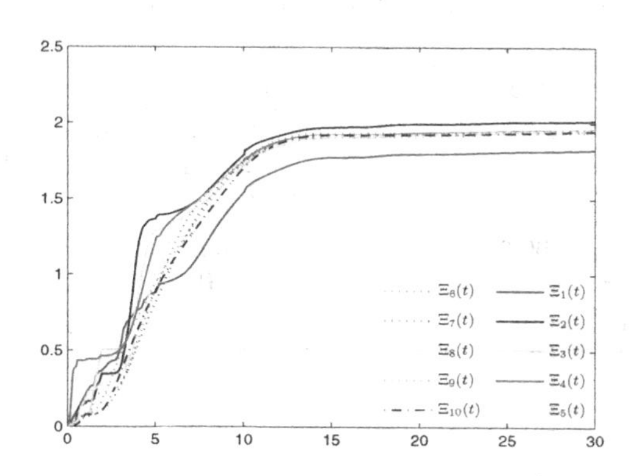

FIGURE 9.8
Responses of adaptive updated $\Xi_i(t)$.

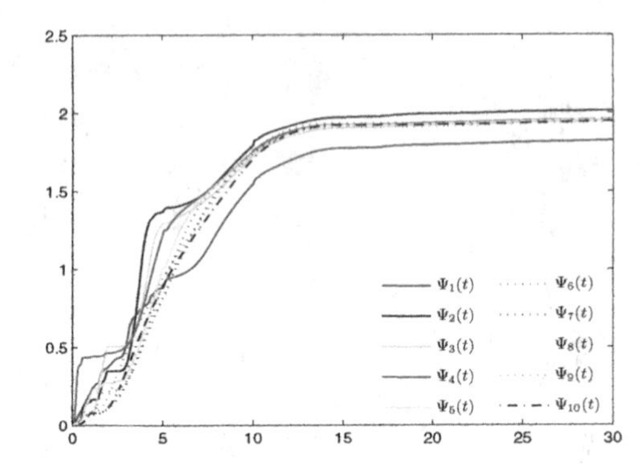

FIGURE 9.9
Responses of adaptive updated $\Psi_i(t)$.

TABLE 9.1
Simulation parameters' values of vehicle dynamics for $i \in \mathscr{V}_N$

Parameter Name	Simulation Values
Desired vehicle spacing	$\Delta_{i-1,i} = 8m$
Specific mass of the air	$\rho_{mi} = 1N/m^3$
Cross-sectional area	$H_i = 2.2m^2$
Drag coefficient	$c_i = 0.35$
Mechanical drag	$d_{mi} = 150N$
Headway time	$h_i = 1s$

TABLE 9.2
Controller and adaptive designed parameters for $i \in \mathscr{V}_N$

Parameter Name	Simulation Values
Controller parameters	$K_i = 3$, $\lambda = 0.75$, $\pi_i = 5$
PID parameters	$K_P = 2$, $K_I = 3$, $K_D = 1$
Initial adaptive values	$\Xi_i(0) = 0, \Psi_i(0) = 0$
Adaptive parameters 1	$\alpha_i = 2, \beta_i = 5$
Adaptive parameters 2	$\sigma_{1,i} = 0.1e^{-10t}$, $\sigma_{2,i} = 0.1e^{-20t}$
Tuning parameters	$\varphi_{1,i} = \varphi_{2,i} = 0.3$, $\rho_i = 0.3$

Considering actuator physical constraints, *adaptive compensation* in combination with the sliding mode technique and Lyapunov synthesis has been proposed. As an extension of recent results, a *MCTH policy* is introduced to improve the traffic

TABLE 9.3
The parameters $m_i(kg)$, $\tau_i(s)$ and $L_i(m)$ for $i \in \mathcal{V}_N$ [43]

i	1	2	3	4	5	6	7	8	9	10
m_i	1500	1600	1550	1650	1500	1400	1550	1550	1700	1600
τ_i	0.1	0.3	0.2	0.4	0.25	0.4	0.3	0.15	0.35	0.2
L_i	4	4.5	5	5	4.5	3.5	4	4	5	4.5

TABLE 9.4
The initial values $x_i(0)(m)$, $v_i(0)(m/s)$ and $a_i(0)(m/s^2)$ for $i \in \mathcal{V}_N \cup \{0\}$

i	0	1	2	3	4	5	6	7	8	9	10
$x_i(0)$	150	135	125.5	112.5	99.5	87	75.5	63.5	51.5	38.5	26
$v_i(0)$	1	4	2	0	5	3	1	2	1	3	2
$a_i(0)$	0	1	5	2	1	3	1	2	1	2	−1

density and remove the common assumption that the initial spacing, velocity and acceleration errors must be zero. It has proved that the proposed control scheme is able to guarantee the *strong string stability* of the whole vehicle platoon and the convergence of the spacing, velocity and acceleration errors. Finally, simulation studies have been presented to illustrate the effectiveness of the proposed adaptive control scheme.

10

Non-Fragile Quantized Consensus for Multi-Agent Systems Based on Incidence Matrix

This chapter concerns the *quantization* effect on a non-fragile control law for Lipschitz nonlinear *multi-agent systems* with *quantization* information under undirect communication graphs. Quantized insensitive consensus control laws are implementable with quantized values of the relative states by using *incidence matrix* instead of adjacency matrix. The designed controllers are non-fragile to additive interval-bounded controller coefficient variations, where a new technique by converting *interval-bounded variation* format to norm-bounded uncertain format is proposed to solve numerical problems caused by *interval-bounded variations* effectively. By taking advantage of tools from non-smooth analysis, explicit convergence results are derived for both uniform and *logarithmic quantizers* with the aid of *linear matrix inequality* (LMI) technique. It is pointed out that finite time practical consensus can be achieved when uniform quantizers are used, while consensus can be achieved under some conditions when a logarithmic quantizer is considered. Sufficient conditions of the existence of the proposed control strategy are also obtained by using matrix inequality techniques.

10.1 Introduction

Consensus is an important problem in the area of cooperative control of *multi-agent systems* [77, 110, 111, 113, 181] and has wide application background in areas such as formation flight, traffic control, networked control and flocking problems [100, 103, 105]. When digital communications are adopted, due to the finite channel capacity [100, 103, 105], and the widespread application of analog-to-digital and digital-to-analog converters in sensors and actuators [210, 211], *quantization* plays an important role in information exchange among agents [174]. Therefore, quantized consensus or consensus with quantized communications becomes interesting and more meaningful, and has drawn the attention of more and more researchers from various perspectives [28, 40, 39, 87, 103, 105, 117, 188]. In [103], the authors considered discrete-time distributed average-consensus with limited communication datarate and time-varying communication topologies for first-order integrator systems, and was then extended to second-order integrator systems in [105]. In [87], average-consensus algorithms based on the assumption that the states of agents were integer-valued was designed for first-order integrator systems. In [28], the state agreement problem for first-order integrator systems with quantized relative information under undirect communication graphs was investigated. In [39], the authors studied the adaptive coordinated tracking problem for continuous-time first-order integrator systems with quantized information. The authors in [117] considered the *quantization* effects on synchronized motion of teams of mobile agents with second-order dynamics. Note that the above results about the *quantization* effect on continuous-time consensus problems mostly dealt with first-order or second-order integrator systems. Recently, the synchronization of high order linear systems with both uniform quantizer and logarithmic quantizer under *undirected graph* was studied in [188]. The authors in [40] extended the results to the case of directed communication graphs. Note that most of the previously mentioned works on the quantized control for *multi-agent systems* mainly focused on the consensus problem for linear systems. However, in practice, it is almost impossible to get an exact mathematical model of a dynamic system due to the complexity of the system, the difficulty of measuring various parameters, environmental noises, etc. Therefore, the quantized consensus of nonlinear *multi-agent systems* is important and challenging in both theory and practice. Up to now, very few works have addressed quantized consensus of high-order nonlinear *multi-agent systems* by using the *incidence matrix*, for instance, [41, 139], and many additional aspects of the topic are yet to be fully investigated.

On the other hand, one common feature in the above-mentioned results is that the researchers are all based on a common assumption that all controllers can be implemented exactly. However, inaccuracies and uncertainties do occur in controller implementation due to finite word length (FWL) and numerical roundoff errors, etc. [61]. Consequently, how to design a controller that is insensitive or *non-fragile* to some amount of error with respect to its coefficients is of practical significance for the implementation of digital controllers because even vanishingly small perturbations in the control coefficients may destabilize the closed-loop systems [89]. As a result, some profound results have been obtained and reported in the literature [134, 184], in which the *non-fragile* synchronisation control for multi-agent systems with respect to additive norm-bounded controller coefficient variations have been investigated. However, since norm-bounded controller coefficient variations cannot exactly reflect the uncertain information due to FWL effects, *non-fragile* control problems with respect to interval-bounded coefficient variations for linear systems have received more and more attention [13, 29, 50, 190, 191]. Many effective methods such as the structured vertex separator method [190], randomized algorithm [29], the sparse structure method [13] and the transfer function sensitivity approach [50, 191] have been proposed to deal with the numerical problem caused by interval-bounded coefficient variations. However, the number of *linear matrix inequality* (LMI) constraints involved in the design conditions by using the first three methods is still large, while the fourth method involves the closed-loop system's transfer function and cannot be applied to multi-agent systems with *quantization* information. Therefore, new design tools and analysis methods are needed.

Inspired by the above pioneering work, this chapter is concerned with *non-fragile* quantized consensus for *Lipschitz* nonlinear multi-agent systems for both the cases of uniform quantizer and logarithmic quantizer under undirect communication graphs. The main contributions of this chapter are threefold.

(1) In contrast to [41, 139], because of the simultaneous existence of input *quantization* and additive interval-bounded controller coefficient variations, the *quantized consensus* problem in this case becomes quite challenging to solve. In addition, measurement quantization with only one case of uniform quantizer is considered in [139], while a different type of nonlinear system is considered in [41], which make the design tools and analysis methods in [41, 139] much different from those in this chapter.

(2) In contrast to the existing results in [13, 29, 50, 190, 191], a new technique by converting *interval-bounded variation* format to norm-bounded uncertain format is proposed to solve the type of additive interval-bounded controller coefficient variations, thereby reducing the computational complexity significantly, and a new lemma (Lemma 2.10 as in Chapter 2) is also proposed.

(3) By using LMI technique, a distributed consensus control is proposed based on the relative states of neighboring agents by using *incidence matrix* instead of adjacency matrix. Using the Lyapunov method and non-smooth analysis, it is shown that the original multi-agent system can reach finite-time practical consensus when a uniform quantizer is used, while the given systems can reach consensus under some conditions when a logarithmic quantizer is considered. Then, sufficient conditions of the existence of the proposed protocols are derived, and the gain of controllers can be computed by solving matrix inequalities.

10.2 Uniform Quantizer and Logarithmic Quantizer

Two classes of quantized sensors are considered: *uniform* and *logarithmic quantizers*. The definitions of uniform quantizer and logarithmic quantizer are introduced as follows. A uniform quantizer is a map $q_u : \mathbf{R} \rightarrow \mathbf{R}$ such that

$$q_u(x) = \Delta\lfloor \frac{x}{\Delta} + \frac{1}{2} \rfloor \tag{10.1}$$

where Δ is a positive number and $\lfloor \alpha \rfloor$ denotes the greatest integer that is less than or equal to α. For a uniform quantizer, the quantization error is always bounded by $\frac{\Delta}{2}$, i.e., $|q_u(x) - x| \leq \frac{\Delta}{2}$ for all $x \in \mathbf{R}$.

A logarithmic quantizer is an odd map q: $\mathbb{R} \to \mathbb{R}$ defined by

$$q_l(x) = \begin{cases} exp\left(\Delta\lfloor \frac{ln(x)}{\Delta} + \frac{1}{2}\rfloor\right), x > 0 \\ 0, \ x = 0 \\ -q_l(-x), x < 0 \end{cases} \tag{10.2}$$

where $\Delta > 0$ is the quantizer parameter and $\lfloor \alpha \rfloor$ denotes the greatest integer that is less than or equal to α. The quantization error for the logarithmic quantizer satisfies

$$|q_l(x) - x| \leq \Delta_l |x|$$

where the parameter $\Delta_l = 1 - e^{-\frac{\Delta}{2}}$.

Remark 10.1 *It can easily generalize the above scalar-valued definitions of quantizers to vector-valued case. For example, for a vector* $v = [v_1, \cdots, v_d]^T \subset \mathbf{R}^d$ *of size* d*, we define* $q_u(v) \triangleq [q_u(v_1), \cdots, q_u(v_d)]^T$ *and* $q_l(v) \triangleq [q_l(v_1), \cdots, q_l(v_d)]^T$*. Then, the following bounds also hold* $|q_u(v) - v| \leqslant \sqrt{d}\frac{\Delta}{2}$, $|q_l(v) - v| \leqslant \Delta_l |v|$ *[28, 40, 117].*

Remark 10.2 *The above two quantizers are most commonly used in practice and have very distinct characteristics. Precise control is usually not achievable with uniformly quantized information, while it is possible to achieve precise consensus under a logarithmic quantizer [40].*

10.3 Problem Formulation

A multi-agent system $\mathscr{G}$ composed of N interconnected nonlinear continuous time subsystems $\mathscr{G}_i, i = 1, \cdots, N$ is considered. Each edge corresponds to an available information link from one subsystem to another. Suppose, dynamics of the i-th agent are described by

$$\dot{x}_i \quad = f(x_i) + Bu_i, i = 1, \cdots, N \tag{10.3}$$

where $x_i \in R^n$ is the state of agent subsystem i, $u_i \in R^p$ is the control input; $f(x_i)$ is continuously differentiable and can be divided into linear and nonlinear parts as follows

$$f(x_i) \quad = Ax_i(t) + A_g g(x_i), i = 1, \cdots, N \tag{10.4}$$

where A and A_g are real constant matrices with appropriate dimensions and the nonlinear function $g(x_i)$ is assumed to satisfy the *Lipschitz* constraint with a *Lipschitz* constant $\gamma > 0$, i.e.,

$$\|g(x) - g(y)\| \quad \leq \gamma \|x - y\|, \forall x, y \in R^n. \tag{10.5}$$

Remark 10.3 *Many nonlinearities can be regarded as* Lipschitz*, at least locally. Even terms like* x^2 *can regarded as* Lipschitz*, provided we know that the operating* *range of x is bounded [146]. Thus, the class of systems covered by this chapter is fairly general. In addition, the function $g(\cdot)$ satisfying (10.5) can be used to represent some practical nonlinearities, for instance, trigonometric nonlinearities occurring in many robotic problems, nonlinear softening spring models frequently used in mechanical systems, nonlinearities which are square or cubic in nature [206], nonlinearities in a single-link flexible joint robot [147, 216], and so on.*

Our objective is to develop *non-fragile* quantized control laws that are implementable with local information, i.e., the ith member can use the information $x_i - x_j$ if ith and jth members are neighbors connected by edge k $(k = 1, \cdots, M)$, where the difference variable is defined as follows

$$z_k := \sum_{i=1}^{N} d_{ik} x_i = \begin{cases} x_i - x_j & \text{if } i\text{th node is the positive end} \\ x_j - x_i & \text{if } i\text{th node is the negative end} \end{cases}$$

By the definition of *incidence matrix* $\mathscr{D}$ in Chapter 2, the variable z can be concisely represented as

$$z = (\mathscr{D}^T \otimes I_n)x$$

where $z = \left[z_1^T, \cdots, z_M^T\right]^T$, $x = \left[x_1^T, \cdots, x_N^T\right]^T$.

The non-fragile *quantized consensus* control law is designed as

$$\begin{aligned} u_i &= -c(K + \Delta_K) \sum_{k=1}^{M} d_{ik} q(z_k), i = 1, \cdots, N \\ \text{for } q(z_k) &= \begin{cases} q_u(z_k), & \text{Uniform Quantizer} \\ q_l(z_k), & \text{Logarithmic Quantizer} \end{cases} \end{aligned} \tag{10.6}$$

where $c > 0$ is the coupling weight among neighboring agents, $K \in R^{p \times n}$ is the feedback gain matrix, d_{ik} is (i,k)-th entry of the *incidence matrix* $\mathscr{D}$ associated with $\mathscr{G}$, and Δ_K represents additive interval-bounded coefficient variations with the following form

$$\Delta_K = [\delta_{lj}]_{p \times n}, j = 1, \cdots, n; l = 1, \cdots, p \tag{10.7}$$

where δ_{lj}, (for $j = 1, \cdots, n; l = 1, \cdots, p$) are used to denote the deviation of the controller coefficients of K with the following form:

$$|\delta_{lj}| \leq \delta_k, \text{ for } j = 1, \cdots, n; l = 1, \cdots, p.$$

with δ_k being the maximum possible deviation of the controller coefficients.

Remark 10.4 *When quantization effect and controller coefficient variations are not considered, a similar problem has been investigated in [110, 181], where a consensus protocol based on the relative states between neighboring agents is given as*

$$u_i = cK \sum_{j=1}^{N} a_{ij}(x_i - x_j), i = 1, \cdots, N$$

where a_{ij} is (i, j)-th entry of the adjacency matrix $\mathscr{A}$ associated with $\mathscr{G}$, defined as

$a_{ii}=0$, $a_{ij}=a_{ji}=1$ if $(i,j)\in\mathscr{E}$ and $a_{ij}=a_{ji}=0$ otherwise. It is noted that the incidence matrix $\mathscr{D}$ instead of adjacency matrix $\mathscr{A}$ is introduced in this chapter. The properties of incidence matrix $\mathscr{D}$ will facilitate the non-fragile quantized control synthesis.

It is noted that Δ_K in (10.7) can further be described as $\Delta_K=\sum\limits_{j=1}^{n}\sum\limits_{l=1}^{p}\delta_{lj}g_le_j^T$, where $e_k\in R^n$ and $g_k\in R^p$ denote the column vectors in which the kth element equals 1 and the others equal 0. In fact, Δ_K can also be rewritten as

$$\Delta_K=E\Re F \tag{10.8}$$

where

$$\begin{aligned}&E=\begin{bmatrix}E_1 & E_2 & \cdots & E_\mu & \cdots & E_{qn}\end{bmatrix}, F^T=\begin{bmatrix}F_1^T & F_2^T & \cdots & F_\mu^T & \cdots & F_{pn}^T\end{bmatrix}\\&\Re=diag\{\delta_{11},\delta_{12},\cdots,\delta_{lj},\cdots,\delta_{pn}\}\end{aligned}$$

with

$$E_\mu=g_l, F_\mu=e_j^T, \text{ for } \mu=(l-1)n+j; l=1,\cdots,p; j=1,\cdots,n.$$

Remark 10.5 *For example, let* $\Delta_K=\begin{bmatrix}\delta_{11} & \delta_{12} & \delta_{13}\\ \delta_{21} & \delta_{22} & \delta_{23}\end{bmatrix}$, *then,*

$$\begin{aligned}&g_1=\begin{bmatrix}1\\0\end{bmatrix}, g_2=\begin{bmatrix}0\\1\end{bmatrix}, e_1^T=\begin{bmatrix}1 & 0 & 0\end{bmatrix}\\&e_2^T=\begin{bmatrix}0 & 1 & 0\end{bmatrix}, e_3^T=\begin{bmatrix}0 & 0 & 1\end{bmatrix}\end{aligned}$$

and E, F and $\Re$ in this case can be given as follows

$$\begin{aligned}&E=\begin{bmatrix}1 & 1 & 1 & 0 & 0 & 0\\ 0 & 0 & 0 & 1 & 1 & 1\end{bmatrix}, F=\begin{bmatrix}I_3\\ I_3\end{bmatrix}\\&\Re=diag\{\delta_{11},\delta_{12},\delta_{13},\delta_{21},\delta_{22},\delta_{23}\}.\end{aligned}$$

Remark 10.6 *It is worth mention that the control laws which do not account for possible coefficient variations in themselves often do not work well with real systems because even vanishingly small perturbations in the control coefficients may destabilize the closed-loop systems [50, 191]. This fact can be found in Figure 3 and Figure 5 in Section 10.5, Numerical Example. Therefore, the controllers designed in [41, 139] are necessary to consider the effect of coefficient variations in controllers themselves. In addition, measurement quantization is considered in [139], while a different type of nonlinear system is considered in [41], which make the design tools and analysis methods in [41, 139] much different from this chapter. The main difference of the procedure proposed in this chapter compared to the procedures developed by [41, 139] is that the problem considered in this chapter is solved very efficiently by the convex optimization technique with the aid of the LMI technique.*

Remark 10.7 *As pointed out in [190], the type of norm-bounded controller coefficient variations investigated in [61, 134, 184] cannot exactly reflect the uncertain information due to FWL effects. Therefore, the type of interval-bounded controller coefficient variations is investigated in this chapter. In addition, by using the transformation as in (10.8), the computational complexity can be reduced significantly compared to previous methods in [13, 29, 50, 190, 191].*

By the *Kronecker product*, the control law (10.6) can be rewritten as:

$$u = -c[I_N \otimes \tilde{K}](\mathscr{D} \otimes I_n)q((\mathscr{D}^T \otimes I_n)x). \tag{10.9}$$

where $\tilde{K} = K + \Delta_K$.

Then, it is not difficult to obtain that the closed-loop network dynamics can be written as follows

$$\dot{x} = (I_N \otimes A)x + (I_N \otimes A_g)G(x) - [cI_N \otimes B\tilde{K}](\mathscr{D} \otimes I_n)q((\mathscr{D}^T \otimes I_n)x) \tag{10.10}$$

where $G(x) = \left[g(x)_1^T, \cdots, g(x)_N^T\right]^T$.

Define $\xi = (\Pi \otimes I_n)x$, where $\Pi = I_N - \frac{1}{N}\mathbf{1}\mathbf{1}^T$ satisfies $\Pi \cdot \Pi = \Pi$, and $\xi = \left[\xi_1^T, \cdots, \xi_N^T\right]^T$. It is easy to see that 0 is a simple eigenvalue of Π with $\mathbf{1}$ as a corresponding eigenvector and 1 is the other eigenvalue with multiplicity $N-1$. From $\mathbf{1}_N^T \mathscr{D} = 0$, one can obtain that

$$\begin{aligned} \mathscr{D}^T \Pi &= \mathscr{D}^T \\ (\mathscr{D}^T \otimes I_n)x &= (\mathscr{D}^T \Pi \otimes I_n)x = (\mathscr{D}^T \otimes I_n)(\Pi \otimes I_n)x = (\mathscr{D}^T \otimes I_n)\xi. \end{aligned}$$

Then, (10.10) can be rewritten as

$$\begin{aligned} \dot{\xi} &= (\Pi \otimes I_n)\dot{x} \\ &= (\Pi \otimes I_n)[(I_N \otimes A)x + (I_N \otimes A_g)G(x) - [cI_N \otimes B\tilde{K}](\mathscr{D} \otimes I_n)q((\mathscr{D}^T \otimes I_n)x)] \\ &= (I_N \otimes A)(\Pi \otimes I_n)x + (\Pi \otimes A_g)G(x) - (c\Pi\mathscr{D} \otimes B\tilde{K})q((\mathscr{D}^T \otimes I_n)x) \\ &= (I_N \otimes A)\xi + (\Pi \otimes A_g)G(x) - [cI_N \otimes B\tilde{K}](\mathscr{D} \otimes I_n)q((\mathscr{D}^T \otimes I_n)\xi). \end{aligned} \tag{10.11}$$

Remark 10.8 *When quantization effect is not considered, it is easy to verify that* $\xi = 0$ *if and only if* $x_1 = x_2 = \cdots = x_N$*, thus the consensus problem can be reduced to the asymptotical stability of* ξ*. When quantization effect is considered, it is possible to achieve precise consensus under a logarithmic quantizer. However, due to the uniform quantization effect, complete consensus connot be ensured, but only practical consensus can be achieved. That is, all the states of agents enter into a ball and never flow out, and the radius of this ball can be estimated. Hereafter, we refer to* ξ *as the consensus error.*

Since the quantizers introduce discontinuities into the closed-loop system, the system (10.10) has a discontinuous right-hand side. Therefore, a Filippov definition is introduced to analyze the property of the solution of given differential equation.

Definition 10.1 *[24] For a vector differential equation $\dot{x} = f(t,x)$, $f(t,x) : \mathbb{R} \times \mathbb{R}^d \to \mathbb{R}^d$ is not necessarily discontinuous. A vector function x is said to be a Filippov solution of the differential equation if x is absolutely continuous and satisfies the differential inclusion $\dot{x} \in \mathscr{F}\{f(x,t)\}$ for almost every time t, where*

$$\mathscr{F}\{f(x,t)\} = \bigcap_{r>0} \bigcap_{\mu(s)=0} \overline{co}\{f(t, \mathscr{B}(x,r) \backslash S)\}$$

with $\overline{co}$ denoting the convex closure, $\mathscr{B}(x,r)$ denoting the Euclidean ball of radius r centered at x and $\bigcap_{\mu(s)=0}$ denoting the intersection over all sets S of Lebesgue measure zero.

With the definition of Filippov solution, the differential inclusion associated with (10.10) is

$$\dot{\xi} \in (I_N \otimes A)\xi + (\Pi \otimes A_g)G(x) - [cI_N \otimes B\tilde{K}](\mathscr{D} \otimes I_n) \cdot \mathscr{F}\{q((\mathscr{D}^T \otimes I_n)\xi)\}. \tag{10.12}$$

To analyze the Filippov solutions to the discontinuous system (10.12), the following lemma is useful.

Lemma 10.1 *[156] Let $x(t)$ be a Filippov solution to $\dot{x} = f(x)$ on an interval containing t and $V(x)$: $R^n \to R$ be continuously differentiable. Then $V(x)$ is absolutely continuous and $\dot{V}(x)$ exists almost everywhere satisfying $\dot{V}(x) \in (\nabla V)^T \mathscr{F}\{f(x)\}$.*

10.4 Non-Fragile Quantized Controller Design

In this section, multi-agent system (10.3) with undirect communication graph using the given consensus protocol (10.9) under quantized relative state information is considered.

10.4.1 Non-Fragile Control with Uniform Quantization

In this subsection, we consider multi-agent systems (10.10) using the given non-fragile quantized consensus protocol (10.6) under uniform quantization (10.1).

To study the non-fragile control of the consensus algorithm to quantized relative information, we introduce a Lyapunov function as follows:

$$V(\xi) = \frac{1}{2}\xi^T(\mathscr{L} \otimes P^{-1})\xi \tag{10.13}$$

where P is a symmetric and positive definite matrix $P = P^T > 0$.

From Lemma 2.1 of Chapter 2, one can infer that $V(\xi)$ in (10.13) is a positive semi-definite function. By the definition of ξ, it is easy to obtain that $(\mathbf{1}^T \otimes I)\xi = 0$. For a connected graph $\mathscr{G}$, as shown in [110], it follows from Lemma 2.1 that

$$V(\xi) \geq \tfrac{1}{2}\lambda_2 \xi^T(I \otimes P^{-1})\xi \geq \tfrac{\lambda_2}{2\lambda_{max}(P)}\|\xi\|^2,$$

on the other hand, one can obtain that

$$V(\xi) \leq \frac{1}{2}\lambda_N \xi^T (I \otimes P^{-1})\xi \leq \frac{\lambda_N}{2\lambda_{min}(P)} \|\xi\|^2.$$

Therefore, we have

$$c_1\|\xi\|^2 \leq V(\xi) \leq c_2\|\xi\|^2 \tag{10.14}$$

where

$$c_1 = \frac{\lambda_2}{2\lambda_{max}(P)}, \; c_2 = \frac{\lambda_N}{2\lambda_{min}(P)}.$$

The following theorem provides an estimate of the region where the solutions converge. It is shown that practical consensus under uniform quantizers can be achieved.

Theorem 10.1 *Let the communication graph $\mathscr{G}$ be undirected and connected. Solving the following matrix inequality*

$$\begin{bmatrix} AP + PA^T - 2\tau BB^T + \mu A_g A_g^T & P & \tau BE\Lambda & \tau \delta_k PF^T \\ * & -\mu\gamma^{-2} I & 0 & 0 \\ * & * & -\tau\Lambda & 0 \\ * & * & * & -\tau\Lambda \end{bmatrix} < 0 \tag{10.15}$$

to get a positive matrix $P = P^T > 0$, a diagonal matrix $\Lambda > 0$, scalars $\tau > 0$ and $\mu > 0$. The parameters in the consensus protocol (10.9) are designed as

$$c \geq \frac{\tau}{\lambda_2}, K = B^T P^{-1} \tag{10.16}$$

where λ_2 is the smallest non-zero eigenvalue of the Laplacian matrix *$\mathscr{L}$. Then, practical consensus can be achieved and for any Filippov solution x to (10.10) there exists a time $T > 0$ such that*

$$\|x - \frac{(\mathbf{1}\mathbf{1}^T \otimes I_n)x}{N}\| \leq \sqrt{\frac{c_2}{c_1}} \frac{\|c\mathscr{L}\| \|P^{-1}\| \|B\| (\|P^{-1}\| \|B\| + \delta_k \|E\| \|F\|) \|\mathscr{D}\| \sqrt{n} M \Delta}{2\alpha c_1 (1-\theta)} \tag{10.17}$$

where c_1 and c_2 are denoted in (10.14), Δ is quantizer parameter, M is the number of edges for graph $\mathscr{G}$ and $\theta > 0$ is a constant which can be chosen as small as possible. Moreover, α is given out as

$$\alpha = -\lambda_{max}(AP + PA^T - 2\tau BB^T + \tau\Omega + \mu A_g A_g^T + \mu^{-1}\gamma^2 P^2)/\lambda_{max}(P) \tag{10.18}$$

with $\Omega = BE\Lambda E^T B^T + \delta_k^2 PF^T \Lambda^{-1} FP$.

Proof 10.1 *Consider the Lyapunov function $V(\xi)$ in (10.13) with symmetric and positive definite matrix $P = P^T > 0$ being a solution of (10.15). From Lemma 10.1,* *for almost every t, there exists* $\zeta \in \mathscr{F}\{q((\mathscr{D}^T \otimes I_n)\xi)\}$ *such that the derivative of* $V(\xi)$ *along the solution of (10.12) satisfies*

$$\begin{aligned}\dot{V} &= \xi^T(\mathscr{L}\otimes P^{-1})[(I_N\otimes A)\xi + (\Pi\otimes A_g)G(x) - (cI_N\otimes B\tilde{K})(\mathscr{D}\otimes I_n)\cdot\zeta] \\ &= \xi^T(\mathscr{L}\otimes P^{-1}A)\xi + \xi^T(\mathscr{L}\otimes P^{-1})(\Pi\otimes A_g)G(x) - \xi^T(c\mathscr{L}\otimes P^{-1}B\tilde{K})(\mathscr{D}\otimes I_n)\zeta \\ &= \xi^T[(\mathscr{L}\otimes P^{-1}A) - (\mathscr{L}\otimes P^{-1})(cI_N\otimes\tilde{K})(\mathscr{L}\otimes I_n)]\xi \\ &\quad + \xi^T(\mathscr{L}\otimes P^{-1})(\Pi\otimes A_g)G(x) + \xi^T(c\mathscr{L}\otimes P^{-1}B\tilde{K})(\mathscr{D}\otimes I_n)((\mathscr{D}^T\otimes I_n)\xi - \zeta) \\ &= \xi^T(\mathscr{L}\otimes P^{-1}A - c\mathscr{L}^2\otimes P^{-1}B\tilde{K})\xi + \xi^T(\mathscr{L}\otimes P^{-1}A_g)G(x) \\ &\quad + \xi^T(c\mathscr{L}\otimes P^{-1}B\tilde{K})(\mathscr{D}\otimes I_n)((\mathscr{D}^T\otimes I_n)\xi - \zeta)\end{aligned}$$

where $\tilde{K} = K + \Delta_K$.

Note that

$$\begin{aligned}\xi^T(\mathscr{L}\otimes P^{-1}A_g)G(x) &= \xi^T(\mathscr{L}\otimes P^{-1}A_g)(\Pi\otimes I_n)G(x) \\ &= \xi^T(\mathscr{L}\otimes P^{-1}A_g)(G(x) - \mathbf{1}\otimes g(\bar{x}) + \mathbf{1}\otimes g(\bar{x}) - \tfrac{1}{N}\mathbf{1}\mathbf{1}^T G(x)) \\ &= \xi^T(\mathscr{L}\otimes P^{-1}A_g)(G(x) - \mathbf{1}\otimes g(\bar{x}))\end{aligned}$$

where $\bar{x} = \frac{1}{N}(\mathbf{1}^T\otimes I)x$ *and we have used* $\mathscr{L}\mathbf{1} = 0$ *to get the last equation.*

In addition, in terms of properties of uniform quantizer and differential inclusion, it can be implied that any $\zeta \in \mathscr{F}\{q((\mathscr{D}^T\otimes I_n)\xi)\}$ *satisfies* $\|\zeta - (\mathscr{D}^T\otimes I_n)\xi\| \le \sqrt{nM}\frac{\Delta}{2}$, *then we have*

$$\begin{aligned}\dot{V} &= \tfrac{1}{2}\xi^T[\Xi\xi + 2(\mathscr{L}\otimes P^{-1}A_g)(G(x) - \mathbf{1}\otimes g(\bar{x}))] \\ &\quad + \xi^T(c\mathscr{L}\otimes P^{-1}B\tilde{K})(\mathscr{D}\otimes I_n)((\mathscr{D}^T\otimes I_n)\xi - \zeta) \\ &\le \tfrac{1}{2}\xi^T[\Xi\xi + 2(\mathscr{L}\otimes P^{-1}A_g)(G(x) - \mathbf{1}\otimes g(\bar{x}))] \\ &\quad + \|\xi\|\|c\mathscr{L}\|\|P^{-1}\|\|B\|(\|P^{-1}\|\|B\| + \delta_k\|E\|\|F\|)\|\mathscr{D}\|\sqrt{nM}\tfrac{\Delta}{2}\end{aligned} \tag{10.19}$$

where

$$\begin{aligned}\Xi = \mathscr{L}\otimes(P^{-1}A + A^TP^{-1}) - 2c\mathscr{L}^2\otimes P^{-1}BB^TP^{-1} \\ - c\mathscr{L}^2\otimes(P^{-1}B\Delta_K + \Delta_K^TB^TP^{-1}),\end{aligned}$$

and we have used the facts $\mathscr{L} = \mathscr{D}\mathscr{D}^T$ *and*

$$\|\tilde{K}\| \le \|K\| + \|\Delta_K\| \le \|P^{-1}\|\|B\| + \delta_k\|E\|\|F\|.$$

Note that (10.19) can be rewritten as

$$\begin{aligned}\dot{V} &\le -\alpha V + \alpha V + \tfrac{1}{2}\xi^T[\Xi\xi + 2(\mathscr{L}\otimes P^{-1}A_g)(G(x) - \mathbf{1}\otimes g(\bar{x}))] \\ &\quad + \|\xi\|\|c\mathscr{L}\|\|P^{-1}\|\|B\|(\|P^{-1}\|\|B\| + \delta_k\|E\|\|F\|)\|\mathscr{D}\|\sqrt{nM}\tfrac{\Delta}{2} \\ &= -\alpha V + \tfrac{1}{2}\xi^T[(\Xi + \alpha\mathscr{L}\otimes P^{-1})\xi + 2(\mathscr{L}\otimes P^{-1}A_g)(G(x) - \mathbf{1}\otimes g(\bar{x}))] \\ &\quad + \|\xi\|\|c\mathscr{L}\|\|P^{-1}\|\|B\|(\|P^{-1}\|\|B\| + \delta_k\|E\|\|F\|)\|\mathscr{D}\|\sqrt{nM}\tfrac{\Delta}{2}.\end{aligned} \tag{10.20}$$

Since graph $\mathscr{G}$ *is connected, there exists an orthogonal matrix* $U \in R^{N\times N}$ *with*

$$U = \begin{bmatrix} \frac{\mathbf{1}}{\sqrt{N}} & T_1^T \end{bmatrix}, U^{-1} = \begin{bmatrix} \frac{\mathbf{1}^T}{\sqrt{N}} \\ T_1 \end{bmatrix}$$

such that $\Gamma = U^{-1}\mathscr{L}U = diag\{0, \lambda_2, \cdots, \lambda_N\}$, *where* $\lambda_2 \leq \lambda_3 \leq \cdots, \leq \lambda_N$ *are the non-zero eigenvalues. As the definition of orthogonal matrix, one can obtain that* $U^TU = UU^T = I$ *and* $U^T = U^{-1}$.

Let $\eta = [\eta_1^T, \cdots, \eta_N^T]^T = (U^T \otimes I_n)\xi$. *By the definition of* η *and* ξ, *it is easy to see that* $\eta_1 = (\frac{\mathbf{1}^T}{\sqrt{N}} \otimes I_n)\xi = 0$. *Using Young's inequality in Lemma 2.7 of Chapter 2 with* $\mu > 0$ *as the auxiliary variable, it then follows that*

$$\begin{aligned}
&2\xi^T(\mathscr{L} \otimes P^{-1}A_g)(G(x) - \mathbf{1} \otimes g(\bar{x})) \\
&= 2\eta^T(\Gamma \otimes P^{-1}A_g)(U^T \otimes I_n)(G(x) - \mathbf{1} \otimes g(\bar{x})) \\
&\leq \mu\eta^T(\Gamma \otimes P^{-1}A_g)(\mathscr{W}^{-1} \otimes I_n)(\Gamma \otimes A_g^T P^{-1})\eta \\
&+ \tfrac{1}{\mu}(G(x) - \mathbf{1} \otimes g(\bar{x}))^T(U \otimes I_n)(\mathscr{W} \otimes I_n)(U^T \otimes I_n)(G(x) - \mathbf{1} \otimes g(\bar{x})) \\
&= \mu\eta^T(\Gamma\mathscr{W}^{-1}\Gamma \otimes P^{-1}A_gA_g^TP^{-1})\eta + \tfrac{1}{\mu}\sum_{i=1}^{N}\omega_i\|g(x_i) - g(\bar{x})\|^2 \\
&\leq \eta^T(\mu\Gamma\mathscr{W}^{-1}\Gamma \otimes P^{-1}A_gA_g^TP^{-1} + \tfrac{\gamma^2}{\mu}\mathscr{W})\eta
\end{aligned} \tag{10.21}$$

where $\mathscr{W} = diag\{\omega_1, \cdots, \omega_N\}$ *is a positive-definite diagonal matrix. By choosing* $\mathscr{W} = diag\{1, \lambda_2, \cdots, \lambda_N\}$ *and letting* $\hat{\eta} = (I \otimes P^{-1})\eta = (U^T \otimes P^{-1})\xi$, *it then follows from Lemma 2.10 in Chapter 2, (10.20) and (10.21) that*

$$\begin{aligned}
&\xi^T[\Xi + \alpha\mathscr{L} \otimes P^{-1}]\xi + 2\xi^T(\mathscr{L} \otimes P^{-1}A_g)(G(x) - \mathbf{1} \otimes g(\bar{x})) \\
&= \sum_{i=2}^{N}\lambda_i\hat{\eta}^T(AP + PA^T - 2c\lambda_iBB^T + c\lambda_i(B(-\Delta_K)P + P(-\Delta_K)^TB^T) \\
&\quad + \alpha P + \mu A_gA_g^T + \mu^{-1}\gamma^2P^2)\hat{\eta} \\
&\leq \sum_{i=2}^{N}\lambda_i\hat{\eta}^T(AP + PA^T - 2\tau BB^T + \tau(BE\Lambda E^TB^T + \delta_k^2PF^T\Lambda^{-1}FP) \\
&\quad + \alpha P + \mu A_gA_g^T + \mu^{-1}\gamma^2P^2)\hat{\eta}
\end{aligned} \tag{10.22}$$

if c *satisfies the constraint in (10.16).*

By using Schur complement lemma, it can be implied by (10.15) that

$$AP + PA^T - 2\tau BB^T + \tau(BE\Lambda E^TB^T + \delta_k^2PF^T\Lambda^{-1}FP) + \mu A_gA_g^T + \mu^{-1}\gamma^2P^2 < 0.$$

Let

$$\begin{aligned}
\alpha \quad = \quad & -\lambda_{max}(AP + PA^T - 2\tau BB^T + \tau(BE\Lambda E^TB^T + \delta_k^2PF^T\Lambda^{-1}FP) + \mu A_gA_g^T \\
& + \mu^{-1}\gamma^2P^2)/\lambda_{max}(P),
\end{aligned}$$

and one can get from (10.22) that

$$\xi^T[\Xi + \alpha\mathscr{L} \otimes P^{-1}]\xi + 2\xi^T(\mathscr{L} \otimes P^{-1}A_g)(G(x) - \mathbf{1} \otimes g(\bar{x})) \leq 0.$$

It then follows from (10.20) that

$$\begin{aligned}
\dot{V} \quad & \leq -\alpha V + \|\xi\|\|\mathscr{L}\|\|P^{-1}\|\|B\|(\|P^{-1}\|\|B\| + \delta_k\|E\|\|F\|)\|\mathscr{D}\|\sqrt{nM}\tfrac{\Delta}{2} \\
& \leq -\alpha c_1\|\xi\|^2 + \|\xi\|\|c\mathscr{L}\|\|P^{-1}\|\|B\|(\|P^{-1}\|\|B\| + \delta_k\|E\|\|F\|)\|\mathscr{D}\|\sqrt{nM}\tfrac{\Delta}{2} \\
& = -\alpha c_1\|\xi\|(\|\xi\| - \tfrac{\|\mathscr{L}\|\|P^{-1}\|\|B\|(\|P^{-1}\|\|B\| + \delta_k\|E\|\|F\|)\|\mathscr{D}\|\sqrt{nM}\Delta}{2\alpha c_1}) \\
& = -\alpha c_1\|\xi\|(\theta\|\xi\| + (1-\theta)\|\xi\| - \tfrac{\|\mathscr{L}\|\|P^{-1}\|\|B\|(\|P^{-1}\|\|B\| + \delta_k\|E\|\|F\|)\|\mathscr{D}\|\sqrt{nM}\Delta}{2\alpha c_1})
\end{aligned}$$

where $0 < \theta < 1$ is a constant which can be chosen as small as possible.

Hence, for

$$\|\xi\| \geq \frac{\|\mathscr{L}\|\|P^{-1}\|\|B\|(\|P^{-1}\|\|B\| + \delta_k\|E\|\|F\|)\|D\|\sqrt{n}M\Delta}{2\alpha c_1(1-\theta)},$$

we have $\dot{V} \leq -\alpha c_1 \theta \|\xi\|^2$.

Note that $c_1\|\xi\|^2 \leq V(\xi) \leq c_2\|\xi\|^2$, then from Theorem 4.18 in [90], one can easily obtain that there is a time instant $t \geq T$ with $T > 0$ such that

$$\|\xi\| \leq \sqrt{\frac{c_2}{c_1}} \frac{\|\mathscr{L}\|\|P^{-1}\|\|B\|(\|P^{-1}\|\|B\|+\delta_k\|E\|\|F\|)\|\mathscr{D}\|\sqrt{n}M\Delta}{2\alpha c_1(1-\theta)}.$$

By definition of ξ, this is equivalent to (10.17). This completes the proof. □

Remark 10.9 *Theorem 10.1 shows that when uniform quantizers are used, exact state consensus cannot be achieved. However, there is a time instant $t \geq T$, the states $x_i, i = 1, \cdots, N$ will enter into a ball as in (10.17).*

Remark 10.10 *It must be pointed out that the proposed protocol is indeed not a fully distributed one because the coupling weight c, presented in Theorem 1, should satisfy the inequality (10.16), while λ_2 is global information in the sense that each agent has to know the entire communication graph to compute it. Therefore, some fully distributed consensus protocols without using λ_2 will be a direction of future research. In addition, the result of Theorem 10.1 applies the tuning parameter τ due to the product terms between τ and Λ or P in (10.15). However, applying a numerical optimization algorithm, such as the program fminsearch in the optimization toolbox of MATLAB, one can obtain the tuning parameter τ [62].*

From the proof of Theorem 10.1, it is easy to see that

$$\begin{aligned}\dot{V} \quad &\leq \quad \sum_{i=2}^{N} \lambda_i \hat{\eta}^T (AP + PA^T - 2\tau BB^T + \tau(BE\Lambda E^T B^T + \delta_k^2 PF^T \Lambda^{-1} FP) \\ &\quad + \alpha P + \mu A_g A_g^T + \mu^{-1}\gamma^2 P^2)\hat{\eta} < 0\end{aligned}$$

when the effect of quantization is not considered. Then the following corollary provides a sufficient condition to design non-fragile consensus control without considering the effect of quantization.

Corollary 10.1 *Let the communication graph $\mathscr{G}$ be undirected and connected. The N agents described by (10.3) reach global consensus under the protocol $u_i = -c(K + \Delta_K) \sum_{k=1}^{M} d_{ik} z_k$ if there exists a solution to (10.15). c and K are also computed by (10.16).*

In addition, when the effect of quantization and controller coefficient variations are not considered simultaneously, it is easy to see that

$$\dot{V} \leq \sum_{i=2}^{N} \lambda_i \hat{\eta}^T (AP + PA^T - 2\tau BB^T + \mu A_g A_g^T + \mu^{-1}\gamma^2 P^2)\hat{\eta} < 0,$$

then the following corollary provides a sufficient condition to design standard consensus control which can make the original high-order identical dynamic systems achieve consensus.

Corollary 10.2 *Let the communication graph $\mathcal{G}$ be undirected and connected. The N agents described by (10.3) reach global consensus under the protocol $u_i = -cK\sum_{k=1}^{M} d_{ik}z_k$ if there exists a solution to the following LMI*

$$\begin{bmatrix} AP + PA^T - 2\tau BB^T + \mu A_g A_g^T & P \\ * & -\mu\gamma^{-2}I \end{bmatrix} < 0 \tag{10.23}$$

to get a matrix $P > 0$, and scalars $\tau > 0$ and $\mu > 0$. In addition, c and K are also computed by (10.16).

10.4.2 Non-Fragile Control with Logarithmic Quantization

In this subsection, we consider the effects of logarithmic quantizers (10.2) on the non-fragile quantized consensus-type scheme (10.6). When the logarithmic quantizer is used, one can show that the proposed consensus scheme can still make all agents achieve consensus.

Theorem 10.2 *Let the communication graph $\mathcal{G}$ be undirected and connected. Solving the matrix inequality (10.15) to get a positive matrix $P = P^T > 0$, a diagonal matrix $\Lambda > 0$, scalars $\tau > 0$ and $\mu > 0$. The parameters in the consensus protocol (10.9) are designed as (10.16). If*

$$\Delta_l < \frac{\alpha c_1}{\|c\mathcal{L}\|\|P^{-1}\|\|B\|(\|P^{-1}\|\|B\| + \delta_k\|E\|\|F\|)\|D\|^2} \tag{10.24}$$

then any Filippov solution x to (10.10) satisfies

$$\|x - \frac{(\mathbf{1}\mathbf{1}^T \otimes I_n)x}{N}\| = 0$$

where α is defined in (10.18).

Proof 10.2 *Consider the Lyapunov function $V(\xi)$ in (10.13) with symmetric and positive definite matrix $P = P^T > 0$ being a solution of (10.15). Then, after some similar process of proof in Theorem 10.1, one can obtain*

$$\begin{aligned} \dot{V} &= \xi^T(\mathcal{L} \otimes P^{-1})[(I_N \otimes A)\xi + (\Pi \otimes A_g)G(x) - (cI_N \otimes B\tilde{K})(\mathcal{D} \otimes I_n)\cdot\zeta] \\ &\le -\alpha V_1 + \|\xi\|^2\|c\mathcal{L}\|\|P^{-1}\|\|B\|(\|P^{-1}\|\|B\| + \delta_k\|E\|\|F\|)\|\mathcal{D}\|^2\Delta_l \\ &\le -\alpha c_1\|\xi\|^2 + \|\xi\|^2\|c\mathcal{L}\|\|P^{-1}\|\|B\|(\|P^{-1}\|\|B\| + \delta_k\|E\|\|F\|)\|\mathcal{D}\|^2\Delta_l \\ &= -\|\xi\|^2(\alpha c_1 - \|c\mathcal{L}\|\|P^{-1}\|\|B\|(\|P^{-1}\|\|B\| + \delta_k\|E\|\|F\|)\|\mathcal{D}\|^2\Delta_l) \end{aligned}$$

where we have used the fact that any $\zeta \in \mathcal{F}\{q((\mathcal{D}^T \otimes I_n)\xi)\}$ satisfies

$$\|\zeta - (\mathcal{D}^T \otimes I_n)\xi\| \le \Delta_l\|(\mathcal{D}^T \otimes I_n)\xi\| = \Delta_l\|\mathcal{D}\|\|\xi\|.$$

Note that if

$$\alpha c_1 > \|c\mathscr{L}\|\|P^{-1}\|\|B\|(\|P^{-1}\|\|B\| + \delta_k\|E\|\|F\|)\|\mathscr{D}\|^2\Delta_l,$$

one can obtain that $\dot{V} < 0$*, which implies that* $\xi = x - \frac{(\mathbf{1}\mathbf{1}^T \otimes I_n)x}{N}$ *converges to zero. This completes the proof.* □

Remark 10.11 *One can find that* ξ *has a bounded error in Theorem 1 while converges to zero in Theorem 2, which means that only practical consensus is achievable with uniformly quantized information while precise consensus can be achieved with logarithmically quantized information. This is consistent with the observation in [39] that precise control is usually not achievable with uniform quantizer because a uniform quantizer has equally spaced quantization levels and the quantization errors are uniformly bounded for all input values, while it is possible to achieve precise control with logarithmical quantizer because only the logarithm of the quantization levels is equally spaced, which enables the quantizer to provide a higher precision when the input is close to the origin.*

10.5 Numerical Example

To check how the proposed non-fragile quantized control law works in a real system, a network of four single-link manipulators with revolute joints actuated by a DC motor is considered. The dynamics of the i-th agent system is described by (10.3) and (10.4) with [147, 216]

$$x_i = \begin{bmatrix} x_{i1} \\ x_{i2} \\ x_{i3} \\ x_{i4} \end{bmatrix}, A = \begin{bmatrix} 0 & 1 & 0 & 0 \\ -48.6 & -1.25 & 48.6 & 0 \\ 0 & 0 & 0 & 10 \\ 1.95 & 0 & -1.95 & 0 \end{bmatrix}, B = \begin{bmatrix} 0 \\ 21.6 \\ 0 \\ 0 \end{bmatrix}$$
$$A_g = I_4, g(x_i) = \begin{bmatrix} 0 & 0 & 0 & -0.333\sin(x_{i1}) \end{bmatrix}^T.$$

Clearly, $g(x_i)$ here satisfies (10.5) with a Lipschitz constant $\gamma = 0.333$.

Let $\Delta_K = \begin{bmatrix} -0.2 & -0.2 & -0.2\sin(2t) & -0.2\sin(2t) \end{bmatrix}$, we can imply that $|\delta_{lj}| \leq 0.2$, which means $\delta_k = 0.2$. When Δ_K is rewritten as (10.8), E, F and $\Re$ can be given as follows

$$E = \begin{bmatrix} 1 & 1 & 1 & 1 \end{bmatrix}; F = I_4; \Re = diag\{-0.2, -0.2, -0.2\sin(2t), -0.2\sin(2t)\}.$$

The communication topology is shown in Figure 10.1. From the signs of each edge and the topology given in Figure 10.1, one can derive the *incidence matrix* $\mathscr{D}$ as follows

$$\mathscr{D} = \begin{bmatrix} -1 & -1 & 0 & 0 \\ 1 & 0 & -1 & 0 \\ 0 & 1 & 1 & -1 \\ 0 & 0 & 0 & 1 \end{bmatrix}.$$

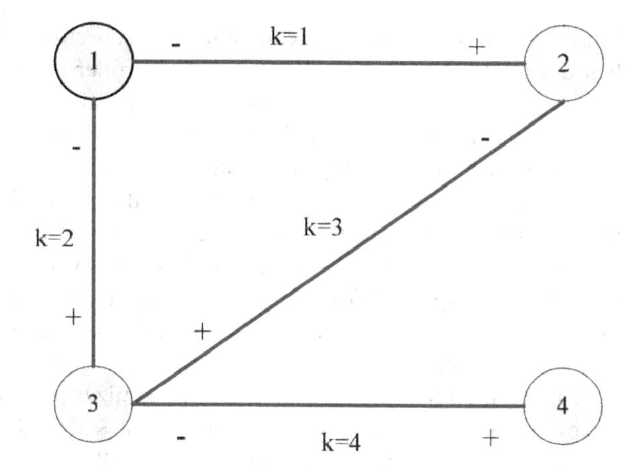

FIGURE 10.1
The framework of communication topology.

Then, from $L = \mathscr{D}\mathscr{D}^T$, one can easily obtain the non-zero eigenvalues of the *Laplacian matrix* are $\lambda_2 = 1, \lambda_3 = 3, \lambda_4 = 4$.

Solving the LMI (10.23) in Corollary 10.2 gives the feedback gain matrix of standard controller as $K = \begin{bmatrix} 3.0229 & 0.1966 & -2.0975 & 5.8545 \end{bmatrix}$ and $c \geq \frac{\tau}{\lambda_2} = \tau = 16.2424$. For the standard controller, let $c = 16.2424$.

In addition, by applying Corollary 10.1 and Remark 10.10, solving the condition in (10.15) by using fminsearch algorithm with the initial value $\tau = 16.2424$ gives the non-fragile controller gain as

$$\begin{aligned} K &= \begin{bmatrix} 75.1155 & 4.7297 & -50.1312 & 127.6364 \end{bmatrix} \\ \tau &= 15.8256, \end{aligned}$$

then, $c \geq 15.8256$. In the following simulation, we always let $c = 15.8256$.

In order to verify the effectiveness of the proposed control strategy, simulations are given with the following *initial conditions*:

$$x_1(0) = \begin{bmatrix} -2.5 \\ 0 \\ 0.5 \\ 0.4 \end{bmatrix}, x_2(0) = \begin{bmatrix} -0.3 \\ 0.5 \\ -0.8 \\ 0.9 \end{bmatrix}, x_3(0) = \begin{bmatrix} -1 \\ 2 \\ 0.5 \\ 0.4 \end{bmatrix}, x_4(0) = \begin{bmatrix} -0.3 \\ 2.5 \\ -0.8 \\ 0.9 \end{bmatrix}.$$

Choosing the initial state of control input u as zero. When quantization effect is not considered, with the same standard controller parameters but with or without *additive* interval-bounded controller coefficient variations Δ_K, the simulation results are shown in Figure 10.2 and Figure 10.3. The state trajectories under non-fragile controller are shown in Figure 10.4. From Figure 10.2 and Figure 10.4, it can be observed that the global consensus is indeed achieved. However, comparing these results with Figure 10.3, one can see that the standard controller performs as well as non-fragile controller when $\delta_k = 0$, whereas the states cannot reach a consensus under standard controller when $\delta_k = 0.2$, which means the standard controller is very sensitive to *additive* controller coefficient variations. Besides, the state trajectories obtained by non-fragile controller is almost not affected by the same variations in Figure 10.4, which shows the effectiveness of the proposed design method. The responses of control input u_i are shown in Figure 10.5. For the cases of uniform quantizer parameter $\Delta = 10$ in (10.1) and logarithmic quantizer parameter $\Delta = 0.2$ in (10.2), the simulation results are shown in Figure 10.6 and Figure 10.7, respectively. Comparing the results of Figure 10.6 and Figure 10.7 to Figure 10.4, we can see that all the states x of all agents achieve practical consensus with uniform quantization information, but more overshoot will be caused by uniform quantizer in Figure 10.6. When *logarithmic quantizers* are used in (10.9), for the case of quantizer parameter $\Delta = 0.2$, the simulation results in Figure 10.7 show that all the states x of all agents reach consensus, from which the efficiency of the quantized controller is demonstrated.

10.6 Conclusion

In this chapter, a non-fragile quantized consensus strategy by using *incidence matrix* for *Lipschitz* nonlinear *multi-agent systems* with quantization information is proposed. Both cases of uniform quantizer and logarithmic quantizer are considered under undirect communication graphs, respectively. It is shown that the original multi-agent system can reach finite-time practical consensus when uniform quantizers are used, while the given systems can reach consensus under some conditions when a logarithmic quantizer is considered. A new technique has been developed to solve *additive* interval-bounded controller coefficient variations where the numerical problem caused by using *interval-bounded variations* can be solved effectively. Sufficient conditions of the existence of the proposed controllers are derived by using matrix inequalities. Future research directions include considering the case of switching directed communication graphs and the practical consensus problem of quantized multi-agent systems with time-varying communication topology and distinct time-varying communication delays.

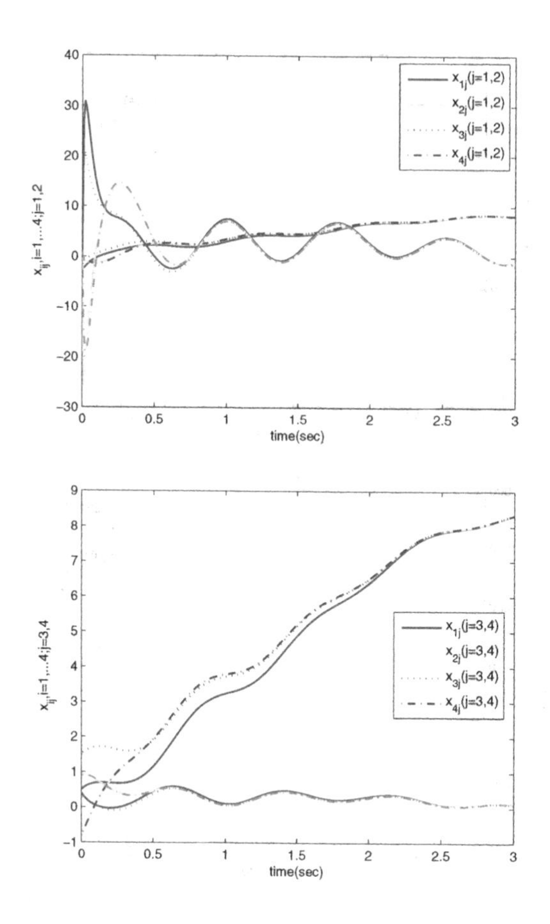

FIGURE 10.2
The state trajectories under standard controller with $\delta_k = 0$ and without quantizers.

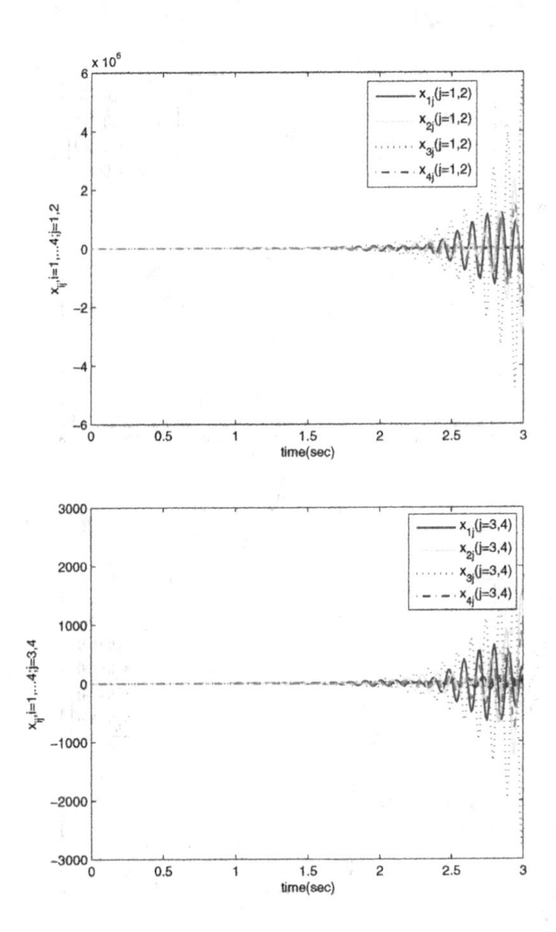

FIGURE 10.3
The state trajectories under standard controller with $\delta_k = 0.2$ and without quantizers.

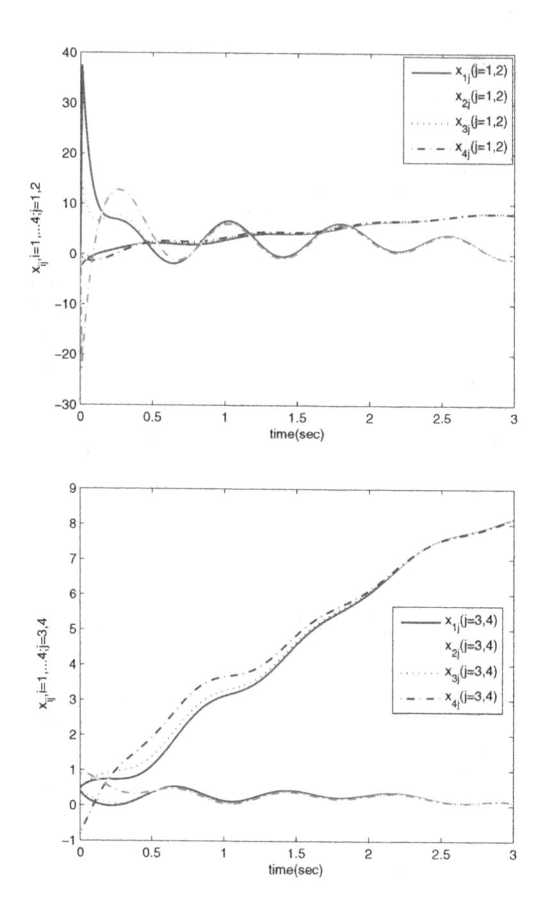

FIGURE 10.4
The state trajectories under non-fragile controller with $\delta_k = 0.2$ and without quantizers.

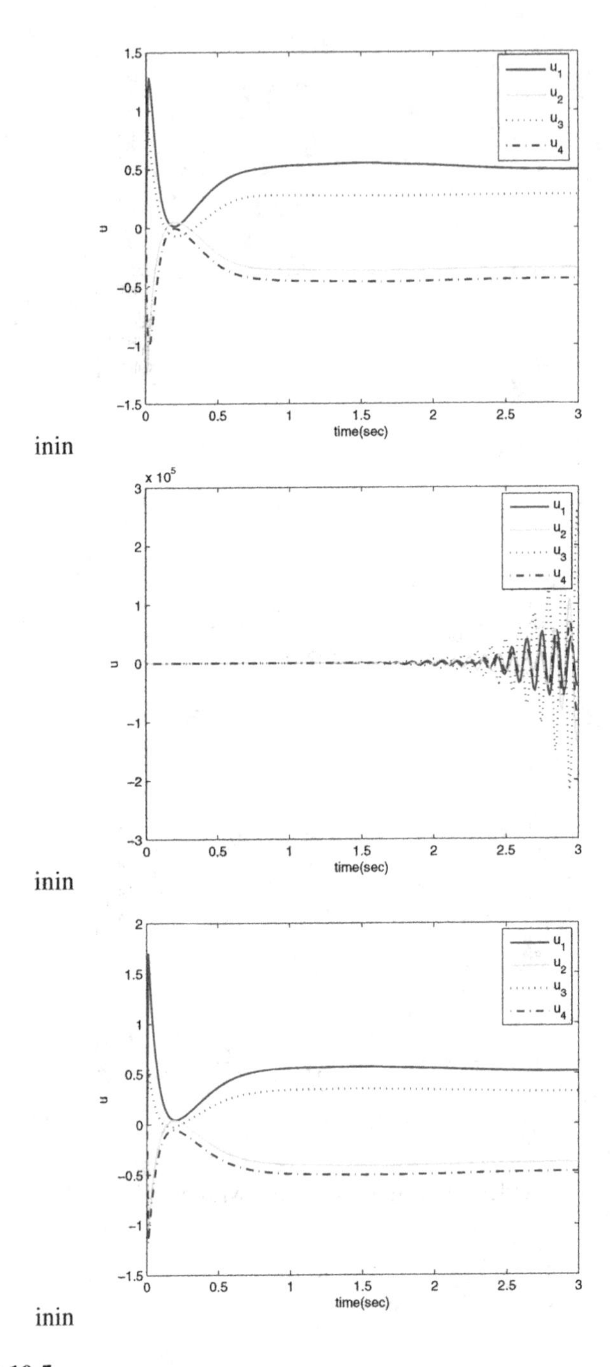

FIGURE 10.5
The responses of control input without quantizers (top: Standard controller; middle: Standard controller with $\delta_k = 0.2$; bottom: Non-fragile controller with $\delta_k = 0.2$).

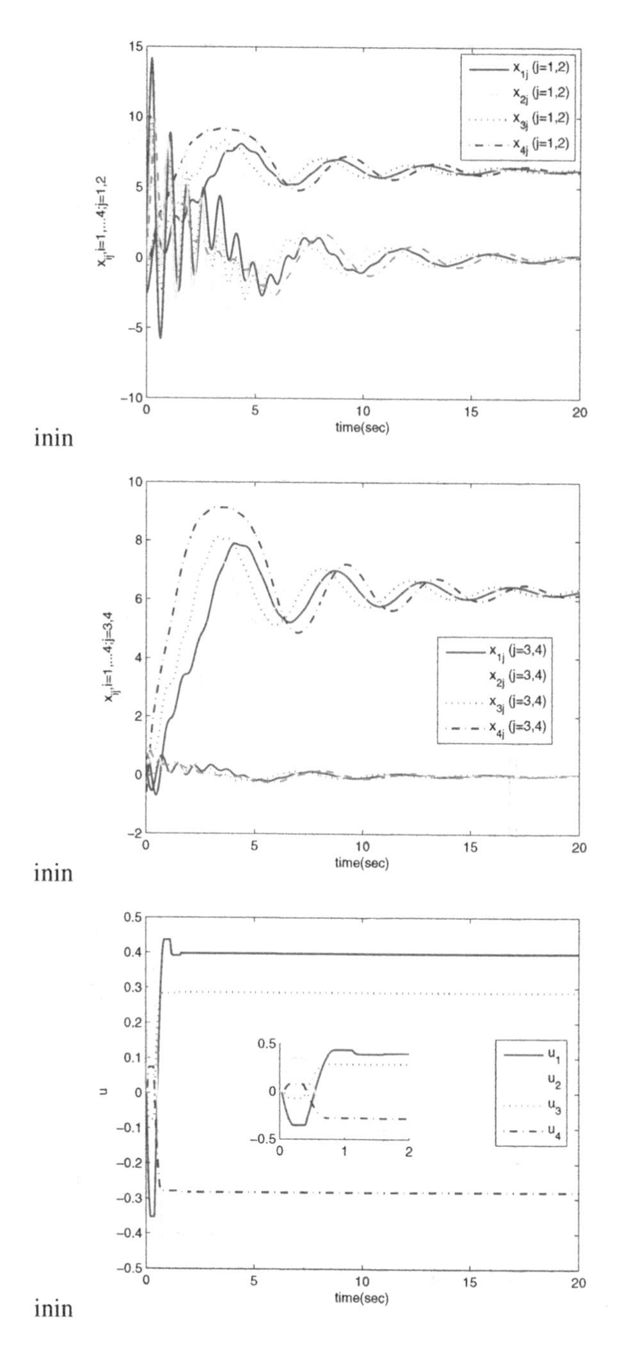

FIGURE 10.6
The state trajectories and control input under non-fragile controller with $\delta_k = 0.2$ and uniform quantizer $\Delta = 10$.

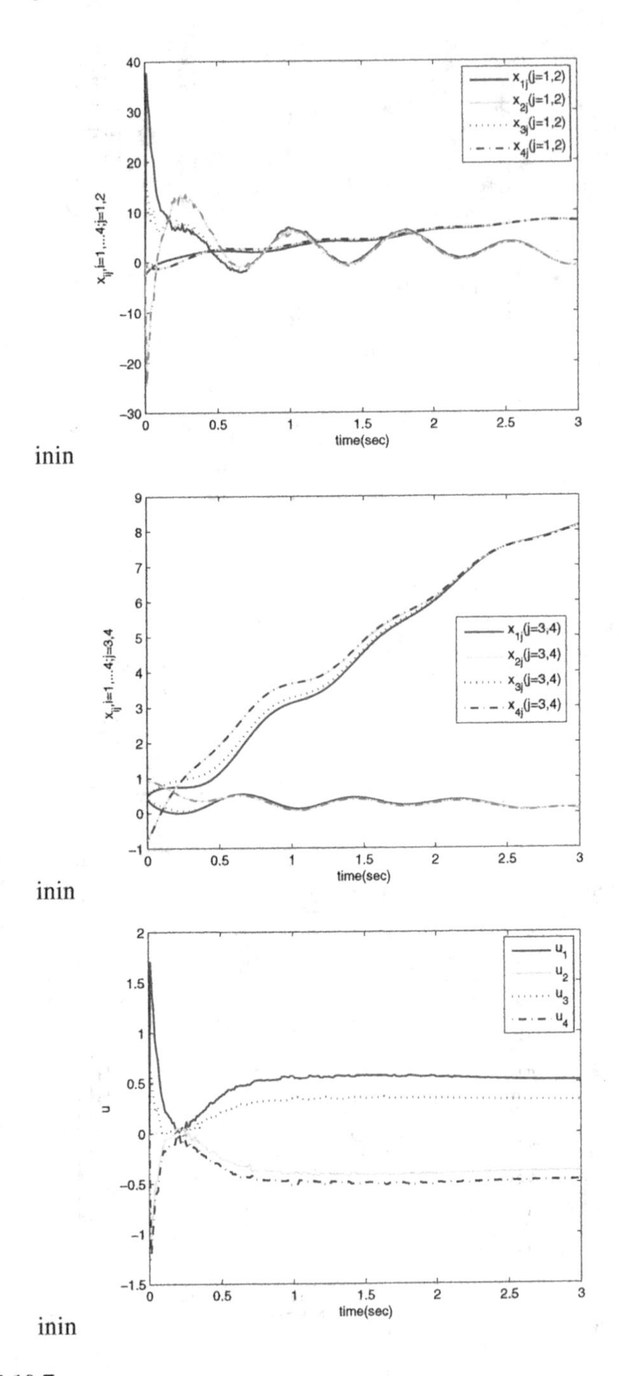

FIGURE 10.7
The state trajectories and control input under non-fragile controller with $\delta_k = 0.2$ and logarithmic quantizer $\Delta = 0.2$.

11

Non-Fragile H_∞ Consensus for Multi-Agent Systems with Interval-Bounded Variations

This chapter studies the distributed *non-fragile H_∞ consensus* problems for linear *multi-agent systems* with *external disturbances* and unknown initial disturbances under switching weighted balanced directed topologies. The designed controllers are insensitive to *multiplicative* controller coefficient variations. Sufficient conditions for the existence of the proposed control strategy are also obtained by using linear matrix inequality (LMI) technique. It is worth mentioning that instead of requiring the coupling strength among neighboring agents to be larger than a threshold value as in previous literature, the coupling strength in this chapter can be determined by solving some LMIs.

11.1 Introduction

Consensus control problem has been an emerging topic in *multi-agent systems*, aiming to design appropriate control protocols using only local information to ensure that the agents reach an agreement on certain quantities of interest [110, 113, 208]. It has been extensively studied by various scholars from different perspectives, for instance, [110, 112, 113, 176, 179, 181, 199] and the references therein. Consensus of *multi-agent systems* with general linear and *Lipschitz* nonlinear dynamics using distributed adaptive protocols was studied in [113], and then the proposed method was extended to *multi-agent systems* with heterogeneous matching uncertainties in [110]. Consensus control of second-order *multi-agent systems* with sampled data was investigated in [199]. In [112, 176, 179, 181], the H_∞ *consensus* control problem for high-order *multi-agent systems* was studied due to the fact that the systems are unavoidably affected by various disturbances such as sensor noises, actuator bias, measurement or calculation errors and the variation of communication links.

It is well known that, in practice, the topological structures of the network may switch among some possible topologies due to, for example, the influence of link failures, field of view constraints of visual sensors, the limit of camera field-of-view, or the presence of communication obstacles, and so on [25, 129]. Thus, it is scientifically meaningful but usually difficult to deal with the distributed consensus over switching topologies due to complexity of the topology and lack of theoretical tools.

It should be emphasized that all the above-mentioned results are based on an implicit assumption that the controller can be realized exactly. However, in practical engineering systems, small variations in controller coefficients may result in huge performance degradation or even instability of the systems [13, 29, 50, 67, 190]. Therefore, how to design a controller insensitive to the variations in its coefficients is of practical significance for the implementation of designed controllers. In addition, it is worth mentioning that the model of *multiplicative* uncertainty is frequently used in many practical systems such as production line, environmental risk assessments and economic systems [50]. However, the existence of the uncertainties in systems and controller itself makes the consensus problem very challenging to solve.

To our knowledge, there is no existing result on *non-fragile H_∞ consensus* control of *multi-agent systems* under both *multiplicative* interval-bounded controller coefficient variations and unknown initial disturbances. This chapter studies the *non-fragile* control for *multi-agent systems* with eliminating the effect of including model and controller *multiplicative* interval-bounded coefficient variations under switching weighted undirected or balanced directed topologies. The main contributions of this chapter are summarized as follows.

(1) The corresponding consensus protocols are designed for linear multi-agent systems. The designed controller is insensitive to *additive* interval-bounded model coefficient variations and *multiplicative* controller coefficient variations simultaneously.

(2) In contrast to artificial selection methods in [112, 176, 179, 181], the coupling strength among neighboring agents can be determined by solving some LMI conditions.

(3) The issue of robustness with respect to system uncertainties and controller coefficient variations simultaneously is considered. Thus the multi-agent systems under consideration are more general than those in [112, 176, 179, 181].

11.2 Problem Formulation

A multi-agent system composed of N interconnected continuous time subsystems subjected to *external disturbances* is considered. The dynamics of the ith agent is described by the following

$$\dot{x}_i(t) \quad = \quad \tilde{A}(t)x_i(t) + \tilde{B}(t)u_i(t) + Dw_i(t), i = 1, \cdots, N \tag{11.1}$$

where

$$\tilde{A}(t) = A + \Delta A(t),\ \tilde{B}(t) = B(I + \Delta B(t)),$$

$x_i(t) \in \mathbf{R}^n, i = 0, \cdots, N$ are the states of agent i, $u_i(t) \in \mathbf{R}^p$ is the control input acting on agent i, and $w_i(t) \in \mathbf{R}^q$ is the *external disturbance* input that belongs to $L_2[0, \infty)$.

$\Delta A(t)$ and $\Delta B(t)$ are the system characteristic matrix uncertainty and input matrix uncertainty, respectively.

In practice, the topological structures of the network may switch among some possible topologies due to, for example, abrupt variations in their structures, field of view constraints of visual sensors, the limit of camera field-of-view, or the presence of communication obstacles, and so on [104, 181]. Therefore, the communication topology considered in this chapter is assumed to be dynamically switching over the connected graph set $\mathscr{G}^{\sigma(t)}$, $\sigma(t) \in \{1,2,\cdots,m\}, m \geq 1$, where $\sigma(t)$ denotes the set of all possible undirected or balanced directed topologies. Consider an infinite sequence of non-overlapping time intervals $[t_k, t_{k+1}), k \in \mathbf{N}$, with $t_0 = 0$, $t_{k+1} - t_k \geq \tau$ and $\tau > 0$, across which the communication topology is fixed. Here, the time sequence $t_1, t_2, \cdots$, is called the switching sequence, at which the communication topology changes. During each time interval $[t_k, t_{k+1})$, the graph $\mathscr{G}$ is time invariant. Then, in order to achieve consensus for linear multi-agent systems with *external disturbances*, the following *non-fragile* distributed consensus protocol is proposed

$$u_i(t) = c\tilde{K} \sum_{j=1}^{N} a_{ij}^{\sigma(t)}(x_j - x_i), i = 1, \cdots, N \tag{11.2}$$

where $\tilde{K} = K(I + \Delta K(t))$, $K \in \mathbf{R}^{p \times n}$ is the feedback gain matrix, $\Delta K(t)$ denotes controller gain uncertainties, $c > 0$ is the coupling weight among neighboring agents, and $a_{ij}^{\sigma(t)}$ is (i,j)-th entry of the *adjacency matrix* of communication topology $\mathscr{G}^{\sigma(t)}$.

In addition, throughout this chapter, the follow assumption is made.

Assumption 11.1 *The time-varying uncertainty $\Delta A(t)$, $\Delta B(t)$ and $\Delta K(t)$ satisfies*

$$\begin{aligned} \Delta A(t) &= [\delta_{akj}(t)]_{n \times n}, |\delta_{akj}(t)| \leq \delta_a, k, j = 1, \cdots, n \\ \Delta B(t) &= [\delta_{bll}(t)]_{p \times p}, |\delta_{bll}(t)| \leq \delta_b, l = 1, \cdots, p \\ \Delta K(t) &= [\delta_{ckj}(t)]_{n \times n}, |\delta_{ckj}(t)| \leq \delta_c, k, j = 1, \cdots, n \end{aligned} \tag{11.3}$$

where $\delta_{akj}(t)$, $\delta_{bll}(t)$ and $\delta_{ckj}(t)$ (for $k, j = 1, \cdots, n; l = 1, \cdots, p$) are used to denote the magnitudes of the deviation of the coefficients of A, B and K. Then, following (11.3) and the definition of the matrix Frobenius norm, the type of interval coefficient variations can be converted to the following norm bounded type.

$$\|\Delta A(t)\| \leq n\delta_a, \ \|\Delta B(t)\| \leq p\delta_b, \ \|\Delta K(t)\| \leq n\delta_c. \tag{11.4}$$

Remark 11.1 *The uncertainty model in (11.3) has been investigated extensively in [13, 29, 50, 67, 190], where this model is used to describe the controller coefficient variations. Note that most of the above-mentioned results studied additive interval-bounded controller coefficient variations [13, 29, 50, 67, 190], whereas there are few results on non-fragile controller design considering multiplicative interval-bounded controller coefficient variations [50, 67]. It should be mentioned that the methods proposed in [13, 29, 50, 67, 190] cannot be directly used to solve multiplicative interval-bounded controller coefficient variations, while the proposed methods in [50, 67] involve the closed-loop system transfer function and cannot be applied to multi-agent systems. Therefore, it is necessary to develop new methods to deal with this challenge, which motivates us for this study.*

It follows from (11.1) that the closed-loop system is

$$\dot{x}_i(t) = \tilde{A}(t)x_i(t) + c\tilde{B}(t)\tilde{K}\sum_{j=1}^{N} a_{ij}^{\sigma(t)}(x_j - x_i) + Dw_i(t). \quad (11.5)$$

By using the *Kronecker product*, it is not difficult to obtain that the closed-loop network dynamics (11.5) can be rewritten as

$$\dot{x}(t) = [I_N \otimes \tilde{A}(t) - c(\mathscr{L}^{\sigma(t)} \otimes \tilde{B}(t)\tilde{K})]x(t) + (I_N \otimes D)w(t) \quad (11.6)$$

where

$$x(t) = \left[x_1(t)^T, \cdots, x_N(t)^T\right]^T.$$

Inspired by [181], we shall use the following assumption on the communication topology throughout this chapter.

Assumption 11.2 *There exists one common positive vector $\zeta = [\zeta_1, \cdots, \zeta_N]^T \in \mathbf{R}^N$, such that $\zeta^T \mathscr{L}^{\sigma(t)} = \mathbf{0}_N^T$ and $\mathbf{1}_N^T \zeta = 1$, where $\mathscr{L}^{\sigma(t)}$ is the* Laplacian matrix *of directed switching graph $\mathscr{G}^{\sigma(t)}$, $\sigma(t) \in 1, 2, \cdots, m$.*

Remark 11.2 *Assumption 11.2 holds if every possible topology $\mathscr{G}^{\sigma(t)}$, $\sigma(t) \in \{1, 2, \cdots, m\}$ is a connected undirected graph or strongly connected and balanced directed graph. Then, a common positive vector ζ can be given as $\zeta = \frac{1}{N}\mathbf{1}_N$.*

Define $e(t) = (\Pi \otimes I_n)x(t)$, where $\Pi = (I_N - \mathbf{1}_N \zeta^T) \in \mathbf{R}^{N\times N}$ satisfies $\Pi \cdot \Pi = \Pi$ and $\Pi \cdot \mathscr{L}^{\sigma(t)} = \mathscr{L}^{\sigma(t)}$. Then, (11.6) can be rewritten as

$$\dot{e}(t) = [I_N \otimes \tilde{A}(t) - c(\mathscr{L}^{\sigma(t)} \otimes \tilde{B}(t)K)]e(t) + (\Pi \otimes D)w(t). \quad (11.7)$$

Before proceeding further, the consensus for (11.1) under protocol $u_i(t)$ in (11.2) is defined.

Definition 11.1 *The non-fragile consensus problem of multi-agent system (11.1) is solved by protocol $u_i(t)$ in (11.2) if, for any initial condition $x(0)$, the states of the closed-loop system (11.5) satisfy*

$$\lim_{t\to\infty} \|x_i(t) - x_j(t)\|_2 = 0, \forall i, j = 1, \cdots, N. \quad (11.8)$$

Remark 11.3 *Construct the following matrix*

$$\Omega = \begin{bmatrix} 1 & \mathbf{0}_{N-1}^T \\ \mathbf{1}_{N-1} & I_{N-1} \end{bmatrix} \in \mathbf{R}^{N\times N}.$$

Then, pre- and post-multiplying Π by Ω^{-1} and Ω, respectively, it follows by using the fact $\mathbf{1}_N^T \zeta = 1$ that

$$\Gamma \triangleq \Omega^{-1}\Pi\Omega = \begin{bmatrix} 0 & \tilde{\zeta}^T \\ \mathbf{0}_{N-1} & I_{N-1} \end{bmatrix} \in \mathbf{R}^{N\times N}$$

where $\tilde{\zeta} = [-\zeta_2, -\zeta_3, \cdots, -\zeta_N]^T \in \mathbf{R}^{N-1}$. It is easy to see that 0 is a simple eigenvalue of Π with $\mathbf{1}$ as a corresponding eigenvector and 1 is the other eigenvalue with multiplicity $N-1$. In addition, it is easy to verify that $e(t) = \mathbf{0}_{Nn}$ if and only if $x_1(t) = x_2(t) = \cdots = x_N(t)$, thus the consensus problem can be reduced to the asymptotical stability of $e(t)$.

To synthesize a *non-fragile* distributed control law which is optimal in the sense of attenuating both exogenous and initial-state disturbances, the following performance index is introduced.

Definition 11.2 *Given the disturbance attenuation level $\gamma > 0$ and the weighting positive definite matrix $R = R^T > 0$, the H_∞ consensus index synthesized with transient performance of the closed-loop system (11.7) is defined as*

$$\mathbf{J} = \int_0^\infty \{e^T(t)e(t) - \gamma^2[w^T(t)w(t) + e(0)^T(I_N \otimes R)e(0)]\}dt \tag{11.9}$$

where the weighting matrix R is a measure of the relative importance of the uncertainty in initial condition versus the uncertainty in the exogenous signal $w(t)$.

Remark 11.4 *Note that if the initial states are zero, $\mathbf{J}$ is equal to standard H_∞ performance index of $e(t)$ over all admissible exogenous signals $w(t)$. The idea for defining such a performance index is borrowed from the theory of H_∞ control with transients [4].*

11.3 Non-Fragile H_∞ Consensus for Multi-Agent Systems

For notational convenience, let

$$\mathbb{A} = A + n\delta_a I;\ \varepsilon_b = 1 + \sqrt{np}\delta_b;\ \varepsilon_c = 1 + n\delta_c. \tag{11.10}$$

The following theorem presents a sufficient condition for the existence of a protocol $u_i(t)$ in (11.2) for the H_∞ consensus synthesized with model and controller coefficient variations and transient performance problem.

Theorem 11.1 *Given the scalars $\delta_a > 0$, $\delta_b > 0$ and $\delta_c > 0$, suppose that Assumptions 11.1-11.2 hold. Then, the distributed non-fragile H_∞ consensus with performance γ and weighting matrix $R > 0$ for multi-agent system (11.1) can be solved by protocol $u_i(t)$ in (11.2) for any given dwell time $\tau > 0$, if there exist a symmetric matrix $P = P^T > 0$, and positive scalars $c > 0$ and $\gamma > 0$ such that the following LMIs hold.*

$$\begin{bmatrix} He\{\mathbb{A}P\} - 2c\varepsilon_b\varepsilon_c\alpha_0 BB^T & D & P \\ * & -\frac{\gamma^2}{\zeta_{max}}I & 0 \\ * & * & -\zeta_{min}I \end{bmatrix} < 0 \tag{11.11}$$

$$\begin{bmatrix} P & I \\ I & \frac{\gamma^2}{\zeta_{max}} R \end{bmatrix} > 0 \tag{11.12}$$

Here,

$$\zeta_{min} = \min_{i=1,\cdots,N} \zeta_i, \zeta_{max} = \max_{i=1,\cdots,N} \zeta_i$$
$$\alpha_0 = \min_{\sigma(t)\in\{1,2,\cdots,m\}} \alpha(\mathscr{L}^{\sigma(t)})$$

where $\alpha(\mathscr{L}^{\sigma(t)})$ is the generalized algebraic connectivity of graphs $\mathscr{G}^{\sigma(t)}$, $\sigma(t) \in \{1,2,\cdots,m\}$. In addition, $\mathbb{A}$, ε_b, and ε_c are defined in (11.10). Then, the controller gain K can be obtained by

$$K = B^T P^{-1} \tag{11.13}$$

In addition, in order to obtain the minimum upper bound of the H_∞ performance in (11.9), we minimize γ subject to (11.11) and (11.12).

Proof 11.1 *Choose a Lyapunov function for system (11.7):*

$$V(t) = e^T(t)(\Xi \otimes P^{-1})e(t) \tag{11.14}$$

where $\Xi = diag\{\zeta_1,\cdots,\zeta_N\}$, with $\zeta = [\zeta_1,\cdots,\zeta_N]^T \in \mathbf{R}^N$ being the common positive left eigenvector of Laplacian matrices $\mathscr{L}^{\sigma(t)}$, $\sigma(t) \in \{1,2,\cdots,m\}$ associated with the zero eigenvalue, satisfying $\mathbf{1}_N^T\zeta = 1$, and P is a positive Lyapunov matrix. It can be shown that $V(t) \geq 0$ and $V(t) = 0$ if and only if $e(t) = 0$.

Then, for each time interval $[t_k,t_{k+1}), k = 0,1,\cdots$, taking the time derivative of $V(t)$ along the trajectories of systems (11.7) yields

$$\begin{aligned} \dot{V}(t) &= 2e^T(t)(\Xi\otimes P^{-1})[(I_N\otimes\tilde{A}(t) - c(L^{\sigma(t)}\otimes \\ &\quad \tilde{B}(t)\tilde{K})e(t) + (\Pi\otimes D)w(t))] \\ &= e^T(t)[\Xi\otimes He\{P^{-1}\tilde{A}(t)\} - cHe\{\Xi\mathscr{L}^{\sigma(t)}\otimes \\ &\quad P^{-1}\tilde{B}(t)\tilde{K})\}]e(t) + 2e^T(\Xi\Pi\otimes P^{-1}D)w(t). \end{aligned} \tag{11.15}$$

Then, let $K = B^TP^{-1}$, it follows from (11.15) by recalling (11.4) and (11.10) that

$$\begin{aligned} \dot{V}(t) &\leq e^T(t)\{\Xi\otimes He\{P^{-1}\mathbb{A}\} - c\varepsilon_b\varepsilon_c He\{\Xi L^{\sigma(t)}\}\otimes \\ &\quad P^{-1}BB^TP^{-1}\}e(t) + 2e^T(t)(\Xi\Pi\otimes P^{-1}D)w(t) \\ &\leq e^T(t)\{\Xi\otimes[He\{P^{-1}\mathbb{A}\} - 2c\varepsilon_b\varepsilon_c\alpha_0P^{-1}B \\ &\quad B^TP^{-1}]\}e(t) + 2e^T(t)(\Xi\Pi\otimes P^{-1}D)w(t) \end{aligned} \tag{11.16}$$

where $\alpha_0 = \min_{\sigma(t)\in\{1,2,\cdots,m\}} \alpha(L^{\sigma(t)})$, and $\alpha(L^{\sigma(t)})$, $\sigma(t) \in \{1,2,\cdots,m\}$ is the generalized algebraic connectivity of graphs $\mathscr{G}^{\sigma(t)}$, $\sigma(t) \in \{1,2,\cdots,m\}$ defined in Definition 10.1.

Next, the performance of system (11.5) with external disturbances and initial state disturbances simultaneously is discussed. Recalling (11.16) and by using the facts that $\Pi^2 = \Pi$ and the largest eigenvalue λ_{max} of Π is 1 (see Remark 11.3 for details), one can obtain that

$$e^T(t)e(t) \leq e^T(t)(\Xi\otimes\frac{1}{\zeta_{min}}I)e(t)$$

$$
\begin{aligned}
&\gamma^{-2}e^T(t)(\Xi^2\Pi^2 \otimes P^{-1}DD^TP^{-1})e(t)\\
&\leq \zeta_{max}\gamma^{-2}e^T(t)(\Xi\Pi \otimes P^{-1}DD^TP^{-1})e(t)\\
&\leq \zeta_{max}\gamma^{-2}\lambda_{max}e^T(t)(\Xi \otimes P^{-1}DD^TP^{-1})e(t)\\
&= e^T(t)(\Xi \otimes \zeta_{max}\gamma^{-2}P^{-1}DD^TP^{-1})e(t)
\end{aligned}
$$

where $\zeta_{min} = \min_{i=1,\cdots,N} \zeta_i$, *and* $\zeta_{max} = \max_{i=1,\cdots,N} \zeta_i$.

Then, by the definition of the H_∞ *performance in (11.9), one can obtain that*

$$
\begin{aligned}
\mathbf{J} &= \int_0^\infty \{e^T(t)e(t) - \gamma^2[w^T(t)w(t) + e(0)^T(I_N \otimes R)e(0)]\\
&\quad +\dot{V}(e(t))\}dt - V(\infty) + V(0)\\
&\leq \int_0^\infty [e^T(t)\{\Xi \otimes (He\{P^{-1}\mathbb{A}\} - 2c\varepsilon_b\varepsilon_c\alpha_0P^{-1}BB^TP^{-1}\\
&\quad +\tfrac{1}{\zeta_{min}}I) + \gamma^{-2}\Xi^2\Pi^2 \otimes P^{-1}DD^TP^{-1}\}e(t)dt -\\
&\quad \int_0^\infty \gamma^2\tilde{w}(t)^T\tilde{w}(t)dt + V(0) - \gamma^2 e(0)^T(I_N \otimes R)e(0)\\
&\leq \int_0^\infty e^T(t)\{\Xi \otimes [He\{P^{-1}\mathbb{A}\} - 2c\varepsilon_b\varepsilon_c\alpha_0P^{-1}BB^TP^{-1}\\
&\quad +\tfrac{1}{\zeta_{min}}I + \gamma^{-2}\zeta_{max}P^{-1}DD^TP^{-1}]\}e(t)dt\\
&\quad +e(0)^T(\Xi \otimes P^{-1})e(0) - e(0)^T(\Xi \otimes \tfrac{\gamma^2}{\zeta_{max}}R)e(0)
\end{aligned}
$$

where $\tilde{w}(t) = [w(t) - \gamma^{-2}(\Xi\Pi \otimes D^TP^{-1})e(t)]$.

Note that $\mathbf{J} < 0$ *if the following inequalities hold:*

$$
\begin{aligned}
&\Xi \otimes \{He\{P^{-1}\mathbb{A}\} - 2c\varepsilon_b\varepsilon_c\alpha_0P^{-1}BB^TP^{-1} + \tfrac{1}{\zeta_{min}}I\\
&+\gamma^{-2}\zeta_{max}P^{-1}DD^TP^{-1}\} < 0,
\end{aligned}
$$

$$
\Xi \otimes (P^{-1} - \tfrac{\gamma^2}{\zeta_{max}}R) < 0.
$$

Since Ξ *is a positive definite diagonal matrix, the above two inequalities are equivalent to the following inequalities.*

$$
\begin{aligned}
&He\{P^{-1}\mathbb{A}\} - 2c\varepsilon_b\varepsilon_c\alpha_0P^{-1}BB^TP^{-1}\\
&+\tfrac{1}{\zeta_{min}}I + \gamma^{-2}\zeta_{max}P^{-1}DD^TP^{-1} < 0,
\end{aligned} \tag{11.17}
$$

$$
P^{-1} - \tfrac{\gamma^2}{\zeta_{max}}R < 0. \tag{11.18}
$$

Then, by using the Schur complement lemma, (11.18) can be equivalently rewritten as (11.12).

Applying congruence transformations to (11.17) by P, we have

$$
He\{\mathbb{A}P\} - 2c\varepsilon_b\varepsilon_c\alpha_0BB^T + \gamma^{-2}\zeta_{max}DD^T + \tfrac{1}{\zeta_{min}}P^2 < 0.
$$

Thus, we can get (11.12) immediately by using the Schur complement lemma. This completes the proof. □

Remark 11.5 *It is worth noting that the coupling weight c in (11.2) can be obtained by solving the LMI conditions (11.11) and (11.12) in Theorem 11.1. This will reduce the conservativeness of the previous results in [110, 112, 113, 176, 179, 181, 199] which require c to be larger than a threshold value.*

Remark 11.6 *The high-order multi-agent system considered in this chapter is very general and our result is applicable to all lower order multi-agent systems that are considered in most of the existing literature (such as the second-order multi-agent systems considered in [199, 200]). In addition, when the system uncertainties and controller coefficient variations are not considered (i.e., $\Delta A(t) = 0$, $\Delta B(t) = 0$ and $\Delta K(t) = 0$), Theorem 11.1 reduced to the result of [181].*

11.4 Numerical Example

In this section, a simulation example is given to validate the effectiveness of the proposed control design method based on simulation of the linearized longitudinal dynamical equations of aircraft. The equation corresponding to the dynamics of the ith aircraft is given by (11.1) with [203].

$$A = \begin{bmatrix} -0.277 & 1 & -0.0002 \\ -17.1 & -0.178 & -12.2 \\ 0 & 0 & -6.67 \end{bmatrix}$$
$$x_i(t) = \begin{bmatrix} x_{i1}(t) \\ x_{i2}(t) \\ x_{i3}(t) \end{bmatrix}, B = D = \begin{bmatrix} 0 \\ 0 \\ 6.67 \end{bmatrix}$$

where $x_{i1}(t), x_{i2}(t), x_{i3}(t)$ denote the angle of attack, the pitch rate and elevator angle, respectively. The control input $u_i(t)$ is the command to the elevator. The switching weighted topology structures of networks with 4 agents is shown in Figure 11.1. It is assumed that the communication network periodically switches according to $\mathcal{G}^1 \to \mathcal{G}^2 \to \mathcal{G}^3 \to \mathcal{G}^1$ every $\tau = 0.1$ second. As stated in Remark 11.2, we can obtain a common positive vector $\zeta = 0.25 \times \mathbf{1}_4$, which implies $\zeta_{max} = \zeta_{min} = 0.25$. Moreover, by using Algorithm 2.1 in Chapter 2, we can obtain $\alpha(L^1) = 0.2610$, $\alpha(L^2) = 2.9493$ and $\alpha(L^3) = 3.2907$, which implies $\alpha_0 = 0.2610$.

By solving the LMI conditions (11.11) and (11.12) in Theorem 11.1 by using the LMI toolbox of MATLAB, one gets a feasible solution as

$$\begin{aligned} P &= \begin{bmatrix} 0.0869 & -0.1218 & -0.0961 \\ -0.1218 & 1.9847 & 1.5372 \\ -0.0961 & 1.5372 & 2.8387 \end{bmatrix}, c = 1.3798 \\ K &= \begin{bmatrix} 0.0908 & -3.1287 & 4.0469 \end{bmatrix}, \gamma = 2.5261. \end{aligned}$$

In order to verify the effectiveness of the proposed method, simulations are carried out with the following parameters:

$$\begin{aligned} &\Delta A(t) = [0.2\sin(t)]_{3\times3},\ \Delta B(t)(t) = 0.2\cos(5t) \\ &\Delta K(t) = [0.2\sin(t)]_{3\times3},\ w_i(t) = -2\sin(2t),\ R = I_3 \end{aligned} \tag{11.19}$$

which implies $\delta_a = 0.2$, $\delta_b = 0.2$, and $\delta_c = 0.2$. In addition, let each coordinate of the initial states be chosen randomly.

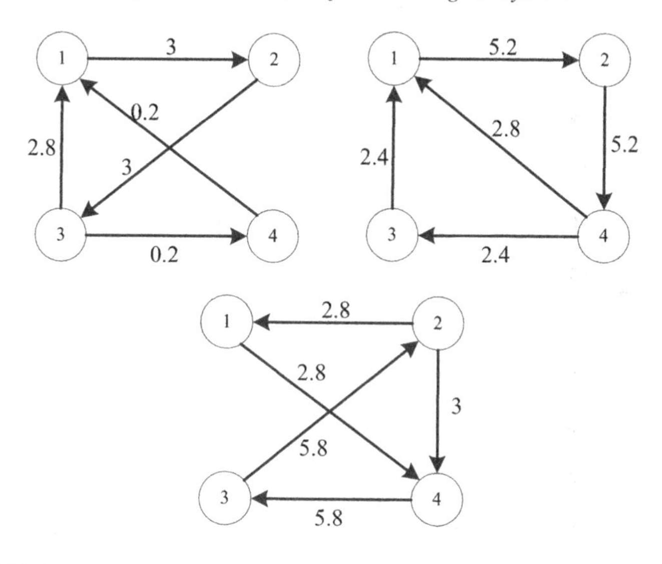

FIGURE 11.1
Three directed switching weighted graphs.

With *non-zero initial* states and *external disturbances* in (11.19), the trajectories of the states and responses of control inputs by using distributed *non-fragile* consensus protocol (11.2) are shown in Figures 11.2-11.4 and Figure 11.5, respectively. It is clear that the state trajectories of the four aircraft under protocol (11.2) reach global consensus despite *multiplicative* controller coefficient variations, which implies that the new protocol can effectively eliminate the effect of *multiplicative* controller coefficient variations. In addition, chattering phenomenon happens in all control input in Figure 11.5 due to switching network.

11.5 Conclusion

This chapter has addressed the *non-fragile H_∞ consensus* problem for linear multi-agent systems with model uncertainties, *multiplicative* controller coefficient variations and *external disturbances*. The considered communication network is a switching weighted balanced directed network. The designed distributed *non-fragile H_∞ consensus* protocols can be used to achieve a satisfactory performance against model uncertainties, *external disturbances* and unknown initial states, and simultaneously eliminate the effect of *multiplicative* controller coefficient variations. It is shown that the original multi-agent systems can reach global consensus by a numerical example. An interesting direction for future study is Markovian switching network topologies.

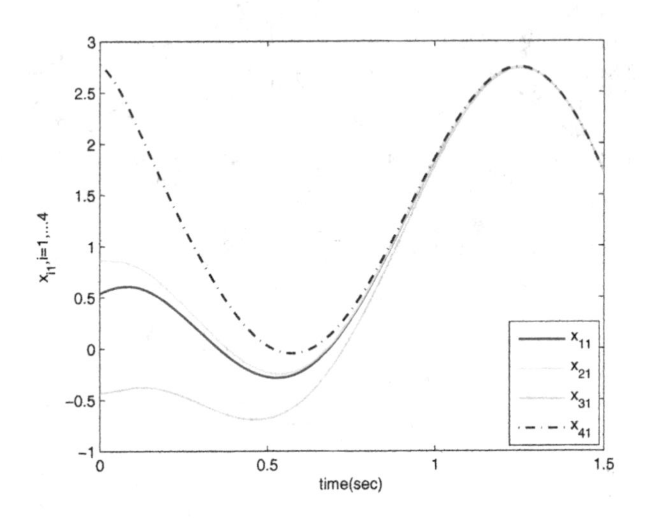

FIGURE 11.2
The states $x_{i1}, i = 1, \cdots, 4$ reach global consensus.

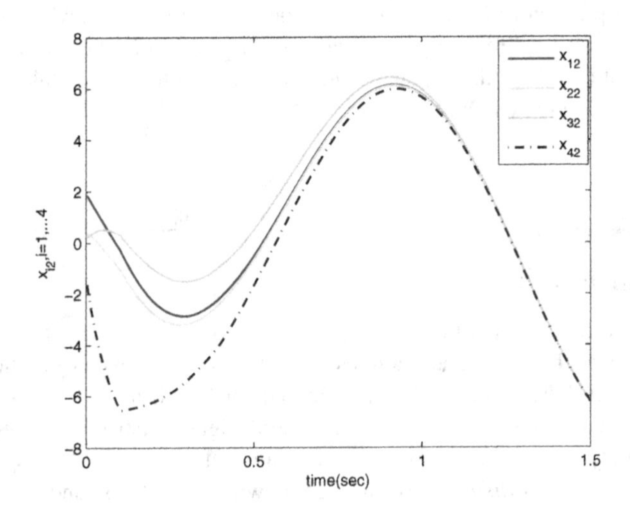

FIGURE 11.3
The states $x_{i2}, i = 1, \cdots, 4$ reach global consensus.

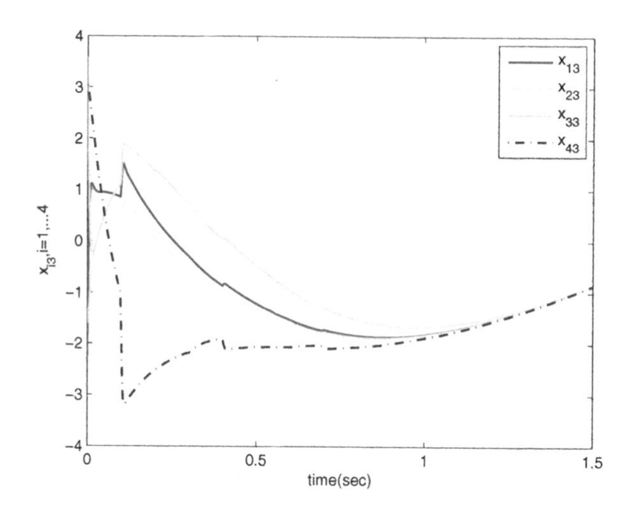

FIGURE 11.4
The states $x_{i3}, i = 1, \cdots, 4$ reach global consensus.

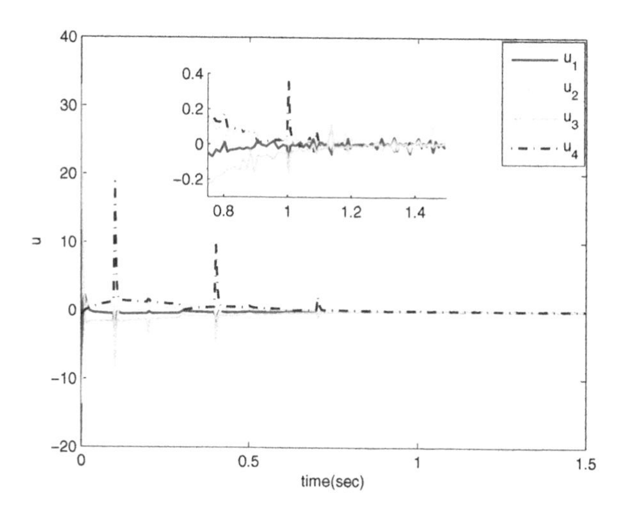

FIGURE 11.5
Distribute control $u_i(t), i = 1, \cdots, 4$.

12

Quantized H_∞ Consensus for Multi-Agent Systems with Quantization Mismatch

This chapter investigates the distributed quantized H_∞ *consensus* problems for general linear and *Lipschitz* nonlinear *multi-agent systems* with input *quantization* mismatch and external disturbances under switching weighted undirected or balanced directed topologies. The designed distributed quantized H_∞ *consensus* protocol can be divided into two parts which are linear and nonlinear parts. The linear part plays a role in achieving a satisfactory performance against interval-bounded model uncertainties, *external disturbances* and unknown initial states. The nonlinear part eliminates the effect of input *quantization*. It should be mentioned that complete consensus instead of practical consensus can be achieved in the presence of uniform *quantization*. In addition, similar to Chapter 11, the coupling strength in this chapter is determined by solving some linear matrix inequalities (LMIs). Sufficient conditions for the existence of the proposed control strategy are also obtained by using the LMI technique.

12.1 Introduction

In recent years, the problems of consensus have become hot topics in the study of cooperative control of *multi-agent systems* due to their broad applications in various areas, such as spacecraft formation flying, sensor networks, flocking, and cooperative surveillance and so on [11, 21, 33, 110, 123, 199]. On the other hand, quantization plays an important role in information exchange among agents and is one of the basis limitations induced by finite channel capacity [39], and finite sampling rates [98, 128, 197, 198]. As a result, *quantized consensus* or consensus with quantized communications becomes interesting and more meaningful, and has been an active research topic in the literature [40, 49, 117, 185, 188]. *Quantized consensus* problem of first-order and second-order *multi-agent systems* with uniform quantizers under static and time-varying communication topologies was considered in [117, 185]. Since many real coupled dynamical systems can be modeled as *multi-agent systems* with high-order dynamics, such as distributed unmanned air vehicles and coupled manipulators, *quantized consensus* control for high-order *multi-agent systems* was further investigated in [40, 49, 188]. Note that the references discussed above do not consider the difference of the quantization parameters between the encoder and decoder side. However, the neighbors of a given agent may change with time, which may result in a *mismatch* between the encoder and decoder [103]. When quantization parameter mismatch is ignored in the controller design, it might induce instability of the closed-loop systems [210]. Therefore, many efforts have been made to investigate how dynamic coding/decoding channels with dynamic uniform quantizers affect the consensus [99, 103, 104]. Nevertheless, the above-mentioned results only focus on agents with first-order integrator dynamics, and the case of high-order *multi-agent systems* remains yet to be a theoretically challenging issue.

On the other hand, the systems are unavoidably affected by various disturbances such as sensor noises, actuator bias, measurement or calculation errors and the variation of communication links, which might degrade convergence performance of the closed-loop multi-agent systems, and sometimes even cause the systems to become unstable [25], which will make a network never reach consensus [136]. Therefore, synthesized consensus protocols should be provided to attenuate the effects of both exogenous disturbances and unknown initial states on state consensus, which has received particular research attention [112, 176, 179, 181]. The H_∞ *consensus* control problem for multi-agent systems with both missing measurements and parameter uncertainties over fixed topology was investigated in [179]. The global H_∞ *consensus* problem of *Lipschitz* nonlinear multi-agent systems was further investigated in [112]. Additionally, a class of distributed protocols to achieve consensus for general linear systems with both *external disturbances* and unknown initial states was constructed in [176, 181]. Nevertheless, it is still an open question and also more challenging to eliminate the quantization effects on consensus for general linear/nonlinear systems with *external disturbances* and/or initial states under switching topologies.

Motivated by the aforementioned observations, this chapter concerns with the distributed quantized H_∞ *consensus* problems for both linear and *Lipschitz* nonlinear multi-agent systems with mismatched quantization parameters and *external disturbances* under switching weighted undirected or balanced directed topologies. The main contributions of this chapter are summarized as follows.

(1) The corresponding consensus protocols are designed for both linear and *Lipschitz* nonlinear multi-agent systems. Each of these protocols consists of two parts which are linear and nonlinear parts. The linear part is used to achieve a satisfactory performance against model uncertainty, *external disturbances* and unknown initial states. The model uncertainty considered in this chapter is an interval-bounded uncertainty. The nonlinear part is used to eliminate the effect of input quantization.

(2) Compared with the previous results in [39, 40, 49, 117, 185, 188], where only practical consensus can be achieved for the case of uniform quantization, it is proved in this chapter that complete consensus can still be reached even if there exists uniform quantization.

(3) The issue of robustness with respect to mismatch of the encoder/decoder is considered. Thus the quantized multi-agent systems under consideration are more general than those in [40, 39, 49, 117, 188, 202].

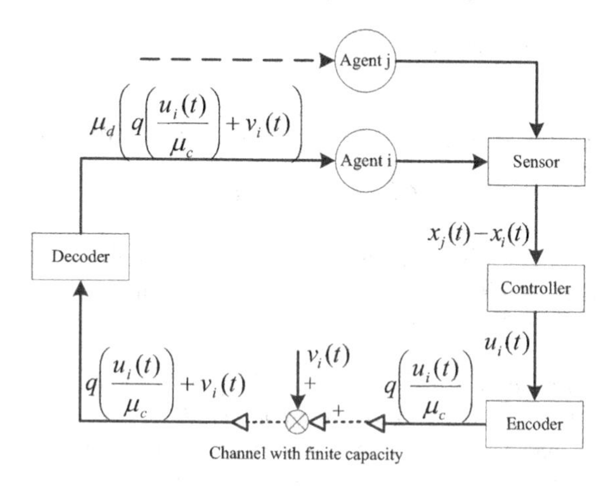

FIGURE 12.1
The framework of quantized distributed consensus protocol.

12.2 Quantized H_∞ Consensus for General Linear Dynamics

A multi-agent system composed of N interconnected continuous time subsystems subjected to *external disturbances* is considered. The dynamics of the ith agent is described by the following

$$\dot{x}_i(t) = \tilde{A}(t)x_i(t)+\tilde{B}(t)Q(u_i(t))+Dw_i(t) \tag{12.1}$$

where $\tilde{A}(t)=A+\Delta A(t)$, $\tilde{B}(t)=B(I+\Delta B(t))$, $x_i(t)\in \mathbf{R}^n, i=0,\cdots,N$ are the states of agent i, $u_i(t)\in \mathbf{R}^p$ is the control input acting on agent i, and $w_i(t)\in \mathbf{R}^q$ is the external disturbance input that belongs to $L_2[0,\infty)$. $\Delta A(t)$, $\Delta B(t)$ and $Q(\cdot)$ are the system characteristic matrix uncertainty, input matrix uncertainty and quantization operator, respectively.

The quantizer operator $Q(\cdot)$ as in [86, 202] is taken as follows.

$$\begin{aligned} Q(u_i(t)) &\triangleq \mu_d(round(u_i(t)/\mu_c)+v_i(t)) \\ &\triangleq \mu_d(q(u_i(t)/\mu_c)+v_i(t)) \end{aligned} \tag{12.2}$$

where the function $q(\cdot)\triangleq round(\cdot)$ rounds toward the nearest integer, and $q(u_i(t)/\mu_c)$ is the information transmitted over the digital communication channel (see Figure 12.1). The positive real parameters $\mu_c>0$ and $\mu_d>0$ are called *quantization adjustable parameters*, which are given at the encoder side and the decoder side, respectively. The term $v_i(t)$ in Figure 12.1 corresponds to a general noise in the channel, and it can model the pure time-delay, packet loss, bit-errors, and so on [86]. In this chapter, the issue of robustness with respect to mismatch of the encoder/decoder is considered. To simplify the presentation, the following assumption is made.

Assumption 12.1 *[86] Perfect channel assumption that* $v_i(t) \equiv 0, \forall t \geq 0, i = 1, \cdots, N$.

The quantizer operator $Q(\cdot)$ in (12.2) can further be rewritten as

$$Q(u_i(t)) = \mu q_\mu(u_i(t)) \tag{12.3}$$

where $q_\mu(u_i(t)) \triangleq \mu_c q(u_i(t)/\mu_c)$, and $\mu = \frac{\mu_d}{\mu_c}$ is defined as the *mismatch ratio*. Then, one can find that $q_\mu(u_i(t))$ is a uniform quantizer which is essentially the same as those reported in [39, 40, 49, 117, 185, 188].

Remark 12.1 *In general, it is assumed that the quantization adjustable parameters at the encoder and at the decoder are the same [39, 40, 117, 188, 202], i.e.,* $\mu_c = \mu_d$. *However, due to uncertainty in the parameters of the quantizers, this may not be always the case [86, 210], i.e.,* $\mu_c \neq \mu_d$. *The robustness of quantized linear systems with respect to this mismatch between encoder/decoder initializations is first considered in [86]. Then, [210] further considered a special case of* $\mu \in [\mu_m, \mu_M]$ *with* $\mu_M \geq \mu_m > 0$, *together with matched external disturbances, i.e.,* $B = D$.

In addition, throughout this chapter, the following assumption is made.

Assumption 12.2 *The time-varying uncertainty* $\Delta A(t)$ *satisfies*

$$\Delta A(t) = [\delta_{kj}(t)]_{n\times n}, k = 1, \cdots, n; j = 1, \cdots, n \tag{12.4}$$

where $\delta_{kj}(t)$, *(for* $k, j = 1, \cdots, n$*) are used to denote the magnitudes of the deviation of the coefficients of* A *with* $|\delta_{kj}(t)| \leq \delta_a$, *for* $k, j = 1, \cdots, n$. *In addition, the time-varying uncertainty* $\Delta B(t)$ *satisfies* $\|\Delta B(t)\|_\infty \leq \delta_b < 1$.

Remark 12.2 *The uncertainty model in (12.4) has been investigated extensively in [13, 29, 50, 190], where this model is used to describe the controller coefficient variations. This interval-bounded uncertainty is obviously less conservative than the norm-bounded uncertainty considered in [183, 202, 210].*

It is noted that $\Delta A(t)$ in (12.4) can be further transformed into

$$\Delta A(t) = \sum_{k=1}^{n}\sum_{j=1}^{n} \delta_{kj}(t) e_k e_j^T$$

which can also be rewritten as

$$\Delta A(t) = E\Re(t)F \tag{12.5}$$

where

$$\begin{aligned} E &= \begin{bmatrix} E_1 & E_2 & \cdots & E_\nu & \cdots & E_{n^2} \end{bmatrix} \\ F^T &= \begin{bmatrix} F_1^T & F_2^T & \cdots & F_\nu^T & \cdots & F_{n^2}^T \end{bmatrix} \\ \Re(t) &= diag\{\delta_{11}(t), \delta_{12}(t), \cdots, \delta_{kj}(t), \cdots, \delta_{nn}(t)\} \end{aligned}$$

with

$$E_\nu = e_k, F_\nu = e_j^T, \text{ for } \nu = (k-1)n + j; k, j = 1, \cdots, n$$

where $e_k \in R^n$ denotes the unit column vector in which the kth element equals 1 and the others equal 0.

It is well known that, in practice, the underlying topology of the mobile agents may switch among some possible topologies due to, for instance, limited sensing radius, temporary sonar equipment failures, or the presence of communication obstacles [181]. Thus, the communication topology is assumed to be dynamically switching over the connected graph set $\mathscr{G}^{\sigma(t)}$, $\sigma(t) \in \{1, 2, \cdots, p\}, p \geq 1$, where $\sigma(t)$ denotes the set of all possible undirected or balanced directed topologies. Consider an infinite sequence of non-overlapping time intervals $[t_k, t_{k+1}), k \in \mathbf{N}$, with $t_0 = 0$, $t_{k+1} - t_k \geq \tau$ and $\tau > 0$, across which the communication topology is fixed. Here, the time sequence $t_1, t_2, \cdots$, is called the switching sequence, at which the communication topology changes. During each time interval $[t_k, t_{k+1})$, the graph $\mathscr{G}$ is time invariant. Then, in order to achieve consensus for linear multi-agent systems with input quantization and *external disturbances*, the following nonlinear distributed consensus protocol (NDCP) is proposed

$$u_i(t) = u_{1i}(t) + u_{2i}(t), i = 1, \cdots, N \tag{12.6}$$

with

$$u_{1i}(t) = cK \sum_{j=1}^{N} a_{ij}^{\sigma(t)} (x_j - x_i), i = 1, \cdots, N \tag{12.7}$$

where $c > 0$ is the coupling weight among neighboring agents, $K \in \mathbf{R}^{p \times n}$ is the feedback gain matrix, $a_{ij}^{\sigma(t)}$ is (i, j)-th entry of the *adjacency matrix* of communication topology $\mathscr{G}^{\sigma(t)}$. In addition, $u_{2i}(t)$ is a nonlinear part and will be designed later.

Remark 12.3 *We note that the controller $u_i(t)$ in (12.6) consists of two parts, namely linear and nonlinear parts. The linear part $u_{1i}(t)$ is constructed to achieve the H_∞ performance against model uncertainties, external disturbances and initial-state disturbances, while the purpose of introducing the nonlinear part $u_{2i}(t)$ into the control law is to handle quantization errors. The framework of quantized distributed consensus protocol is shown in Figure 12.1.*

Let the quantization error $\Delta u_i(t)$ be defined as

$$\Delta u_i(t) \triangleq q_\mu(u_i(t)) - u_i(t), \tag{12.8}$$

then,

$$q_\mu(u_i(t)) = \Delta u_i(t) + u_i(t) = \Delta u_i(t) + u_{1i}(t) + u_{2i}(t). \tag{12.9}$$

In addition, based on the condition of (12.3) and the definition of $\Delta u_i(t)$, each component of $\Delta u_i(t)$ at time t is bounded by half of the quantization level μ_c, i.e.,

$$\|\Delta u_i(t)\|_\infty \leq \frac{\mu_c}{2}. \tag{12.10}$$

Then, it follows from (12.1) and (12.9) that

$$\begin{aligned}\dot{x}_i(t) &= \tilde{A}(t)x_i(t) + c\mu\tilde{B}(t)K\sum_{j=1}^{N} a_{ij}^{\sigma(t)}(x_j - x_i) \\ &\quad + \mu\tilde{B}(t)(\Delta u_i(t) + u_{2i}(t)) + Dw_i(t).\end{aligned} \tag{12.11}$$

By using the *Kronecker product*, it is not difficult to obtain that the closed-loop network dynamics (12.11) can be rewritten as

$$\begin{aligned}\dot{x}(t) &= [I_N \otimes \tilde{A}(t) - c\mu(\mathscr{L}^{\sigma(t)} \otimes \tilde{B}(t)K)]x(t) \\ &\quad + \mu(I_N \otimes \tilde{B}(t))(u_2(t) + \Delta u(t)) + (I_N \otimes D)w(t)\end{aligned} \tag{12.12}$$

where

$$\begin{aligned}x^T(t) &= \left[x_1^T(t), \cdots, x_N^T(t)\right] \\ u_2^T(t) &= \left[u_{21}^T(t), \cdots, u_{2N}^T(t)\right] \\ \Delta u^T(t) &= \left[\Delta u_1^T(t), \cdots, \Delta u_N^T(t)\right] \\ w^T(t) &= \left[w_1^T(t), \cdots, w_N^T(t)\right].\end{aligned}$$

For designing NDCP protocol $u_i(t)$ in (12.6)-(12.7), the following assumption is made.

Assumption 12.3 *There exists one common positive vector* $\zeta = [\zeta_1, \cdots, \zeta_N]^T \in \mathbf{R}^N$, *such that* $\zeta^T \mathscr{L}^{\sigma(t)} = \mathbf{0}_N^T$ *and* $\mathbf{1}_N^T \zeta = 1$, *where* $\mathscr{L}^{\sigma(t)}$ *is the Laplacian matrix of strongly connected graph* $\mathscr{G}^{\sigma(t)}$, $\sigma(t) \in 1, 2, \cdots, p$.

Remark 12.4 *Assumption 12.3 holds if every possible topology* $\mathscr{G}^{\sigma(t)}$, $\sigma(t) \in \{1, 2, \cdots, p\}$ *is a connected undirected graph or strongly connected and balanced directed graph.*

Define

$$e(t) = (\Pi \otimes I_n)x(t) \tag{12.13a}$$

where

$$\Pi = (I_N - \mathbf{1}_N \zeta^T) \in \mathbf{R}^{N \times N} \tag{12.13b}$$

which can imply that $\Pi \cdot \Pi = \Pi$ and $\Pi \cdot \mathscr{L}^{\sigma(t)} = \mathscr{L}^{\sigma(t)}$. Then, (12.12) can be rewritten as

$$\begin{aligned}\dot{e}(t) &= [I_N \otimes \tilde{A}(t) - c\mu(\mathscr{L}^{\sigma(t)} \otimes \tilde{B}(t)K)]e(t) \\ &\quad + \mu(\Pi \otimes \tilde{B}(t))(u_2(t) + \Delta u(t)) + (\Pi \otimes D)w(t).\end{aligned} \tag{12.14}$$

Before proceeding further, the consensus for (12.1) under NDCP protocol $u_i(t)$ in (12.6)-(12.7) is defined.

Definition 12.1 *The quantized consensus problem of multi-agent system (12.1) is solved by NDCP protocol* $u_i(t)$ *in (12.6)-(12.7) if, for any initial condition* $x(0)$, *the states of the closed-loop system (12.11) satisfy*

$$\lim_{t \to \infty} \|x_i(t) - x_j(t)\|_2 = 0, \forall i, j = 1, \cdots, N. \tag{12.15}$$

Remark 12.5 *Construct the following matrix*

$$\Omega = \begin{bmatrix} 1 & \mathbf{0}_{N-1}^T \\ \mathbf{1}_{N-1} & I_{N-1} \end{bmatrix} \in \mathbf{R}^{N\times N}.$$

Then, pre- and post-multiplying Π by Ω^{-1} and Ω, respectively, it follows by using the fact $\mathbf{1}_N^T\zeta = 1$ that

$$\Gamma = \Omega^{-1}\Pi\Omega = \begin{bmatrix} 0 & \tilde{\zeta}^T \\ \mathbf{0}_{N-1} & I_{N-1} \end{bmatrix} \in \mathbf{R}^{N\times N}$$

where $\tilde{\zeta} = [-\zeta_2, -\zeta_3, \cdots, -\zeta_N]^T \in \mathbf{R}^{N-1}$. It is easy to see that 0 is a simple eigenvalue of Π with $\mathbf{1}$ as a corresponding eigenvector and 1 is the other eigenvalue with multiplicity $N-1$. In addition, it is easy to verify that $e(t) = \mathbf{0}_{Nn}$ if and only if $x_1(t) = x_2(t) = \cdots = x_N(t)$, thus the consensus problem can be reduced to the asymptotical stability of $e(t)$.

To synthesize a quantized control law which is optimal in the sense of attenuating both exogenous and initial-state disturbances, the following performance index is introduced.

Definition 12.2 *Given the disturbance attenuation level $\gamma > 0$ and the weighting positive definite matrix $R = R^T > 0$, the H_∞ consensus index synthesized with the transient performance of the closed-loop system (12.14) is defined as*

$$\mathbf{J} \;=\; \int_0^\infty \{e^T(t)e(t) - \gamma^2[w^T(t)w(t) + e^T(0)(I_N \otimes R)e(0)]\}dt \qquad (12.16)$$

where the weighting matrix R is a measure of the relative importance of the uncertainty in initial condition versus the uncertainty in the exogenous signal $w(t)$.

Remark 12.6 *Note that if the initial states are zero, $\mathbf{J}$ is equal to standard H_∞ performance index of $e(t)$ over all admissible exogenous signals $w(t)$. The idea for defining such a performance index is borrowed from the theory of H_∞ control with transients [4].*

The following theorem presents a sufficient condition for the existence of an NDCP protocol $u_i(t)$ in (12.6)-(12.7) for the H_∞ consensus synthesized with input quantization and transient performance problem.

Theorem 12.1 *Suppose that Assumptions 12.1-12.3 hold. Then, the distributed quantized H_∞ consensus with performance γ and weighting matrix $R > 0$ for multi-agent system (12.1) can be solved by NDCP protocol $u_i(t)$ in (12.6)-(12.7) for any given dwell time $\tau > 0$ and mismatch ratio $\mu > 0$, if there exist a symmetric matrix $P = P^T > 0$, a diagonal matrix $\Lambda > 0$, and positive scalars $c > 0$ and $\gamma > 0$ such that the following LMIs hold.*

$$\begin{bmatrix} \Phi & D & E\Lambda & \delta_a PF^T \\ * & -\frac{\gamma^2}{\zeta_{max}}I & 0 & 0 \\ * & * & -\Lambda & 0 \\ * & * & * & -\Lambda \end{bmatrix} < 0 \qquad (12.17)$$

$$\begin{bmatrix} P & I \\ I & \frac{\gamma^2}{\zeta_{max}}R \end{bmatrix} > 0 \tag{12.18}$$

Here, $\Phi = He\{AP\} - 2c\mu(1+\delta_b)\alpha_0 BB^T + \frac{P^2}{\zeta_{min}}$, $\zeta_{min} = \min_{i=1,\cdots,N} \zeta_i$, $\zeta_{max} = \max_{i=1,\cdots,N} \zeta_i$, $\alpha_0 = \min_{\sigma(t)\in\{1,2,\cdots,p\}} \alpha(\mathscr{L}^{\sigma(t)})$ *and* $\alpha(\mathscr{L}^{\sigma(t)})$ *is the generalized algebraic connectivity of graphs* $\mathscr{G}^{\sigma(t)}$, $\sigma(t) \in \{1,2,\cdots,p\}$. *Then, the controller gain* K *can be obtained by*

$$K = B^T P^{-1} \tag{12.19}$$

and the nonlinear part of NDCP protocol $u_i(t)$ *in (12.6) is designed as*

$$\begin{aligned} u_{2i}(t) &= -\tfrac{\mu_c(1+\delta_b)}{2(1-\delta_b)} sign(\rho_i^T(x(t))) \\ \rho(x(t)) &= x^T(t)(\Pi^T \Xi \Pi \otimes P^{-1}B) \end{aligned} \tag{12.20}$$

where $\rho_i(x(t))$ *is the ith component of* $\rho(x(t))$, $sign(\rho)$ *is the sign of a scalar* ρ *with* $sign(0) = 0$, *and* Π *and* Ξ *are defined in (12.13b) and (12.21) below, respectively. In addition, in order to obtain the minimum upper bound of the* H_∞ *performance in (12.16), we minimize* γ *subject to (12.17) and (12.18).*

Proof 12.1 *Consider the following Lyapunov function candidate:*

$$V(e(t)) = e^T(t)(\Xi \otimes P^{-1})e(t) \tag{12.21}$$

where $\Xi = diag\{\zeta_1,\cdots,\zeta_N\}$, *with* $\zeta = [\zeta_1,\cdots,\zeta_N]^T \in \mathbf{R}^N$ *being the common positive left eigenvector of Laplacian matrices* $\mathscr{L}^{\sigma(t)}$, $\sigma(t) \in \{1,2,\cdots,p\}$ *associated with the zero eigenvalue, satisfying* $\mathbf{1}_N^T\zeta = 1$, *and* P *is a positive Lyapunov matrix. It will be shown that* $V(e(t)) \geq 0$ *and* $V(e(t)) = 0$ *if and only if* $e(t) = 0$.

Since $e(t) = (\Pi \otimes I_n)x(t)$, *one can obtain that*

$$\begin{aligned} \rho(x(t)) &= x^T(t)(\Pi^T \Xi \Pi \otimes P^{-1}B) \\ &= x^T(t)(\Pi^T \otimes I_n)(\Xi\Pi \otimes P^{-1}B) \\ &= e^T(t)(\Xi\Pi \otimes P^{-1}B) \\ &\triangleq \hat{\rho}(e(t)). \end{aligned} \tag{12.22}$$

Then, for each time interval $[t_k, t_{k+1}), k = 0,1,\cdots$, *taking the time derivative of* $V(e(t))$ *along the trajectories of systems (12.14) yields*

$$\begin{aligned} \dot{V}(e(t)) &= 2e^T(t)(\Xi \otimes P^{-1})[(I_N \otimes \tilde{A}(t) - c\mu(\mathscr{L}^{\sigma(t)} \otimes \tilde{B}(t)K)e(t) \\ &\quad + \mu(\Pi \otimes \tilde{B}(t))(u_2(t) + \Delta u(t)) + (\Pi \otimes D)w(t))] \\ &= e^T(t)[\Xi \otimes He\{P^{-1}\tilde{A}(t)\} - c\mu He\{\Upsilon_1(t)\}]e(t) \\ &\quad + 2e^T(t)(\Xi\Pi \otimes P^{-1}D)w(t) + \Upsilon_2(t) \end{aligned} \tag{12.23}$$

where

$$\begin{aligned} \Upsilon_1(t) &= \Xi\mathscr{L}^{\sigma(t)} \otimes P^{-1}\tilde{B}(t)K \\ \Upsilon_2(t) &= 2\mu\hat{\rho}(e(t))[I_N \otimes (I_p + \Delta B(t))](u_2(t) + \Delta u(t)). \end{aligned}$$

By using the condition (12.10) and relations of $|\alpha^T\beta| \leq \|\alpha\|_1\|\beta\|_\infty$ and $\|X\alpha\|_\infty \leq \|X\|_\infty\|\alpha\|_\infty$ for $\alpha, \beta \in \mathbf{R}^n$ and appropriate dimension matrix X, we have that

$$\begin{aligned}
\Upsilon_2(t) &= 2\mu\hat{\rho}(e(t))u_2(t) + 2\mu\hat{\rho}(e(t))(I_N \otimes \Delta B(t))u_2(t) \\
&\quad +2\mu\hat{\rho}(e(t))[I_N \otimes (I_p + \Delta B(t))]\Delta u(t) \\
&\leq 2\mu\hat{\rho}(e(t))u_2(t) + 2\mu\|\hat{\rho}(e(t))\|_1\|\Delta B(t)\|_\infty\|u_2(t)\|_\infty \\
&\quad +2\mu\|\hat{\rho}(e(t))\|_1\|(I_p + \Delta B(t))\|_\infty\|\Delta u(t)\|_\infty \\
&\leq 2\mu\hat{\rho}(e(t))u_2(t) + 2\mu\delta_b\|\hat{\rho}(e(t))\|_1\|u_2(t)\|_\infty \\
&\quad +\mu_c\mu(1+\delta_b)\|\hat{\rho}(e(t))\|_1 \\
&= 2\mu\delta_b\hat{\rho}(e(t))u_2(t) + 2\mu\delta_b\|\hat{\rho}(e(t))\|_1\|u_2(t)\|_\infty \\
&\quad +2\mu(1-\delta_b)\hat{\rho}(e(t))u_2(t) + \mu_c\mu(1+\delta_b)\|\hat{\rho}(e(t))\|_1.
\end{aligned} \tag{12.24}$$

Recalling (12.22), (12.20) can be rewritten as

$$u_{2i}(t) = -\frac{\mu_c(1+\delta_b)}{2(1-\delta_b)} sign(\hat{\rho}_i^T(e(t))). \tag{12.25}$$

Applying $u_{2i}(t)$ in (12.25) to (12.24), one can obtain that

$$\begin{aligned}
&2\mu\delta_b\hat{\rho}(e(t))u_2(t) + 2\mu\delta_b\|\hat{\rho}(e(t))\|_1\|u_2(t)\|_\infty = 0 \\
&2\mu(1-\delta_b)\hat{\rho}(e(t))u_2(t) + \mu_c\mu(1+\delta_b)\|\hat{\rho}(e(t))\|_1 = 0
\end{aligned}$$

which means $\Upsilon_2(t) = 0$. Then, substituting (12.19) into (12.23) gives

$$\begin{aligned}
\dot{V}\ &(e(t)) \\
=\ & e^T(t)[\Xi \otimes He\{P^{-1}\tilde{A}(t)\} + 2e^T(t)(\Xi\Pi \otimes P^{-1}D)w(t) \\
& -c\mu He\{\Xi\mathscr{L}^{\sigma(t)} \otimes P^{-1}\tilde{B}(t)B^TP^{-1}\}]e(t) \\
\leq\ & e^T(t)\{\Xi \otimes He\{P^{-1}\tilde{A}(t)\} + 2e^T(t)(\Xi\Pi \otimes P^{-1}D)w(t) \\
& -c\mu(1+\delta_b)He\{\Xi\mathscr{L}^{\sigma(t)}\} \otimes P^{-1}BB^TP^{-1})\}e(t) \\
\leq\ & e^T(t)\{\Xi \otimes [He\{P^{-1}\tilde{A}(t)\} - \mathscr{B}]\}e(t) \\
& +2e^T(t)(\Xi\Pi \otimes P^{-1}D)w(t)
\end{aligned} \tag{12.26}$$

where

$$\begin{aligned}
\mathscr{B} &= 2c\mu(1+\delta_b)\alpha_0 P^{-1}BB^TP^{-1} \\
\alpha_0 &= \min_{\sigma(t)\in\{1,2,\cdots,p\}} \alpha(\mathscr{L}^{\sigma(t)})
\end{aligned}$$

and $\alpha(\mathscr{L}^{\sigma(t)})$, $\sigma(t) \in \{1,2,\cdots,p\}$ is the generalized algebraic connectivity of graphs $\mathscr{G}^{\sigma(t)}$, $\sigma(t) \in \{1,2,\cdots,p\}$ defined in Definition 2.3 of Chapter 2.

Next, the performance of system (12.11) with external disturbances and initial state disturbances simultaneously is discussed. Recalling (12.26) and by using the facts $\Pi^2 = \Pi$ and the largest eigenvalue λ_{max} of Π is 1 (see Remark 12.5 for details), one can obtain that

$$e^T(t)e(t) \leq e^T(t)(\Xi \otimes \frac{1}{\zeta_{min}}I)e(t)$$

$$\begin{aligned}
&\gamma^{-2}e^T(t)(\Xi^2\Pi^2 \otimes P^{-1}DD^TP^{-1})e(t) \\
&\leq \zeta_{max}\gamma^{-2}e^T(t)(\Xi\Pi \otimes P^{-1}DD^TP^{-1})e(t) \\
&\leq \zeta_{max}\gamma^{-2}\lambda_{max}e^T(t)(\Xi \otimes P^{-1}DD^TP^{-1})e(t) \\
&= e^T(t)(\Xi \otimes \zeta_{max}\gamma^{-2}P^{-1}DD^TP^{-1})e(t)
\end{aligned}$$

where $\zeta_{min} = \min_{i=1,\cdots,N} \zeta_i$, $\zeta_{max} = \max_{i=1,\cdots,N} \zeta_i$.

Then, by the definition of the H_∞ *performance in (12.16), one can obtain that*

$$\begin{aligned}
\mathbf{J} &= \int_0^\infty \{e^T(t)e(t) - \gamma^2[w^T(t)w(t) + e^T(0)(I_N \otimes R)e(0)] \\
&\quad + \dot{V}(e(t))\}dt - V(e(\infty)) + V(e(0)) \\
&\le \int_0^\infty [e^T(t)\{\Xi \otimes (He\{P^{-1}\tilde{A}(t)\} - \mathscr{B} + \tfrac{1}{\zeta_{min}} I) \\
&\quad + \gamma^{-2}\Xi^2\Pi^2 \otimes P^{-1}DD^TP^{-1}\}e(t)dt \\
&\quad - \int_0^\infty \gamma^2[w(t) - \gamma^{-2}(\Xi\Pi \otimes D^TP^{-1})e(t)]^T \\
&\quad [w(t) - \gamma^{-2}(\Xi\Pi \otimes D^TP^{-1})e(t)]dt \\
&\quad + V(e(0)) - \gamma^2 e^T(0)(I_N \otimes R)e(0) \\
&\le \int_0^\infty e^T(t)\{\Xi \otimes [He\{P^{-1}\tilde{A}(t)\} - \mathscr{B} + \tfrac{1}{\zeta_{min}} I] \\
&\quad + \Xi \otimes \gamma^{-2}\zeta_{max} P^{-1}DD^TP^{-1}\}e(t)dt \\
&\quad + e^T(0)(\Xi \otimes P^{-1})e(0) - e^T(0)(\Xi \otimes \tfrac{\gamma^2}{\zeta_{max}} R)e(0)
\end{aligned}$$

where $\mathscr{B}$ *is defined in (12.26), and* $\zeta_{min} = \min_{i=1,\cdots,N} \zeta_i$, $\zeta_{max} = \max_{i=1,\cdots,N} \zeta_i$.

Note that $\mathbf{J} < 0$ *if the following inequalities hold:*

$$\begin{aligned}
\Xi \otimes \{He\{P^{-1}\tilde{A}(t)\} &- 2c(1+\delta_b)\alpha_0 P^{-1}BB^TP^{-1} \\
&+ \tfrac{1}{\zeta_{min}} I + \tfrac{\zeta_{max}}{\gamma^2} P^{-1}DD^TP^{-1}\} < 0
\end{aligned}$$

$$\Xi \otimes (P^{-1} - \tfrac{\gamma^2}{\zeta_{max}} R) < 0.$$

Since Ξ *is a positive definite diagonal matrix, the above two inequalities are equivalent to the following inequalities.*

$$\begin{aligned}
He\{P^{-1}\tilde{A}(t)\} - 2c(1+\delta_b)\alpha_0 P^{-1}BB^TP^{-1} &+ \tfrac{1}{\zeta_{min}} I \\
&+ \tfrac{\zeta_{max}}{\gamma^2} P^{-1}DD^TP^{-1} < 0
\end{aligned} \tag{12.27}$$

$$P^{-1} - \tfrac{\gamma^2}{\zeta_{max}} R < 0. \tag{12.28}$$

Then, by using the Schur complement lemma, (12.28) can be equivalently rewritten as (12.17).

Performing a congruence transformation to (12.27) by P, we obtain

$$He\{\tilde{A}(t)P\} - 2c\mu(1+\delta_b)\alpha_0 BB^T + \gamma^{-2}\zeta_{max} DD^T + \tfrac{1}{\zeta_{min}} P^2 < 0. \tag{12.29}$$

Recalling definition of $\tilde{A}(t)$ *and using Lemma 2.10 in Chapter 2, one can obtain that*

$$\begin{aligned}
He\{\tilde{A}(t)P\} &= He\{(A + E\Re(t)F)P\} \\
&\le He\{AP\} + E\Lambda E^T + \delta_a^2 PF^T\Lambda^{-1}FP.
\end{aligned} \tag{12.30}$$

Thus, we can get (12.18) immediately from (12.29) and (12.30) by using the Schur complement lemma. This completes the proof. □

Remark 12.7 *It is worth noting that the coupling weight c can be obtained by solving the LMI conditions (12.17) and (12.18) in Theorem 12.1, which will reduce the conservativeness of the previous results in [112, 176, 181, 180] which require c to be larger than a threshold value.*

Remark 12.8 *NDCP protocol $u_i(t)$ in (12.6)-(12.7) presents a nonlinear controller, under which the agents with dynamics given by (12.1) with uniform input quantization and external disturbances can reach complete consensus. It should be mentioned that due to the uniform quantization effect, complete consensus cannot be achieved, and only practical consensus is achieved in [39, 40, 49, 117, 185, 188], i.e., all the states of agents enter into a ball.*

When uncertainties of A and B are not considered, i.e., $\Delta A(t) = 0$, $\Delta B(t) = 0$, then the following corollary provides a sufficient condition to design consensus control without considering model uncertainties.

Corollary 12.1 *Suppose that Assumptions 12.1 and 12.3 hold. Then, the distributed quantized H_∞ consensus with performance γ and weighting matrix $R > 0$ for multi-agent systems (12.1) can be solved by NDCP protocol $u_i(t)$ in (12.6)-(12.7) for any given dwell time $\tau > 0$ and mismatch ratio $\mu > 0$, if there exist a symmetric matrix $P = P^T > 0$, and positive scalars $c > 0$ and $\gamma > 0$ such that (12.18) and*

$$\begin{bmatrix} He\{AP\} - 2c\mu\alpha_0 BB^T & D & P \\ * & -\frac{\gamma^2}{\zeta_{max}} I & 0 \\ * & * & -\zeta_{min} I \end{bmatrix} < 0 \tag{12.31}$$

hold, where ζ_{min}, ζ_{max} and α_0 are defined in Theorem 12.1. Then, the two parts of NDCP protocol $u_i(t)$ in (12.6) are designed as in (12.19) and

$$u_{2i}(t) = -\frac{\mu_c}{2} sign(\rho_i^T(x(t))) \tag{12.32}$$

where $\rho_i(x(t))$ is the same in Theorem 12.1.

Remark 12.9 *The high-order multi-agent system considered in this chapter is very general and our result is applicable to all lower order multi-agent systems that are considered in most of the existing literature (such as the second-order multi-agent systems considered in [199, 200]). The first numerical example in Section 12.4, Numerical Examples, is given to demonstrate this fact.*

12.3 Quantized H_∞ Consensus for *Lipschitz* Nonlinearity

In this section, we study the *quantized consensus* problem for a group of N identical nonlinear agents, which are described by

$$\dot{x}_i(t) = \tilde{A}(t)x_i(t) + A_f f(x_i(t)) + \tilde{B}(t)Q(u_i(t)) + Dw_i(t) \tag{12.33}$$

where $\tilde{A}(t)$ and $\tilde{B}(t)$ are the same as in (12.1). The nonlinear function $f(x_i(t))$ is assumed to satisfy the *Lipschitz* condition with a *Lipschitz* constant $\psi > 0$, i.e.,

$$\|f(x(t)) - f(y(t))\|_2 \leq \psi\|x(t) - y(t)\|_2, \forall x(t), y(t) \in \mathbf{R}^n. \tag{12.34}$$

The following theorem presents a sufficient condition for achieving quantized H_∞ consensus of (12.33) by using NDCP protocol $u_i(t)$ in (12.6)-(12.7).

Theorem 12.2 *Suppose that Assumptions 12.1-12.3 hold. Then, the distributed quantized H_∞ consensus with performance γ and weighting matrix $R > 0$ for the nonlinear multi-agent systems (12.33) can be solved by NDCP protocol $u_i(t)$ in (12.6)-(12.7) for any given dwell time $\tau > 0$ and mismatch ratio $\mu > 0$, if there exist a symmetric matrix $P = P^T > 0$, a diagonal matrix $\Lambda > 0$, and positive scalars $c > 0$ and $\gamma > 0$ such that the LMI conditions (12.18) with $\Phi = He\{AP\} - 2c\mu(1+\delta_b)\alpha_0 BB^T + \psi^2 A_f A_f^T + \frac{\zeta_{min}+1}{\zeta_{min}} P^2$ and (12.17) hold, where ψ is a Lipschitz constant. The two parts of the distributed protocol in (12.6) are designed as the same in (12.19) and (12.20).*

Proof 12.2 *Using NDCP protocol $u_i(t)$ in (12.6)-(12.7) for (12.33), we obtain the closed-loop network dynamics as*

$$\begin{aligned} \dot{x}_i(t) &= \tilde{A}(t)x_i(t) + A_f f(x_i(t)) + \mu\tilde{B}(t)(\Delta u_i(t) + u_{2i}(t)) \\ &\quad + c\mu\tilde{B}(t)K\sum_{j=1}^{N} a_{ij}^{\sigma(t)}(x_j(t) - x_i(t)) + Dw_i(t). \end{aligned} \tag{12.35}$$

Define $e(t) = (\Pi \otimes I_n)x(t)$, where $\Pi = (I_N - \mathbf{1}_N\zeta^T) \in \mathbf{R}^{N\times N}$ satisfies $\Pi \cdot \Pi = \Pi$ and $\Pi \cdot \mathscr{L}^{\sigma(t)} = \mathscr{L}^{\sigma(t)}$. It is easy to obtain from (12.35) that $e(t)$ satisfies the following dynamics:

$$\begin{aligned} \dot{e}(t) &= [I_N \otimes \tilde{A}(t) - c\mu(\mathscr{L}^{\sigma(t)} \otimes \tilde{B}(t)K)]e(t) \\ &\quad + (\Pi \otimes D)w(t) + (\Pi \otimes A_f)F(x(t)) \\ &\quad + \mu(\Pi \otimes \tilde{B}(t))(u_2(t) + \Delta u(t)) \end{aligned} \tag{12.36}$$

where

$$F(x(t)) = [f^T(x_1(t)), f^T(x_2(t)), \cdots, f^T(x_N(t))]^T.$$

By following steps similar to those in Theorem 12.1, one can obtain the time derivative of $V(t)$, which is defined in (12.21), along the trajectory of (12.36) as

$$\begin{aligned} \dot{V}(e(t)) &= e^T(t)[\Xi \otimes He\{P^{-1}\tilde{A}(t)\} \\ &\quad + 2e^T(t)(\Xi\Pi \otimes P^{-1}A_f)F(x(t)) \\ &\quad - c\mu He\{\Xi\mathscr{L}^{\sigma(t)} \otimes P^{-1}\tilde{B}(t)K)\}]e(t) \\ &\quad + 2\mu e^T(t)(\Xi\Pi \otimes P^{-1}\tilde{B}(t))(u_2(t) + \Delta u(t)) \\ &\quad + 2e^T(t)(\Xi\Pi \otimes P^{-1}D)w(t). \end{aligned} \tag{12.37}$$

Since $e(t) = (\Pi \otimes I_n)x(t)$, that is

$$e_i(t) = x_i(t) - \sum_{j=1}^{N} \zeta_j x_j(t),$$

it follows from $\sum_{j=1}^{N} \zeta_j = 1$ *that*

$$\sum_{i=1}^{N} \zeta_i e_i(t) = \sum_{i=1}^{N} \zeta_i x_i(t) - \sum_{i=1}^{N} \zeta_i \sum_{j=1}^{N} \zeta_j x_j(t) = 0. \tag{12.38}$$

In light of (12.38), it then follows that

$$2\sum_{i=1}^{N} \zeta_i e_i^T(t) P^{-1} A_f (f(\bar{x}(t)) - \sum_{j=1}^{N} \zeta_j f(x_j(t))) = 0 \tag{12.39}$$

where $\bar{x}(t) = \sum_{i=1}^{N} \zeta_i x_i(t)$*. Therefore, it follows from (12.39) that*

$$\begin{aligned}
&2e^T(t)(\Xi\Pi \otimes P^{-1}A_f)F(x(t)) \\
&= 2e^T(t)(\Xi \otimes P^{-1}A_f)(\Pi \otimes I_n)F(x(t)) \\
&= 2e^T(t)(\Xi \otimes P^{-1}A_f)(F(x(t)) - \mathbf{1}_N \zeta^T F(x(t))) \\
&= 2e^T(t)(\Xi \otimes P^{-1}A_f)(F(x(t)) - \mathbf{1}_N f(\bar{x}(t)) \\
&\quad + \mathbf{1}_N f(\bar{x}(t)) - \mathbf{1}_N \zeta^T F(x(t))) \\
&= 2\sum_{i=1}^{N} \zeta_i e_i^T(t) P^{-1} A_f (f(x_i(t)) - f(\bar{x}(t)) \\
&\quad + f(\bar{x}(t)) - \sum_{j=1}^{N} \zeta_j f(x_j(t))) \\
&= 2\sum_{i=1}^{N} \zeta_i e_i^T(t) P^{-1} A_f (f(x_i(t)) - f(\bar{x}(t))) \\
&\le 2\sum_{i=1}^{N} \zeta_i (\psi \|A_f^T P^{-1} e_i(t)\| \|e_i(t)\|) \\
&\le \sum_{i=1}^{N} \zeta_i e_i^T(t)(\psi^2 P^{-1} A_f A_f^T P^{-1} + I) e_i(t) \\
&= e^T(t)[\Xi \otimes (\psi^2 P^{-1} A_f A_f^T P^{-1} + I)] e(t).
\end{aligned}$$

Then, the rest of the proof is straightforward by following similar steps in proving Theorem 12.1, which is omitted here for brevity. Therefore, the proof is complete. □

Remark 12.10 *As shown in the proof of Theorem 12.1 and Theorem 12.2, in order to perfectly reject the effect of input quantization, we design* $u_{2i}(t)$ *as in (12.20), which provides an effective and robust means of controlling multi-agent systems with model uncertainties and input quantization.*

In addition, when the effect of input quantization is not considered, the nonlinear multi-agent system (12.33) is reduced to

$$\dot{x}_i(t) = \tilde{A}(t)x_i(t) + A_f f(x_i) + \tilde{B}(t)u_i(t) + Dw_i(t). \tag{12.40}$$

Then, a linear distributed consensus protocol (LDCP) as in (12.7) is constructed to make all the states of (12.40) achieve complete consensus. Then, the following corollary provides a sufficient condition to design linear consensus control which can make the original high-order identical dynamic systems achieve consensus.

Corollary 12.2 *Suppose that Assumptions 12.2 and 12.3 hold. Then, the distributed H_∞ consensus with performance γ and weighting matrix $R > 0$ for the nonlinear multi-agent systems (12.40) can be solved by LDCP protocol in (12.7) for any given dwell time $\tau > 0$, if there exist a symmetric matrix $P = P^T > 0$, a diagonal matrix $\Lambda > 0$, and positive scalars $c > 0$ and $\gamma > 0$ such that the LMIs (12.18) and (12.17) hold with*

$$\Phi = He\{AP\} - 2c(1+\delta_b)\alpha_0 BB^T + \psi^2 A_f A_f^T + \frac{\zeta_{min}+1}{\zeta_{min}} P^2,$$

where ψ is a Lipschitz constant. The controller gain of LDCP protocol in (12.7) can be designed as the same in (12.19).

Remark 12.11 *When quantized effect and Lipschitz nonlinear uncertainties are not considered, a similar problem was studied in [181] with the assumptions that $\Delta A(t) = 0$ and $\Delta B(t) = 0$. For this case, LDCP protocol in (12.7) recovers the result in [181].*

12.4 Numerical Examples

In this section, two numerical examples are included to show the effectiveness of the proposed control strategies.

12.4.1 Example 1

This numerical example is given to demonstrate the capability of our methodology in also covering and handling a network of second-order agents with *external disturbances* and undirect switching topologies

$$\begin{cases} \dot{p}_i(t) &= v_i(t) \\ \dot{v}_i(t) &= Q(u_i(t)) + w_i(t), i = 1, 2, \cdots, 8, \end{cases}$$

where $p_i(t), v_i(t) \in \mathbf{R}$ denote the position and the velocity of agent i, respectively, $u_i(t) \in \mathbf{R}$ is agent i's control input, and $w_i(t) \in \mathbf{R}$ is the external disturbance. For motion in the 2-D plane, for instance, one would take $x_i(t) = [p_i(t)^T, v_i(t)^T]^T$, and the parameters of local node dynamics are given by

$$\begin{aligned} A &= \begin{bmatrix} 0 & 1 \\ 0 & 0 \end{bmatrix} B = \begin{bmatrix} 0 \\ 1 \end{bmatrix}, D = \begin{bmatrix} 0 \\ 1 \end{bmatrix} \\ \Delta A(t) &= 0, \Delta B(t) = 0. \end{aligned}$$

Figure 12.2 depicts three different undirected weighted topologies $\mathcal{G}^{\sigma(t)}$, $\sigma(t) \in \{1, 2, 3\}$ with $N = 8$ agents. The switching mode starts at $\mathcal{G}^1$ and the order is $\mathcal{G}^1 \to \mathcal{G}^2 \to \mathcal{G}^3 \to \mathcal{G}^1$ and the topology of the multi-agent system switches every $\tau = 0.1s$

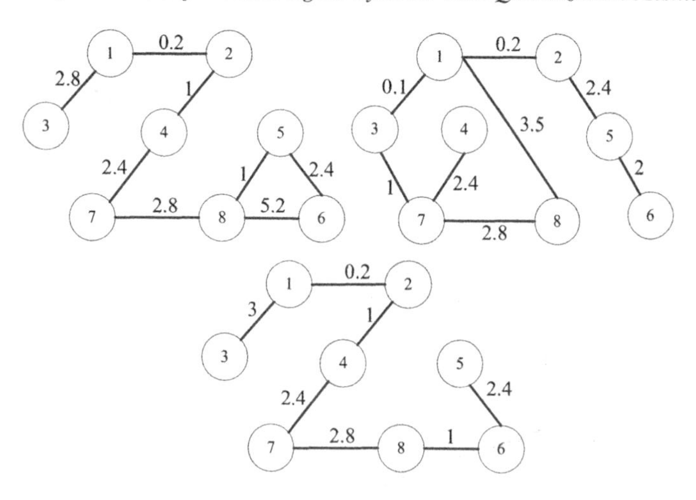

FIGURE 12.2
Three undirected switching weighted graphs.

to the next state. Using Algorithm 2.1 in Chapter 2, we can obtain a common positive vector $\zeta = 0.1250 \times \mathbf{1}_8$, which implies $\zeta_{max} = \zeta_{min} = 0.1250$. Moreover, by using Algorithm 2.1 in Chapter 2, we can obtain $\alpha(L^1) = 0.1066$, $\alpha(L^2) = 0.0952$ and $\alpha(L^3) = 0.1036$, which implies $\alpha_0 = 0.0952$. Then, taking $\mu_c = 0.2$, $\mu_d = 0.1$ (i.e., a mismatch ratio of $\mu = 0.5$) and the external disturbance $w_i(t) = \sin(10t)$, and solving the LMI conditions (12.18) and (12.31) in Corollary 12.1 with $R = 2I_2$, one can get that

$$
\begin{aligned}
P &= \begin{bmatrix} 0.0880 & -0.1183 \\ -0.1183 & 0.7924 \end{bmatrix}, c = 84.6401 \\
K &= \begin{bmatrix} 2.1224 & 1.5788 \end{bmatrix}, \gamma = 1.3809.
\end{aligned}
$$

Corollary 12.1 can be validated through simulations. First, let each coordinate of the initial positions and velocities be chosen randomly. Figure 12.3 illustrates the evolutions of the positions and velocities of the eight agents. It is easy to see that all states of the eight agents can achieve complete consensus, which shows that the designed nonlinear distributed protocol can guarantee the complete consensus of the second-order agents even when there exist encoder and decoder quantization mismatch, *external disturbances* and *non-zero initial* states.

12.4.2 Example 2

This subsection discusses the performance of the proposed NDCP protocol $u_i(t)$ in (12.6)-(12.7) designed in Theorem 12.2 in comparison with LDCP protocol $u_{1i}(t)$ in (12.7) designed in Corollary 12.2 when the effect of input quantization is not considered. Then, a numerical example is presented to show the effectiveness of our proposed method based on simulation of the linearized longitudinal dynamical equations of aircraft. The equation corresponding to the dynamics of the ith aircraft

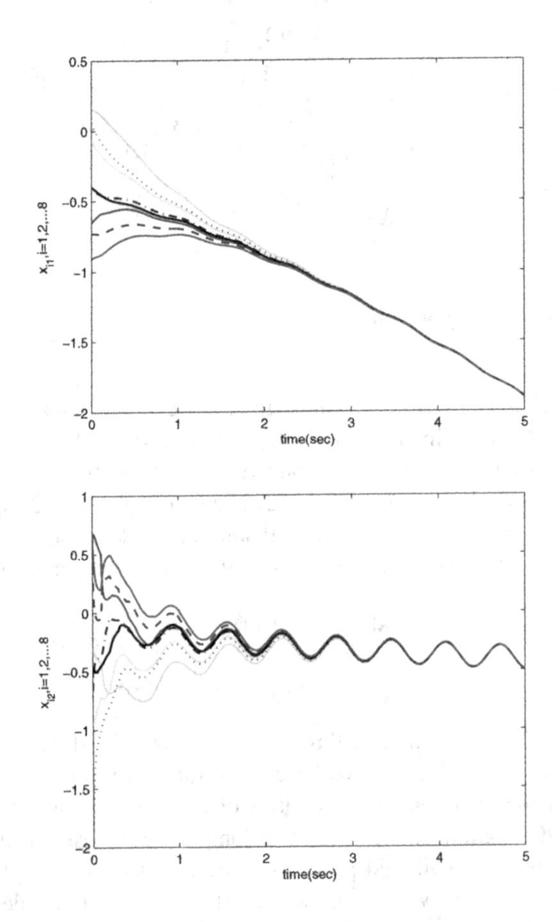

FIGURE 12.3
Evolutions of $p_i(t)$ and $v_i(t), i = 1, 2, \cdots, 8$ with input quantization.

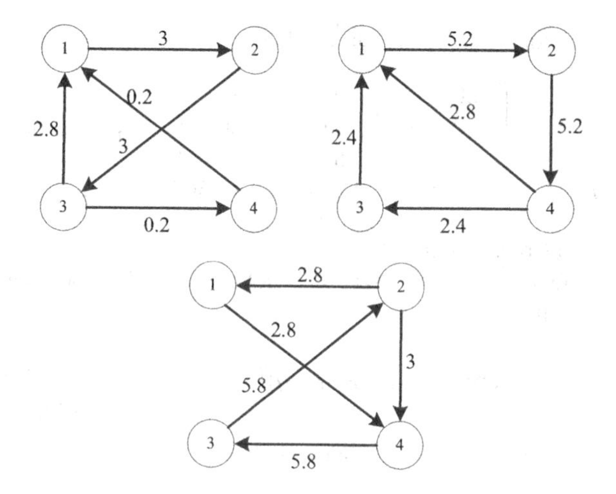

FIGURE 12.4
Three directed switching weighted graphs.

is given by (12.1) with [203].

$$x_i(t) = \begin{bmatrix} x_{i1}(t) \\ x_{i2}(t) \\ x_{i3}(t) \end{bmatrix}, A = \begin{bmatrix} -0.277 & 1 & -0.0002 \\ -17.1 & -0.178 & -12.2 \\ 0 & 0 & -6.67 \end{bmatrix}$$
$$B^T = D^T = \begin{bmatrix} 0 & 0 & 6.67 \end{bmatrix}$$

where $x_{i1}(t), x_{i2}(t), x_{i3}(t)$ denote the angle of attack, the pitch rate and elevator angle, respectively. The control input $u_i(t)$ is the command to the elevator. The switching weighted topology structures of networks with 4 agents is shown in Figure 12.4. It is assumed that the communication networks periodically switch according to $\mathscr{G}^1 \to \mathscr{G}^2 \to \mathscr{G}^3 \to \mathscr{G}^1$ every $\tau = 0.1s$. Using Algorithm 2.1 in Chapter 2, we can obtain a common positive vector $\zeta = 0.25 \times \mathbf{1}_4$, which implies $\zeta_{max} = \zeta_{min} = 0.25$. Moreover, by using Algorithm 2.1 in Chapter 2, we can obtain $\alpha(L^1) = 0.2610$, $\alpha(L^2) = 2.9493$ and $\alpha(L^3) = 3.2907$, which implies $\alpha_0 = 0.2610$.

In order to verify the effectiveness of the proposed method, simulations are carried out with the following parameters and *initial conditions*:

$$\begin{aligned}
A_f &= I_3, \Delta A(t) = [0.12\sin(t)]_{3\times 3}, R = 25I_3 \\
x_1(0) &= [\,0.2\ -0.2\ 0.5\,]^T, x_2(0) = [\,0\ 0.2\ -0.5\,]^T \\
x_3(0) &= [\,0.1\ 0.1\ 0.3\,]^T, x_4(0) = [\,0.2\ 0.1\ -2.5\,]^T \\
f(x_i(t)) &= \begin{bmatrix} 0 & 0.2\sin(x_{i2}(t)) & 0 \end{bmatrix}^T, \Delta B(t) = 0.3\cos(5t) \\
w_i(t) &= 0.01\sin(10t), \mu_c = 0.6, \mu_d = 0.01
\end{aligned}$$

which implies $\delta_a = 0.12$, $\delta_b = 0.3$, $\psi = 0.2$ and $\mu = 0.0167$.

By solving the LMI conditions (12.18) with

$$\Phi = He\{AP\} - 2c\mu(1+\delta_b)\alpha_0 BB^T + \psi^2 A_f A_f^T + \frac{\zeta_{min}+1}{\zeta_{min}} P^2$$

and (12.17) in Theorem 12.2, one can get that

$$\begin{aligned} P &= \begin{bmatrix} 0.0282 & -0.1287 & -0.0084 \\ -0.1287 & 1.4462 & 1.6445 \\ -0.0084 & 1.6445 & 7.2909 \end{bmatrix}, c = 1305.8 \\ K &= \begin{bmatrix} -12.8175 & -2.9108 & 1.5567 \end{bmatrix}, \gamma = 3.5930. \end{aligned}$$

Then, $u_{2i}(t)$ can also be obtained by (12.20).

When the effect of input quantization is not considered, by solving the LMI conditions (12.18) with

$$\Phi = He\{AP\} - 2c(1+\delta_b)\alpha_0 BB^T + \psi^2 A_f A_f^T + \frac{\zeta_{min}+1}{\zeta_{min}} P^2$$

and (12.17) in Corollary 12.2, one can get that

$$\begin{aligned} P &= \begin{bmatrix} 0.0273 & -0.1285 & -0.0012 \\ -0.1285 & 1.5144 & 1.9133 \\ -0.0012 & 1.9133 & 44.4518 \end{bmatrix}, c = 781.8067 \\ K &= \begin{bmatrix} -1.6199 & -0.3457 & 0.1649 \end{bmatrix}, \gamma = 10.6808. \end{aligned}$$

With *non-zero initial* states and external disturbances, the trajectories of the states by using LDCP protocol $u_{1i}(t)$ with or without input quantization are shown in Figure 12.5 and Figure 12.6, respectively. The trajectories of the states by using NDCP protocol $u_i(t)$ are shown in Figure 12.7. It is clear that LDCP protocol without input quantization and NDCP protocol have similar convergence rates for consensus, which implies that the new protocol can effectively eliminate the effect of input quantization. However, input quantization for LDCP protocol results in high overshoot and low convergence rate of consensus in Figure 12.6, while NDCP protocol has a relatively small overshoot and fast convergence rate by using $u_{2i}(t)$ to compensate for the quantization errors. It is, therefore, necessary to use $u_{2i}(t)$ to eliminate the effect of input quantization.

12.5 Conclusion

In this chapter, a new distributed quantized H_∞ *consensus* strategy for both linear and *Lipschitz* nonlinear multi-agent systems with input quantization mismatch and external disturbances is proposed. The considered communication networks are switching weighted undirected or balanced directed networks. The designed distributed quantized H_∞ *consensus* protocols can be used to achieve a satisfactory performance against interval-bounded model uncertainties, external disturbances and unknown initial states, and simultaneously to eliminate the effect of input quantization and model uncertainties. It is shown that the original multi-agent systems can reach complete consensus even if there exists uniform quantization. In addition, extending the

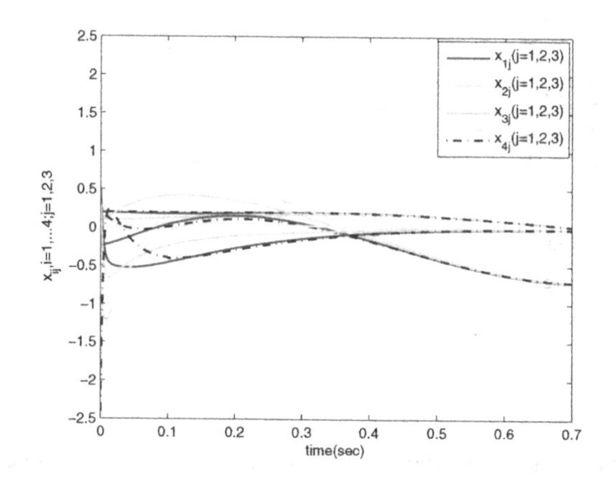

FIGURE 12.5
State trajectories of all agents under LDCP protocol $u_{1i}(t)$ without input quantization.

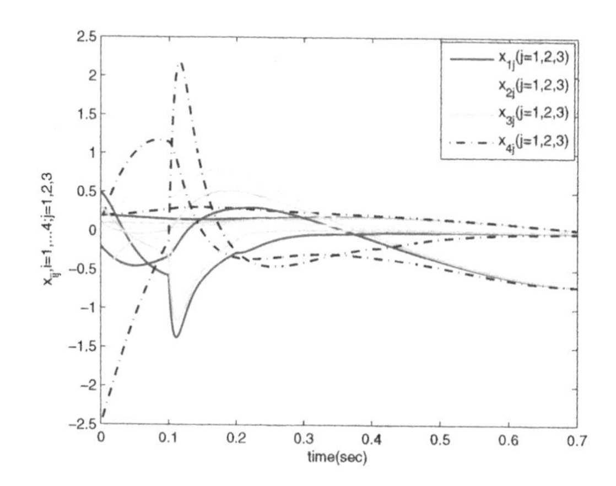

FIGURE 12.6
State trajectories of all agents under LDCP protocol $u_{1i}(t)$ with input quantization.

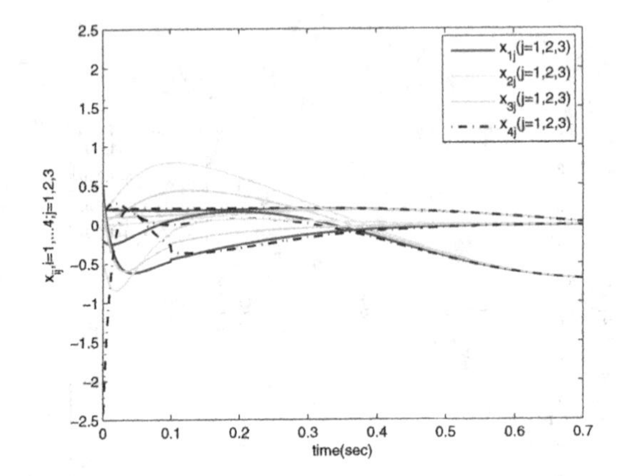

FIGURE 12.7
State trajectories of all agents under NDCP protocol $u_i(t)$ with input quantization.

results of this chapter to the case where the communication topology is not necessarily connected or has noise relative measurement is an important yet challenging topic for future research.

Bibliography

[1] R. Agaev and P. Chebotarev. On the spectra of nonsymmetric laplacian matrices. *Linear Algebra Appl.*, 399:157–178, 2005.

[2] A. Ali, G. Garcia, and P. Martinet. The flatbed platoon towing model for safe and dense platooning on highways. *IEEE Intell. Transp. Syst. Mag.*, 7(1):58–68, 2015.

[3] M. Arcak. Passivity as a design tool for group coordination. *IEEE Trans. Autom. Control*, 52(8):1380–1390, 2007.

[4] D. V. Balandin and M. M. Kogan. LMI-based H_∞-optimal control with transients. *Int. J. Control*, 83(8):1664–1673, 2010.

[5] S. Banerjee, K. Halder, S. Dasgupta, S. Mukhopadhyay, K. Ghosh, and A. Gupta. An interval approach for robust control of a large PHWR with PID controllers. *IEEE Trans. Nucl. Sci.*, 62(1):281–292, 2015.

[6] C. P. Bechlioulis, D. V. Dimarogonas, and K. J. Kyriakopoulos. Robust control of large vehicular platoons with prescribed transient and steady state performance. In *Proc. of 53rd IEEE Conference on Decision and Control*, pages 3689–3694. IEEE, 2014.

[7] R. Bencatel, M. Faied, J. Sousa, and A. R. Girard. Formation control with collision avoidance. In *Proc. IEEE Conf. Decision and Control and European Control Conf.*, pages 591–596, 2011.

[8] M. D. Bernardo, P. Falcone, A. Salvi, and S. Santini. Design, analysis, and experimental validation of a distributed protocol for platooning in the presence of time-varying heterogeneous delays. *IEEE Trans. Control Syst. Technol.*, 24(2):413–427, 2016.

[9] S. Bhat and D. Bernstein. Continuous finite-time stabilization of the translational and rotational double integrators. *IEEE Trans. Autom. Control*, 43(11):678–682, 1998.

[10] C. Canudas de Wit and B. Brogliato. Stability issues for vehicle platooning in automated highway systems. In *Proc. of IEEE International Conference on Control Applications*, volume 2, pages 1377–1382, 1999.

[11] A. Carron, M. Todescato, R. Carli, and L. Schenato. An asynchronous consensus-based algorithm for estimation from noisy relative measurements. *IEEE Trans. Control Network Syst.*, 1(3):283–295, 2014.

[12] W. W. Che and Yang G. H. Non-fragile dynamic output feedback H_∞ control for discrete-time systems. *Int. J. Control Autom. Syst.*, 9(5):993–997, 2011.

[13] W. W. Che, G. H. Yang, and X. Z. Jin. Non-fragile H_∞ filter design with sparse structure for linear discrete-time systems. *J. Franklin Inst.*, 351(1):225–240, 2014.

[14] H. Chehardoli and A. Ghasemi. Adaptive centralized/decentralized control and identification of 1-d heterogeneous vehicular platoons based on constant time headway policy. *IEEE Trans. Intell. Transp. Syst.*, 19(10):3376–3386, 2018.

[15] F. Chen, R. Jiang, C. Wen, and R. Su. Self-repairing control of a helicopter with input time delay via adaptive global sliding mode control and quantum logic. *Inf. Sci.*, 316:123–131, 2015.

[16] F. Chen, Z. Wang, G. Tao, and B. Jiang. Robust adaptive fault-tolerant control for hypersonic flight vehicles with multiple faults. *J. Aerosp. Eng.*, 28(4):04014111.1–04014111.10, 2015.

[17] F. Chen, K. Zhang, B. Jiang, and C. Wen. Adaptive sliding mode observer-based robust fault reconstruction for a helicopter with actuator fault. *Asian J. Control*, 18(4):1558–1565, 2016.

[18] G. Chen, Y. Song, and F. L. Lewis. Distributed fault-tolerant control of networked uncertain Euler-Lagrange systems under actuator faults. *IEEE Trans. Cybern.*, 47(7):1706–1718, 2017.

[19] M. Chen, S. S. Ge, and B. Ren. Adaptive tracking control of uncertain MIMO nonlinear systems with input constraints. *Automatica*, 47(3):452–465, 2011.

[20] M. Chen, G. Tao, and B. Jiang. Dynamic surface control using neural networks for a class of uncertain nonlinear systems with input saturation. *IEEE Trans. Neural Networks Learn. Syst.*, 26(9):2086–2097, 2015.

[21] Z. Chen. Pattern synchronization of nonlinear heterogeneous multi-agent networks with jointly connected topologies. *IEEE Trans. Control Network Syst.*, 1(4):349–359, 2014.

[22] D. F. Chichka, J. L. Speyer, and C. G. Park. Peak-seeking control with application to formation flight. In *Proc. 1999 IEEE Conf. Dec. Control*, volume 3, pages 2463–2470, 1999.

[23] M. L. Corradini, G. Orlando, and G. Parlangeli. A VSC approach for the robust stabilization of nonlinear plants with uncertain nonsmooth actuator nonlinearities-a unified framework. *IEEE Trans. Autom. Control*, 49(5):807–813, 2004.

[24] J. Cortes. Discontinuous dynamical systems. *IEEE Control Syst.*, 28(3):36–73, 2008.

[25] Y. Cui and Y. Jia. $L_2 - L_\infty$ consensus control for high-order multi-agent systems with switching topologies and time-varying delays. *IET Control Theory Appl.*, 6(12):1933–1940, 2012.

[26] M. di Bernardo, P. Falcone, A. Salvi, and S. Santini. Design, analysis, and experimental validation of a distributed protocol for platooning in the presence of time-varying heterogeneous delays. *IEEE Trans. Control Syst. Technol.*, 24(2):413–427, 2016.

[27] M. di Bernardo, A. Salvi, and S. Santini. Distributed consensus strategy for platooning of vehicles in the presence of time-varying heterogeneous communication delays. *IEEE Trans. Intell. Transp. Syst.*, 16(1):102–112, 2015.

[28] D. V. Dimarogonas and K. H. Johansson. Stability analysis for multi-agent systems using the incidence matrix: quantized communication and formation control. *Automatica*, 46(4):695–700, 2010.

[29] D. W. Ding, X. L. Li, and Y. Y. Wang. Nonfragile filtering for discrete-time linear systems in finite-frequency domain. *Int. J. Control*, 86(4):664–673, 2013.

[30] S. Ding, J. Wang, and W. X. Zheng. Second-order sliding mode control for nonlinear uncertain systems bounded by positive functions. *IEEE Trans. Ind. Electron.*, 62(9):5899–5909, 2015.

[31] Z. T. Ding. Adaptive consensus output regulation of a class of nonlinear systems with unknown high-frequency gain. *Automatica*, 51:348–355, 2015.

[32] H. Dong, S. Gao, and B. Ning. Cooperative control synthesis and stability analysis of multiple trains under moving signaling systems. *IEEE Trans. Control Syst. Technol.*, 17(10):2730–2738, 2016.

[33] X. Dong, J. Xi, G. Lu, and Y. Zhong. Formation control for high-order linear time-invariant multi-agent systems with time delays. *IEEE Trans. Control Network Syst.*, 1(3):232–240, 2014.

[34] W. B. Dunbar and D. S. Caveney. Distributed receding horizon control of vehicle platoons: Stability and string stability. *IEEE Trans. Autom. Control*, 56(3):620–633, 2012.

[35] K. Esfandiari, F. Abdollahi, and H. A. Talebi. Adaptive control of uncertain nonaffine nonlinear systems with input saturation using neural networks. *IEEE Trans. Neural Networks Learn. Syst.*, 26(10):2311–2322, 2015.

[36] M. Faieghi, A. Jalali, and S. K. Mashhadi. Robust adaptive cruise control of high speed trains. *ISA Trans.*, 53(2):533–541, 2014.

[37] Y. X. Feng, B. T. Hu, H. Hao, Y. C. Gao, Z. W. Li, and J. R. Tan. Design of distributed cyber-physical systems for connected and automated vehicles with implementing methodologies. *IEEE Trans. Ind. Inf.*, 14(9):4200–4211, 2018.

[38] P. Fernandes and U. Nunes. Multiplatooning leaders positioning and cooperative behavior algorithms of communicant automated vehicles for high traffic capacity. *IEEE Trans. Intell. Transp. Syst.*, 16(3):1172–1187, 2015.

[39] J. J. Fu and J. Z. Wang. Adaptive coordinated tracking of multi-agent systems with quantized information. *Syst. Control Lett.*, 74:115–125, 2014.

[40] J. J. Fu and J. Z. Wang. Output consensus of heterogeneous linear systems with quantized information. *J. Franklin Inst.*, 351(3):1400–1418, 2014.

[41] J. J. Fu and J. Z. Wang. Adaptive motion coordination of passive systems under quantization effect. *Int. J. Robust Nonlinear Control*, 25(11):1638–1653, 2015.

[42] F. Gao, X. S. Hu, S. B. E. Li, K. Q. Li, and Q. Sun. Distributed adaptive sliding mode control of vehicular platoon with uncertain interaction topology. *IEEE Trans. Ind. Electron.*, 65(8):6352–6361, 2018.

[43] A. Ghasemi, R. Kazemi, and S. Azadi. Stable decentralized control of a platoon of vehicles with heterogeneous information feedback. *IEEE Trans. Veh. Technol.*, 62(9):4299–4308, 2013.

[44] A. Ghasemi, R. Kazemi, and S. Azadi. Exact stability of a platoon of vehicles by considering time delay and lag. *J. Mech. Sci. Technol.*, 29(2):799–805, 2015.

[45] G. Guo and W. Yue. Hierarchical platoon control with heterogeneous information feedback. *IET Control Theory Appl.*, 5(15):1766–1781, 2011.

[46] G. Guo and W. Yue. Autonomous platoon control allowing range-limited sensors. *IEEE Trans. Veh. Technol.*, 61(7):2901–2912, 2012.

[47] G. Guo and W. Yue. Sampled-data cooperative adaptive cruise control of vehicles with sensor failures. *IEEE Trans. Intell. Transp. Syst.*, 15(6):2404–2418, 2014.

[48] X. G. Guo, J. L. Wang, and F. Liao. Non-fragile H_∞ consensus of linear multi-agent systems with interval-bounded variations. In *Proc. 2015 23rd Mediterranean Conference on Control and Automation (MED)*, pages 1039–1044, 2015.

[49] X. G. Guo, J. L. Wang, F. Liao, and D. Wang. Quantized H_∞ consensus of multi-agent systems with quantization mismatch under switching weighted topologies. *IEEE Trans. Control Network Syst.*, 4(2):202–212, 2017.

[50] X. G. Guo, Q. Gao, and Z. Xu. Synthesis of low-coefficient sensitivity controllers with respect to multiplicative controller coefficient variations. *IET Control Theory Appl.*, 9(1):120–128, 2015.

[51] X. G. Guo, J. L. Wang, and F. Liao. Adaptive quantised H_∞ observer-based output feedback control for non-linear systems with input and output quantisation. *IET Control Theory Appl.*, 11(2):263–272, 2016.

[52] X. G. Guo, J. L. Wang, and F. Liao. Adaptive fuzzy fault-tolerant control for multiple high-speed trains with proportional and integral-based sliding mode. *IET Control Theory Appl.*, 11(8):1234–1244, 2017.

[53] X. G. Guo, J. L. Wang, and F. Liao. Non-fragile quantized H_∞ output feedback control for nonlinear systems with quantized inputs and outputs. *J. Franklin Inst.*, 354(1):415–438, 2017.

[54] X. G. Guo, J. L. Wang, F. Liao, S. Suresh, and S. Narasimalu. Quantized insensitive consensus of lipschitz nonlinear multi-agent systems using the incidence matrix. *J. Franklin Inst.*, 352(11):4845–4863, 2015.

[55] X. G. Guo, J. L. Wang, F. Liao, and R. S. H. Teo. Distributed adaptive integrated sliding mode controller synthesis for string stability of vehicle platoons. *IEEE Trans. Intell. Transp. Syst.*, 17(9):2419–2429, 2016.

[56] X. G. Guo, J. L. Wang, F. Liao, and R. S. H. Teo. CNN-based distributed adaptive control for vehicle-following platoon with input saturation. *IEEE Trans. Intell. Transp. Syst.*, 19(10):3121–3132, 2018.

[57] X. G. Guo, J. L. Wang, F. Liao, and R. S. H. Teo. Distributed adaptive control for vehicular platoon with unknown dead-zone inputs and velocityacceleration disturbances. *Int. J. Robust Nonlinear Control*, 27(16):2961–2981, 2017.

[58] X. G. Guo, J. L. Wang, F. Liao, and R. S. H. Teo. Distributed adaptive sliding mode control strategy for vehicle-following systems with nonlinear acceleration uncertainties. *IEEE Trans. Veh. Technol.*, 66(2):981–991, 2017.

[59] X. G. Guo, J. L. Wang, F. Liao, and R. S. H. Teo. Neuroadaptive quantized PID sliding-mode control for heterogeneous vehicular platoon with unknown actuator deadzone. *Int. J. Robust Nonlinear Control*, 29(1):188–208, 2019.

[60] X. G. Guo, J. L. Wang, F. Liao, and W. D. Xiao. Adaptive platoon control for nonlinear vehicular systems with asymmetric input deadzone and inter-vehicular spacing constraints. In *Proc. 56th IEEE Conference on Decision and Control*, pages 393–398, 2017.

[61] X. G. Guo and G. H. Yang. Non-fragile H_∞ filter design for delta operator formulated systems with circular region role constraints: an LMI optimization approach. *Acta Autom. Sin.*, 35(9):1209–1215, 2009.

[62] X. G. Guo and G. H. Yang. Reliable H_∞ filter design for discrete-time systems with sector-bounded nonlinearities: an LMI optimization approach. *Acta Autom. Sin.*, 35(10):1347–1351, 2009.

[63] X. G. Guo and G. H. Yang. Reliable H_∞ filter design for a class of discrete-time nonlinear systems with time-varying delay. *Optimal Control Appl. Methods*, 31(4):303–322, 2010.

[64] X. G. Guo and G. H. Yang. H_∞ filter design for delta operator formulated systems with low sensitivity to filter coefficient variations. *IET Control Theory Appl.*, 5(15):1677–1688, 2011.

[65] X. G. Guo and G. H. Yang. Delay-dependent reliable H_∞ filtering for sector-bounded nonlinear continuous-time systems with time-varying state delays and sensor failures. *Int. J. Syst. Sci.*, 43(1):117–131, 2012.

[66] X. G. Guo and G. H. Yang. H_∞ output tracking control for delta operator systems with insensitivity to controller coefficient variations. *Int. J. Syst. Sci.*, 44(4):652–662, 2013.

[67] X. G. Guo and G. H. Yang. A sequential linear programming matrix method to insensitive H_∞ output feedback for linear discrete-time systems. *J. Dyn. Syst. Meas. Contr.*, 136(1):014506.1–014506.7, 2014.

[68] X. G. Guo, G. H. Yang, and W. W. Che. Insensitive dynamic output feedback control with mixed-H_∞ norm sensitivity minimization. *J. Franklin Inst.*, 350(1):72–91, 2013.

[69] J. Guzinski, M. Diguet, Z. Krzeminski, A. Lewicki, and H. Abu-Rub. Application of speed and load torque observers in high-speed train drive for diagnostic purposes. *IEEE Trans. Ind. Electron.*, 56(1):248–256, 2009.

[70] H. Deng and M. Krstić. Stochastic nonlinear stabilization, part i: A backstepping design. *Syst. Control Lett.*, 32(3):143–150, 1997.

[71] M. T. Hamayun, C. Edwards, and H. Alwi. A fault tolerant control allocation scheme with output integral sliding modes. *Automatica*, 49(6):1830–1837, 2013.

[72] H. Hao and P. Barooah. On achieving size-independent stability margin of vehicular lattice formations with distributed control. *IEEE Trans. Autom. Control*, 57(10):2688–2694, 2012.

[73] H. Hao and P. Barooah. Stability and robustness of large platoons of vehicles with double-integrator models and nearest neighbor interaction. *Int. J. Robust Nonlinear Control*, 23(18):2097–2122, 2013.

[74] L. Y. Hao, J. H. Park, and D. Ye. Integral sliding mode fault-tolerant control for uncertain linear systems over networks with signals quantization. *IEEE Trans. Neural Networks Learn. Syst.*, 28(9):2088–2100, 2017.

[75] L. Y. Hao and G. H. Yang. Fault-tolerant control via sliding-mode output feedback for uncertain linear systems with quantisation. *IET Control Theory Appl*, 7(16):1992–2006, 2013.

[76] W. He, Y. Dong, and C. Sun. Adaptive neural impedance control of a robotic manipulator with input saturation. *IEEE Trans. Syst. Man Cybern.: Syst.*, 46(3):334–344, 2016.

[77] H. F. Hong, W. W. Yu, J. J. Fu, and X. H. Yu. Finite-time connectivity-preserving consensus for second-order nonlinear multi-agent systems. *IEEE Trans. Control Network Syst.*, 6(1):236–248, 2019.

[78] R. Horn and C. Johnson. *Matrix analysis*. Cambridge University Press, New York, 1990.

[79] C. Hu, B. Yao, and Q. Wang. Adaptive robust precision motion control of systems with unknown input dead-zones: A case study with comparative experiments. *IEEE Trans. Ind. Electron.*, 58(6):2454–2464, 2011.

[80] C. L. Hwang and C. Y. Kuo. A stable adaptive fuzzy sliding mode control for affine nonlinear systems with application to four-bar linkage systems. *IEEE Trans. Fuzzy Syst.*, 9(2):238–252, 2001.

[81] H. Ji, Z. Hou, and R. Zhang. Adaptive iterative learning control for high-speed trains with unknown speed delays and input saturations. *IEEE Trans. Autom. Sci. Eng.*, 13(1):260–273, 2016.

[82] D. Jia and N. Dong. Platoon based cooperative driving model with consideration of realistic inter-vehicle communication. *Transp. Res. Part C: Emerg. Technol.*, 68:245–264, 2016.

[83] X. Jin. Adaptive fault tolerant control for a class of input and state constrained MIMO nonlinear systems. *Int. J. Robust Nonlinear Control*, 26(2):286–302, 2016.

[84] X. Z. Jin and G. H. Yang. Adaptive sliding mode fault-tolerant control for nonlinearly chaotic systems against network faults and time-delays. *J. Franklin Inst.*, 350(5):1206–1220, 2013.

[85] X. Z. Jin, G. H. Yang, and W. W. Che. Adaptive pinning control of deteriorated nonlinear coupling networks with circuit realization. *IEEE Trans. Neural Networks Learn. Syst.*, 23(9):1345–1355, 2012.

[86] T. Kameneva and D. Nešić. Robustness of quantized control systems with mismatch between coder/decoder initializations. *Automatica*, 45(3):817–822, 2009.

[87] A. Kashyap, T. Basar, and R. Srikant. Quantized consensus. *Automatica*, 43(7):1192–1203, 2007.

[88] B. Kaviarasan, R. Sakthivel, and Y. Shi. Reliable dissipative control of high-speed train with probabilistic time-varying delays. *IEEE Trans. Intell. Transp. Syst.*, 47(6):3940–3951, 2016.

[89] L. H. Keel and S. P. Bhattacharyya. Robust, fragile, or optimal? *IEEE Trans. Autom. Control*, 42(8):1098–1105, 1997.

[90] H. K. Khalil. *Nonlinear systems*. Prentice Hall, NJ, 2002.

[91] S. Khoo, L. Xie, and Z. Man. Robust finite-time consensus tracking algorithm for multirobot systems. *IEEE/ASME Trans. Mechatron.*, 14(2):219–228, 2009.

[92] S. Klinge and R.H. Middleton. Time headway requirements for string stability of homogeneous linear unidirectionally connected systems. In *Proc. 48th IEEE Conf. Decision and Control and 28th Chinese Control Conf.*, pages 1192–1197, 2009.

[93] J. W. Kwon and D. Chwa. Adaptive bidirectional platoon control using a coupled sliding mode control method. *IEEE Trans. Intell. Transp. Syst.*, 15(5):2040–2048, 2014.

[94] G. Lafferriere, A. Williams, J. Caughman, and J. J. P. Veerman. Decentralized control of vehicle formations. *Syst. Control Lett.*, 54(9):899–910, 2005.

[95] G. Lai, Z. Liu, Y. Zhang, C. L. Chen, and S. Xie. Asymmetric actuator backlash compensation in quantized adaptive control of uncertain networked nonlinear systems. *IEEE Trans. Neural Netw. Learn. Syst.*, 28(2):294–307, 2017.

[96] I. Lestas and G. Vinnicombe. Scalability in heterogeneous vehicle platoons. In *Proc. American Control Conference*, pages 4678–4683, 2007.

[97] F. W. Lewis, S. Jagannathan, and A. Yesildirak. *Neural network control of robot manipulators and non-linear systems*. Taylor & Francis, New York, 1998.

[98] D. Li and Z. Lin. Reaching consensus in unbalanced networks with coarse information communication. *Int. J. Robust Nonlinear Control*, 26(10):2153–2168, 2016.

[99] D. Li, Q. Liu, X. Wang, and Z. Yin. Quantized consensus over directed networks with switching topologies. *Syst. Control Lett.*, 165:13–22, 2014.

[100] L. Li, D. W. Ho, and J. Lu. A unified approach to practical consensus with quantized data and time delay. *IEEE Trans. Circuits Syst. Regul. Pap.*, 60(10):2668–2678, 2013.

[101] S. Li, L. Yang, and Z. Gao. Coordinated cruise control for high-speed train movements based on a multi-agent model. *Transp. Res. Part C: Emerg. Technol.*, 56:281–292, 2015.

[102] S. H. Li, H. Sun, J. Yang, and X. Yu. Continuous finite-time output regulation for disturbed systems under mismatching condition. *IEEE Trans. Autom. Control*, 60(1):277–282, 2015.

[103] T. Li, M. Fu, L. Xie, and J. F. Zhang. Distributed consensus with limited communication data rate. *IEEE Trans. Autom. Control*, 56(2):279–292, 2011.

[104] T. Li and L. Xie. Distributed consensus over digital networks with limited bandwidth and time-varying topologies. *Automatica*, 47(9):2006–2015, 2011.

[105] T. Li and L. Xie. Distributed coordination of multi-agent systems with quantized-observer based encoding-decoding. *IEEE Trans. Autom. Control*, 57(12):3023–3037, 2012.

[106] T. S. Li, D. Wang, G. Feng, and S. C. Tong. A DSC approach to robust adaptive NN tracking control for strict-feedback nonlinear systems. *IEEE Trans. Syst., Man, Cybern. B, Cybern.*, 40(4):915–927, 2010.

[107] X. J. Li and G. H. Yang. FLS-based adaptive synchronization control of complex dynamical networks with nonlinear couplings and state-dependent uncertainties. *IEEE Trans. Cybern.*, 46(1):171–180, 2016.

[108] X. J. Li and G. H. Yang. Adaptive fault-tolerant synchronization control of a class of complex dynamical networks with general input distribution matrices and actuator faults. *IEEE Trans. Neural Networks Learn. Syst.*, 28(3):559–569, 2017.

[109] Y. Li, S. Sui, and S. Tong. Adaptive fuzzy control design for stochastic nonlinear switched systems with arbitrary switchings and unmodeled dynamics. *IEEE Trans. Cybern.*, 47(2):403–414, 2017.

[110] Z. K. Li, Z. S. Duan, and F. L. Lewis. Distributed robust consensus control of multi-agent systems with heterogeneous matching uncertainties. *Automatica*, 50(3):883–889, 2014.

[111] Z. K. Li, Z. S. Duan, W. Ren, and G. Feng. Containment control of linear multi-agent systems with multiple leaders of bounded inputs using distributed continuous controllers. *Int. J. Robust Nonlinear Control*, 25(13):2101–2121, 2015.

[112] Z. K. Li, X. D. Liu, M. Y. Fu, and L. H. Xie. Global H_∞ consensus of multi-agent systems with Lipschitz non-linear dynamics. *IET Control Theory Appl.*, 6(13):2041–2048, 2012.

[113] Z. K. Li, W. Ren, X. D. Liu, and M. Y. Fu. Consensus of multi-agent systems with general linear and Lipschitz nonlinear dynamics using distributed adaptive protocols. *IEEE Trans. Autom. Control*, 58(7):1786–1791, 2013.

[114] Z. K. Li, W. Ren, X. D. Liu, and L. H. Xie. Distributed consensus of linear multi-agent systems with adaptive dynamic protocols. *Automatica*, 49(7):1986–1995, 2013.

[115] Y. W. Liang, C. C. Chen, and S. D. Xu. Study of reliable design using T-S fuzzy modeling and integral sliding mode control schemes. *Int. J. Fuzzy Syst.*, 333(1):103–112, 2013.

[116] C. M. Lin and C. H. Chen. Car-following control using recurrent cerebellar model articulation controller. *IEEE Trans. Veh. Technol.*, 56(6):3660–3673, 2007.

[117] H. Liu, M. Cao, and C. De Persis. Quantization effects on synchronized motion of teams of mobile agents with second-order dynamics. *Syst. Control Lett.*, 61(12):1157–1167, 2012.

[118] P. Liu, A. Kurt, and U. Ozguner. Distributed model predictive control for cooperative and flexible vehicle platooning. *IEEE Trans. Control Syst. Technol.*, DOI: 10.1109/TCST.2018.2808911,2018.

[119] S. Liu, L. H. Xie, and D. E. Quevedo. Event-triggered quantized communication based distributed convex optimization. *IEEE Trans. Control Network Syst.*, 5(1):167–178, 2018.

[120] Y. Liu, H. Gao, B. Xu, G. Liu, and H. Cheng. Autonomous coordinated control of a platoon of vehicles with multiple disturbances. *IET Control Theory Appl.*, 8(18):2325–2335, 2014.

[121] Y. J. Liu, Y. Gao, S. Tong, and C. P. Chen. A unified approach to adaptive neural control for nonlinear discrete-time systems with nonlinear dead-zone input. *IEEE Trans. Neural Networks Learn. Syst.*, 27(1):139–150, 2016.

[122] Z. Liu, G. Lai, Y. Zhang, X. Chen, and C. L. P. Chen. Adaptive neural control for a class of nonlinear time-varying delay systems with unknown hysteresis. *IEEE Trans. Neural Networks Learn. Syst.*, 25(12):2129–2140, 2014.

[123] S. Martin, A. Girard, A. Fazeli, and A. Jadbabaie. Multi-agent flocking under general communication rule. *IEEE Trans. Control Network Syst.*, 1(2):155–166, 2014.

[124] R. Middleton and J. Braslavsky. String instability in classes of linear time invariant formation control with limited communication range. *IEEE Trans. Autom. Control*, 7(55):1519–1530, 2010.

[125] U. Montanaro, M. Tufo, G. Fiengo, and S. Santini. A novel cooperative adaptive cruise control approach: Theory and hardware in the loop experimental validation. In *Proc. 2014 22nd Mediterranean Conference of Control and Automation (MED)*, pages 37–42, 2014.

[126] G. J. Naus, R. P. Vugts, J. Ploeg, M. J. Van De Molengraft, and M. Steinbuch. String-stable CACC design and experimental validation: A frequency-domain approach. *IEEE Trans. Veh. Technol.*, 59(9):4268–4279, 2010.

[127] M. B. R. Neila and D. Tarak. Adaptive terminal sliding mode control for rigid robotic manipulators. *Int. J. Autom. Comput.*, 8(2):215–220, 2011.

[128] M. Nokleby, W. U. Bajwa, R. Calderbank, and B. Aazhang. Toward resource-optimal consensus over the wireless medium. *IEEE J. Sel. Top. Sign. Proces.*, 7(2):284–295, 2013.

[129] R. Olfati-Saber and R. M. Murray. Consensus problems in networks of agents with switching toplogy and time-delays. *IEEE Trans. Autom. Control*, 49(9):1520–1533, 2004.

[130] M. Pachter, J. J. D'Azzo, and A. W. Proud. Tight formation flight control. *AIAA J. Guid., Nav., Control*, 24(2):246–254, 2012.

[131] A. Pant, P. Seiler, and K. Hedrick. Mesh stability of look-ahead interconnected systems. *IEEE Trans. Autom. Control*, 47(2):403–407, 2002.

[132] B. S. Park, S. J. Yoo, J. B. Park, and Y. H. Choi. Adaptive neural sliding mode control of nonholonomic wheeled mobile robots with model uncertainty. *IEEE Trans. Control Syst. Technol.*, 17(1):207–214, 2009.

[133] J. H. Park, S. H. Kim, and C. J. Moon. Adaptive neural control for strict-feedback nonlinear systems without backstepping. *IEEE Trans. Neural Networks*, 20(7):1204–1209, 2009.

[134] M. J. Park, O. M. Kwon, J. H. Park, S. M. Lee, and E. J. Cha. Synchronization of discrete-time complex dynamical networks with interval time-varying delays via non-fragile controller with randomly occurring perturbation. *J. Franklin Inst.*, 351(10):4850–4871, 2014.

[135] J. C. Patra and A. C. Kot. Nonlinear dynamic system identification using chebyshev functional link artificial neural networks. *IEEE Trans. Syst. Man Cybern. Part B Cybern.*, 32(4):505–511, 2002.

[136] S. Patterson and B. Bamieh. Consensus and coherence in fractal networks. *IEEE Trans. Control Network Syst.*, 1(4):338–348, 2014.

[137] K. Peng and Y. Yang. Leader-following consensus problem with a varying-velocity leader and time-varying delays. *Physica A*, 388(2-3):193–208, 2009.

[138] Y. F. Peng. Adaptive intelligent backstepping longitudinal control of vehicle-platoons using output recurrent cerebellar model articulation controller. *Expert Syst. Appl.*, 37(3):2016–2027, 2010.

[139] C. D. Persis and B. Jayawardhana. Coordination of passive systems under quantized measurements. *SIAM J. Control Optim.*, 50(6):3155–3177, 2012.

[140] A. A. Peters, R. H. Middleton, and O. Mason. Leader tracking in homogeneous vehicle platoons with broadcast delays. *Automatica*, 50(1):64–74, 2014.

[141] I. R. Petersen. A stabilization algorithm for a class of uncertain linear systems. *Syst. Control Lett.*, 8(4):351–357, 1987.

[142] J. Ploeg, E. Semsar-Kazerooni, G. Lijster, N. Van De Wouw, and H. Nijmeijer. Graceful degradation of cooperative adaptive cruise control. *IEEE Trans. Intell. Transp. Syst.*, 16(1):488–497, 2015.

[143] J. Ploeg, N. Van De Wouw, and H. Nijmeijer. $\mathscr{L}_p$ string stability of cascaded systems: Application to vehicle platooning. *IEEE Trans. Control Syst. Technol.*, 22(2):786–793, 2014.

[144] M. M. Polycarpou and P. A. Ioannou. A robust adaptive nonlinear control design. *Automatica*, 32(3):423–427, 1996.

[145] J. Qin, H. Gao, and W. X. Zheng. Exponential synchronization of complex networks of linear systems and nonlinear oscillators: A unified analysis. *IEEE Trans. Neural Networks and Learn.*, 26(3):510–521, 2015.

[146] R. Rajamani. Observers for Lipschitz nonlinear systems. *IEEE Trans. Autom. Control*, 43(3):397–401, 1998.

[147] R. Rajamani and Y. M. Cho. Existence and design of observers for nonlinear systems: relation to distance to unobservability. *Int. J. Control*, 69(5):717–731, 1998.

[148] B. Ren, S. S. Ge, K. P. Tee, and T. H. Lee. Adaptive neural control for output feedback nonlinear systems using a barrier Lyapunov function. *IEEE Trans. Neural Networks*, 21(8):1339–1345, 2010.

[149] W. Ren and R. Beard. Consensus problems in networks of agents with switching toplogy and time-delays. *IEEE Trans. Autom. Control*, 50(5):655–661, 2005.

[150] J. Rogge and D. Aeyels. Vehicle platoons through ring coupling. *IEEE Trans. Autom. Control*, 53(6):1370–1377, 2008.

[151] J. Ryan, C. E. Hanson, and J. F. Parle. String stability of a linear formation flight control system. In *Proc. AIAA Guid., Nav. Control Conf. Exhibit*, pages AIAA–1CAIAA–12, 2002.

[152] Li S., Yang L., and Gao Z. Adaptive coordinated control of multiple high-speed trains with input saturation. *Nonlinear Dyn.*, 83(4):2157–2169, 2016.

[153] S. Santini, A. Salvi, A. S. Valente, and A. Pescape. A consensus-based approach for platooning with inter-vehicular communications and its validation in realistic scenarios. *IEEE Trans. Veh. Technol.*, 66(3):1985–1999, 2017.

[154] P. Seiler, A. Pant, and K. Hedrick. Disturbance propagation in vehicle strings. *IEEE Trans. Autom. Control*, 49(10):1835–1842, 2004.

[155] G. S. Seyboth, J. Wu, J. Qin, and C. Yu. Collective circular motion of unicycle type vehicles with nonidentical constant velocities. *IEEE Trans. Control Network Syst.*, 1(2):167–176, 2014.

[156] D. Shevitz and B. Paden. Lyapunov stability theory of nonsmooth systems. *IEEE Trans. Autom. Control*, 39(9):1910–1914, 1994.

[157] R. Sipahi, F. M. Atay, and S. I. Niculescu. Stability analysis of a constant time-headway driving strategy with driver memory effects modeled by distributed delays. *IFAC-PapersOnLine*, 48(12):376–381, 2015.

[158] R. S. Smith and F. Y. Hadaegh. Control topologies for deep space formation flying spacecraft. In *Proc. Amer. Control Conf.*, volume 4, pages 2836–2841, 2002.

[159] Q. Song and Y. D. Song. Data-based fault-tolerant control of high-speed trains with traction/braking notch nonlinearities and actuator failures. *IEEE Trans. Neural Networks*, 22(12):2250–2261, 2011.

[160] Q. Song, Y. D. Song, and W. Cai. Adaptive backstepping control of train systems with traction/braking dynamics and uncertain resistive forces. *Veh. Syst. Dyn.*, 49(9):1441–1454, 2011.

[161] Y. Song, Y. Wang, and C. Wen. Adaptive fault-tolerant PI tracking control with guaranteed transient and steady-state performance. *IEEE Trans. Autom. Control*, 62(1):481–487, 2017.

[162] S. S. Stanković, M. J. Stanojević, and D. D. Šiljak. Decentralized overlapping control of a platoon of vehicles. *IEEE Trans. Control Syst. Technol.*, 8(5):816–832, 2000.

[163] G. Sun, X. Ren, Q. Chen, and D. Li. A modified dynamic surface approach for control of nonlinear systems with unknown input dead zone. *Int. J. Robust Nonlinear Control*, 25(8):1145–1167, 2015.

[164] D. Swaroop, J. K. Hedrick, C. C. Chien, and P. Ioannou. A comparison of spacing and headway control laws for automatically controlled vehicles. *Veh. Syst. Dyn.*, 23(8):597–625, 1994.

[165] D. Swaroop, J. K. Hedrick, and S. B. Choi. Direct adaptive longitudinal control of vehicle platoons. *IEEE Trans. Veh. Technol.*, 50(1):150–161, 2001.

[166] H. A. Talebi, K. Khorasani, and S. Tafazoli. A recurrent neural-network-based sensor and actuator fault detection and isolation for nonlinear systems with application to the satellite's attitude control subsystem. *IEEE Trans. Neural Networks*, 20(1):45–60, 2009.

[167] C. W. Tao, J. S. Taur, Y. H. Chang, and C. W. Chang. A novel fuzzy-sliding and fuzzy-integral-sliding controller for the twin-rotor multi-input/multi-output system. *IEEE Trans. Fuzzy Syst.*, 18(5):893–905, 2010.

[168] T. Tao and H. Xu. Adaptive fault-tolerant cruise control for a class of high-speed trains with unknown actuator failure and control input saturation. *Math. Prob. Eng.*, 2014:481315.1–481315.13, 2014.

[169] R. Teo, D. M. Stipanović, and C. J. Tomlin. Decentralized spacing control of a string of multiple vehicles over lossy datalinks. *IEEE Trans. Control Syst. Technol.*, 18(2):469–473, 2010.

[170] R. Teo and C. J. Tomlin. Computing danger zones for provably safe closely spaced parallel approaches. *AIAA J. Guid., Nav., Control*, 26(3):434–442, 2003.

[171] M. D. Tran and H. J. Kang. Adaptive terminal sliding mode control of uncertain robotic manipulators based on local approximation of a dynamic system. *Neurocomputing*, 228:231–240, 2017.

[172] C. K. Verginis, C. P. Bechlioulis, D. V. Dimarogonas, and K. J. Kyriakopoulos. Decentralized 2-d control of vehicular platoons under limited visual feedback. In *Proc. of IEEE/RSJ International Conference on Intelligent Robots and Systems (IROS)*, pages 3566–3571. IEEE, 2015.

[173] P. Vinken, E. Hoffman, and K. Zeghal. Influence of speed and altitude profile on the dynamics of in-trail following aircraft. In *Proc. AIAA Guid., Nav. Control Conf. Exhibit*, pages 1–11, 2000.

[174] C. L. Wang, C. Y. Wen, Q. L. Hu, W. Wang, and X. Y. Zhang. Distributed adaptive containment control for a class of nonlinear multiagent systems with input quantization. *IEEE Trans. neural networks Learn. Syst.*, 29(6):2419–2428, 2018.

[175] H. Wang, X. Liu, and K. Liu. Adaptive neural data-based compensation control of non-linear systems with dynamic uncertainties and input saturation. *IET Control Theory Appl.*, 9(7):1058–1065, 2015.

[176] J. Wang, Z. Duan, Y. Zhao, G. Qin, and Y. Yan. H_∞ and H_2 control of multi-agent systems with transient performance improvement. *Int. J. Control*, 86(12):2131–2145, 2013.

[177] X. Wang and G. H. Yang. Distributed fault-tolerant control for a class of cooperative uncertain systems with actuator failures and switching topologies. *Inf. Sci.*, 370 (Issue C):650–666, 2016.

[178] Y. Wang, Y. Song, H. Gao, and F. L. Lewis. Distributed fault-tolerant control of virtually and physically interconnected systems with application to high-speed trains under traction/braking failures. *IEEE Trans. Intell. Transp. Syst.*, 17(2):535–545, 2016.

[179] Z. Wang, D. Ding, H. Dong, and H. Shu. H_∞ consensus control for multi-agent systems with missing measurements: the finite-horizon case. *Syst. Control Lett.*, 62(10):827–836, 2013.

[180] G. H. Wen, Z. S. Duan, G. R. Chen, and W. W. Yu. Consensus tracking of multi-agent systems with Lipschitz-type node dynamics and switching topologies. *IEEE Trans. Circuits Syst. Regul. Pap.*, 61(2):499–511, 2014.

[181] G. H. Wen, G. Q. Hu, W. W. Yu, and G. R. Chen. Distributed H_∞ consensus of higher order multiagent systems with switching topologies. *IEEE Trans. Circuits Syst. Express Briefs*, 61(5):359–363, 2014.

[182] G. X. Wen, C. P. Chen, Y. J. Liu, and Z. Liu. Neural-network-based adaptive leader-following consensus control for second-order non-linear multi-agent systems. *IET Control Theory Appl.*, 9(13):1927–1934, 2015.

[183] Y. Q. Wu, H. Y. Su, R. Q. Lu, Z. G. Wu, and Z. Shu. Passivity-based non-fragile control for Markovian jump systems with aperiodic sampling. *Syst. Control Lett.*, 84:35–43, 2015.

[184] Z. G. Wu, J. H. Park, H. Su, and J. Chu. Non-fragile synchronisation control for complex networks with missing data. *Int. J. Control*, 86(3):555–566, 2013.

[185] E. Xargay, R. Choe, N. Hovakimyan, and I. Kaminer. Multi-leader coordination algorithm for networks with switching topology and quantized information. *Automatica*, 50(3):841–851, 2014.

[186] L. Xiao and F. Gao. Practical string stability of platoon of adaptive cruise control vehicles. *IEEE Trans. Intell. Transp. Syst.*, 12(4):1184–1194, 2011.

[187] S. D. Xu, C. C. Chen, and Z. L. Wu. Study of nonsingular fast terminal sliding-mode fault-tolerant control. *IEEE Trans. Ind. Electron.*, 62(6):3906–3913, 2015.

[188] Y. M. Xu and J. Z. Wang. The synchronization of linear systems under quantized measurements. *Syst. Control Lett.*, 62(10):972–980, 2013.

[189] C. Yang and Y. Sun. Mixed H_2/H_∞ cruise controller design for high speed train. *Int. J. Control*, 74(9):905–920, 2001.

[190] G. H. Yang and W. W. Che. Non-fragile H_∞ filter design for linear continuous-time systems. *Automatica*, 44(11):2849–2856, 2008.

[191] G. H. Yang and X. G. Guo. Insensitive H_∞ filter design for continuous-time systems with respect to filter coefficient variations. *Automatica*, 46(11):1860–1869, 2010.

[192] G. H. Yang, X. G. Guo, W. W. Che, and W. Guan. *Linear systems: non-fragile control and filtering*. CRC Press, 2013.

[193] Y. Yang, C. Hua, and X. Guan. Adaptive fuzzy finite-time coordination control for networked nonlinear bilateral teleoperation system. *IEEE Trans. Fuzzy Syst.*, 22(3):631–641, 2014.

[194] Y. D. Yang and Y. P. Sun. Mixed H_2/H_∞ cruise controller design for high speed train. *Int. J. Control*, (9):905–920, 2001.

[195] D. Ye, M. M. Chen, and H. J. Yang. Distributed adaptive event-triggered fault-tolerant consensus of multiagent systems with general linear dynamics. *IEEE Trans. Cybern.*, 49(3):757–767, 2019.

[196] D. Ye, N. N. Diao, and X. G. Zhao. Fault-tolerant controller design for general polynomial-fuzzy-model-based systems. *IEEE Trans. Fuzzy Syst.*, 26(2):1046–1051, 2018.

[197] Z. X. Yin, Y. R. Huang, X. Y. Geng, and D. Q. Li. Consensus control for directed networks under quantized information exchange. *Asian J. Control*, 18(1):1–8, 2016.

[198] K. You and L. Xie. Minimum data rate for mean square stabilizability of linear systems with Markovian packet losses. *IEEE Trans. Autom. Control*, 56(4):772–785, 2011.

[199] W. Yu, L. Zhou, X. Yu, J. Lu, and R. Lu. Consensus in multi-agent systems with second-order dynamics and sampled data. *IEEE Trans. Ind. Inf.*, 9(4):2137–2146, 2013.

[200] W. W. Yu, G. R. Chen, M. Cao, and J. Kurths. Second-order consensus for multiagent systems with directed topologies and nonlinear dynamics. *IEEE Trans. Syst. Man Cybern. Part B Cybern.*, 40(3):881–891, 2010.

[201] W. Yue, G. Guo, L. Wang, and W. Wang. Nonlinear platoon control of Arduino cars with range-limited sensors. *Int. J. Control*, 88(5):1037–1050, 2015.

[202] S. W. Yun, Y. J. Choi, and P. Park. H_2 control of continuous-time uncertain linear systems with input quantization and matched disturbances. *Automatica*, 45(10):2435–2439, 2009.

[203] J. H. Zhang and Y. Q. Xia. Design of static output feedback sliding mode control for uncertain linear systems. *IEEE Trans. Ind. Inf.*, 57(6):2161–2170, 2010.

[204] L. Zhang, J. Sun, and G. Orosz. Hierarchical design of connected cruise control in the presence of information delays and uncertain vehicle dynamics. *IEEE Trans. Control Syst. Technol.*, 26(1):139–150, 2018.

[205] T. P. Zhang and S. S. Ge. Adaptive dynamic surface control of nonlinear systems with unknown dead zone in pure feedback form. *Automatica*, 44(7):1895–1903, 2008.

[206] H. Zhao, C. Zhang, G. Wang, and G. Xing. H_∞ estimation for a class of Lipschitz nonlinear discrete-time systems with time delay. *Abstr. Appl. Anal.*, 2011:970978.1–970978.22, 2011.

[207] X. Zhao, Y. H. Chen, and H. Zhao. Robust approximate constraint-following control for autonomous vehicle platoon systems. *Asian J. Control*, 20(6):1–13, 2018.

[208] Y. L. Zhao, T. Wang, and W. J. Bi. Consensus protocol for multi-agent systems with undirected topologies and binary-valued communications. *IEEE Trans. Autom. Control*, 64(1):206–221, 2019.

[209] B. C. Zheng and J. H. Park. Adaptive integral sliding mode control with bounded L_2 gain performance of uncertain quantised control systems. *IET Control Theory Appl.*, 9(15):2273–2282, 2015.

[210] B. C. Zheng and G. H. Yang. H_2 control of linear uncertain systems considering input quantization with encoder/decoder mismatch. *ISA Trans.*, 52(5):577–582, 2013.

[211] B. C. Zheng, G. H. Yang, and T. Li. Quantised feedback sliding mode control of linear uncertain systems. *IET Control Theory Appl.*, 8(7):479–487, 2014.

[212] Y. Zheng, S. Li, J. Wang, D. Cao, and K. Li. Stability and scalability of homogeneous vehicular platoon: Study on the influence of information flow topologies. *IEEE Trans. Intell. Transp. Syst.*, 17(1):14–26, 2016.

[213] Y. Zheng, S. E. Li, K. Li, and L. Y. Wang. Stability margin improvement of vehicular platoon considering undirected topology and asymmetric control. *IEEE Trans. Control Syst. Technol.*, 24(4):1253–1265, 2016.

[214] J. Zhou and H. Peng. Range policy of adaptive cruise control vehicle for improved flow stability and string stability. *IEEE Trans. Intell. Transp. Syst.*, 6(2):229–237, 2005.

[215] K. Zhou, J. C. Doyle, and K. Glover. *Robust and optimal control.* Prentice Hall, Upper Saddle River, New Jersey, 1996.

[216] F. Zhu and Z. Han. A note on observers for Lipschitz nonlinear systems. *IEEE Trans. Autom. Control*, 47(10):1751–1754, 2002.

[217] A. M. Zou and K. D. Kumar. Neural network-based adaptive output feedback formation control for multi-agent systems. *Nonlinear Dyn.*, 70(2):1283–1296, 2012.

Index

9781032338316